FUNDAMENTALS OF CONVEX ANALYSIS

THEORY AND DECISION LIBRARY

General Editors: W. Leinfellner (*Vienna*) and G. Eberlein (*Munich*)

Series A: Philosophy and Methodology of the Social Sciences

Series B: Mathematical and Statistical Methods

Series C: Game Theory, Mathematical Programming and Operations Research

Series D: System Theory, Knowledge Engineering and Problem Solving

SERIES B: MATHEMATICAL AND STATISTICAL METHODS

VOLUME 24

Scope: The series focuses on the application of methods and ideas of logic, mathematics and statistics to the social sciences. In particular, formal treatment of social phenomena, the analysis of decision making, information theory and problems of inference will be central themes of this part of the library. Besides theoretical results, empirical investigations and the testing of theoretical models of real world problems will be subjects of interest. In addition to emphasizing interdisciplinary communication, the series will seek to support the rapid dissemination of recent results.

The titles published in this series are listed at the end of this volume.

FUNDAMENTALS OF CONVEX ANALYSIS

Duality, Separation, Representation, and Resolution

by

MICHAEL J. PANIK
Department of Economics,
University of Hartford,
West Hartford, Connecticut, U.S.A.

SPRINGER-SCIENCE+BUSINESS MEDIA, B.V.

Library of Congress Cataloging-in-Publication Data

Panik, Michael J.
Fundamentals of convex analysis : duality, separation, representation, and resolution / by Michael J. Panik.
p. cm. -- (Theory and decision library. Series B, Mathematical and statistical methods ; v. 24)
Includes index.
ISBN 978-90-481-4271-2 ISBN 978-94-015-8124-0 (eBook)
DOI 10.1007/978-94-015-8124-0
1. Convex functions. 2. Convex sets. I. Title. II. Series.
QA331.5.P36 1993
515'.8--dc20 93-17082

ISBN 978-90-481-4271-2

Printed on acid-free paper

Originally published by Kluwer Academic Publishers in 1993
Softcover reprint of the hardcover 1st edition 1993

To the Memory of
Paul N. Cusano
Father-in-Law and
Father-in-Deed

Contents

Preface

0.1. An Overview

My objective in writing this book is to offer to students of economics, management science, engineering, and mathematics an in-depth look at some of the fundamental features of a particular subset of general nonlinear mathematical analysis called ***convex analysis.*** While the spectrum of topics constituting convex analysis is extremely wide, the principle themes which will be explored herein are those of ***duality, separation, representation, and resolution.*** In this regard, these broad topic areas might even be referred to as the mathematical foundations or basic building blocks of convex analysis. Indeed, one could not reasonable expect to address, with even a modicum of success, the theoretical aspects of matrix games, optimization, and general equilibrium analysis without them.

A theme which occupies a key position in the area of convex analysis is that of ***duality.*** This property asserts that convex structures have a dual description, *i.e.*, corresponding to each convex object A in a space of finite dimension there is a dual object B belonging to the same space. In this regard, if B is given, then A can be uniquely generated. For instance, given a finite cone $\mathfrak{C}$, the cone polar to $\mathfrak{C}$ is $\mathfrak{C}^+ = \{\mathbf{y} \mid \mathbf{y}'\mathbf{x} \geq 0, \mathbf{x} \in \mathfrak{C}\}$. By taking the polar of $\mathfrak{C}^+$ we recover $\mathfrak{C}$ itself so that we may write $\mathfrak{C} \leftrightarrow \mathfrak{C}^+$ or $\mathfrak{C} = (\mathfrak{C}^+)^+$. This specification of duality has its roots in the notion of ***separation***, *e.g.*, under a given set of assumptions it is possible to separate (disjoint) convex sets by a hyperplane. In fact, most of the duality theorems encountered in this work are proved by employing virtually the same set of separation arguments.

Also playing an important role in convex analysis is the idea of ***representation***, *i.e.*, there can exist proper subsets of elements of a convex set which possess all the information about the original convex set itself. For example, knowing only the collection of extreme points of a convex polytope allows us to fully determine the entire set by forming the convex hull of the set of extreme points. In a similar vein, the theme of ***resolution***, drawing upon a finite basis argument, serves to decompose the elements within a given convex set into the convex hulls (or conical hull as appropriate) of a pair of disjoint finite point sets. That is, a point in a particular convex set can, at times, be represented as the (vector) sum of points from two different convex structures. For instance, a convex polyhedron is resolvable into the sum of a polytope and a polyhedral cone.

Chapter 1 lays out the mathematical prerequisites. It is assumed that the reader has had some minimal exposure to set theory as well as linear algebra and matrices. This chapter begins with a review of essential concepts such as: the Euclidian norm; linear dependence and independence; spanning set, basis, and dimension; solutions sets for simultaneous linear systems (treating both the homogeneous and nonhomogeneous cases along with the detection of inconsistencies); and linear subspaces and their duals. It then moves into the realm of point-set-theory and defines in considerable detail notions such as; neighborhoods; open and closed sets; points of closure and accumulations; convergence (in norm); compactness criteria; a linear variety or affine set; affine hull; and affine independence.

In chapter 2 we consider the structure of convex point sets. Specifically, the definition of a convex combination and convex set proper are given along with the essential properties of the latter. Also included are: Helly's theorem (1921); Berge's theorem (1959, 1963); the concept of a convex hull; Carathéodory's theorem (1970); and the relative interior of a convex set. Convex sets and their properties are at the heart of convex analysis and will be used extensively in the remainder of the text.

Chapter 3 begins with a detailed discussion of hyperplanes and their associated open and closed half-planes. After defining the characteristics of weak, proper, strict, and strong separation, a set of theorems is advanced which treats a variety of types of separation between a point and a set and between two convex sets. In particular, these theorems posit conditions for the "existence" of separating hyperplanes. Of special interest is the particular methodology used to execute most proofs of separation theorems. As will be readily demonstrated in this chapter, the usual ***modus operandi*** for dealing with the separation of two convex sets $\mathcal{S}_1, \mathcal{S}_2$ consists of redefining the problem as one involving the separation of the origin from the convex set $\mathcal{S}_1 - \mathcal{S}_2$.

An additional set of theorems pertaining to the existence of supporting or tangent hyperplanes is also included. Such theorems are important because they allow us to characterize or "represent" a convex set $\mathcal{S}$ in terms of its associated (finite) set of supporting hyperplanes at boundary points on $\mathcal{S}$. A representation of this type can be thought of as being of the ***external*** or outer variety.

The concept of separation holds a preeminent position in convex analysis in that it serves as a major input in deriving certain ***theorems of the alternative.*** A few of these theorems (involving disjoint alternatives framed in terms of linear equalities and/or inequalities) are introduced in chapter 3 as an application of the separation concept. In particular, the theorems of Farkas (1902) and Gordan (1873) along with one of Gale's theorems (1960) (dealing with the existence of a nonnegative solution to a system of linear inequalities) are all developed and illustrated with the use of a strong separation theorem. As the reader will soon note, Farkas' theorem will be used extensively throughout this work. In fact, it is actually introduced quite early in chapter 1 as the basis for the specification of a criterion used in detecting whether or not a simultaneous linear equation system is inconsistent.

Chapter 4 deals with the concepts of convex cones and finite cones, with the latter also classified as convex. These structures serve the essential function of geometrically illustrating the solution sets for homogeneous systems of linear equalities/inequalities. Also introduced are negative, orthogonal, dual, polar, normal, support, and barrier cones along with the process of determining the dimension of a finite cone. The concept of a ray or half-line is used extensively as an element in the construction of finite cones; its dual and polar serve to specify closed half-spaces. Properties of convex and finite cones are fully explored and the duality property of finite cones is amplified to theorem status, *i.e.*, a duality theorem for finite cones is proven using Farkas' theorem (in fact, it is demonstrated that the duality theorem is equivalent to Farkas' theorem) and then directly related to the concept of strong separation.

In order to explore the various (equivalent) ways of representing or generating a finite cone, the concepts of a conical combination and conical hull are introduced along with Carathéodory's theorem for cones. Also introduced are the notions of: extreme vectors (as well as extreme half-lines and half-spaces); semi-positively independent set of vectors; extreme supporting hyperplanes and half-planes; and extreme solutions of homogeneous linear inequalities. All of this definitional material lends support to the specification of a set of theorems dealing with the structure of finite cones, *e.g.*, we explore; Minkowski's theorem (1910) (the intersection of finitely many half-spaces is a conical combination of finitely many generators); a second theorem which states essentially that a finite cone can be generated by using only its set of extreme vectors; and Weyl's theorem (1935, 1950) (the set of conical combinations of a finite set of vectors corresponds to the intersection of a finite number of extreme supporting half-spaces). In fact, the theorems of Minkowski and Weyl serve to establish the so-called ***sum cone and intersection cone equivalence.*** It is also observed that Weyl's theorem implies the duality theorem for finite cones (*i.e.*, Farkas' theorem) we well as Minkowski's theorem. Moreover, the theorems of Minkowski and Farkas in combination render an "indirect" proof of Weyl's theorem.

Chapter 4 ends with a discussion of a couple of separation theorems for convex cones. These theorems are then used to obtain an representation theorem for a closed convex cone (*i.e.*, any such cone is the intersection of the set of homogeneous closed half-spaces which contain it). In addition, we establish the notion that Farkas' theorem can be cast in terms of finite cones and then interpreted as a separation theorem for a cone and an individual vector or a cone and an open half-space.

An important question often encountered in convex analysis is whether or not certain dual homogeneous linear systems possess a solution. This is the subject matter of chapter 5. Here we consider pairs of finite systems of homogeneous linear equalities and/or inequalities in which the variables are either nonnegative or unrestricted in sign. Moreover, these systems are structured in a fashion such that there is a one-to-one correspondence between unrestricted variables in one system and equations in the other and between nonnegative variables in one system and inequalities in the other. Under the aforementioned correspondence one can pass from a given system of homogeneous linear inequalities and/or equalities involving nonnegative and/or unrestricted variables to a second such system and conversely.

The chapter begins with a lemma by Tucker (1956) for dual homogeneous linear relations exhibiting a special positivity property and then moves into the analysis of a battery of Tucker's existence theorems (1956) for similar pairs of dual systems. An additional set of existence theorems is addressed which provides the foundation for the development of the concept of ***complementary slackness*** in pairs of dual systems and in a specialized self-dual system. Much of the material appearing in this chapter lends itself to applications in the area of linear programming (especially where questions of the existence and uniqueness of solutions as well as their feasibility are concerned).

The material developed in chapters 3-5 serves as the cornerstone for the treatment of ***theorems of the alternative*** for linear systems presented in chapter 6. Such theorems involve two mutually exclusive systems of linear inequalities and/or equalities denoted as, say, (I) and (II). They then assert that either system (I) has a solution or system (II) has a solution, but never both. In addition, a ***transposition theorem***, which is a special type of theorem of the alternative, considers the disjoint alternatives of solvability or contradiction given that in one system a vector is a linear combination of vectors from the other. In fact, a transposition theorem can be viewed as the algebraic counterpart of a separation theorem.

A whole host of important theorems of the alternative are discussed and, as is appropriate, interpreted geometrically. Specifically, the theorems included are those of: Slater (1951); Tucker (1956); Motzkin (1936); Gordan (1873); Steimke (1915); Farkas (1902); Gale (1960); von Neumann (1944); and Mangasarian (1969). Moreover, these theorems cover both homogeneous as well as nonhomogeneous systems and consider solutions which may be characterized as positive, nonnegative, semi-positive, or restricted to a convex combination. The material offered in this chapter lends itself to a wealth of applications in the areas of game theory and mathematical programming (*i.e.*, the specification of first-order optimality conditions in the presence of constraints; nondifferentiable optimization; constraint qualifications, etc.).

The principal focus of chapter 7 is the determination of what are called ***basic solutions*** (as well as ***basic feasible solutions***) to systems of nonhomogeneous linear equalities. After defining a basic solution, a step-wise procedure for obtaining a set of basic variables is outlined and accompanied by the process of "swapping" one basis vector for another so as to obtain a different basic feasible solution. Here too a step-by-step summary algorithm is reported along with several detailed examples which serve to illustrate the salient features of the calculations involved. Also included is a discussion on the circumstances under which we can find at least one basic (feasible) solution to a linear equation system.

Having developed the concept of a basic feasible solution to a linear equation system, this chapter next explores the structure of the solution set for the same. Let this set be denoted as $\mathcal{S} = \{\mathbf{x} \mid \mathbf{A}\mathbf{x} = \mathbf{b},\ \mathbf{x} \geqq \mathbf{0},\ \mathbf{x} \in \mathbf{R}^n\}$. After specifying the form of the homogeneous system associated with $\mathcal{S}$, a ***resolution theorem*** is established which essentially states that any vector in $\mathcal{S}$ is expressible as a convex combination of the set of basic feasible solutions plus a homogeneous solution associated with the linear system. We next examine extreme homogeneous solutions and then generalize the preceding resolution theorem to take account of such solutions. Both resolution theorems are structured so that $\mathcal{S}$ may or may not be bounded.

The next topic in Chapter 7 deals with the development, via Farkas' theorem, of a set of conditions which indicate when a mixed simultaneous system of linear inequalities and equalities has no solution. After this comes a section on complementary slackness in pairs of dual systems. In this regard, sets of weak and strong complementary slackness conditions are derived for general dual homogeneous systems; along with these, similar conditions, in the form of a set of complementary inequalities, are derived for a specialized self-dual system.

Chapter 8 begins with a discussion of extreme points and extreme direction for general convex sets. It addresses the conditions which underlie the existence of extreme points as well as the representation of a convex set in terms of its collection of extreme points (*i.e.*, we examine a theorem which states that, under certain conditions, extreme points form a minimal subset whose convex hull equals the set itself). Next considered are the concepts of: recession and extreme directions of a convex set; recession and extreme half-lines; and the convex cone generated by a convex set. The extension of the convex hull and affine hull concepts to sets which consist of both points and recession directions is also included. This enables us to offer expanded versions of the preceding representation theorem (a convex set can be expressed as the convex hull of its set of extreme points and extreme directions) and Carathédory's theorem (also framed in terms of points and directions).

We next study what are called "faces" of ***polyhedral convex sets*** (*e.g.*, extreme points, facets, and edges are all types of faces). Before doing so, however, it is important to determine what it actually means for a convex set to be characterized as "polyhedral." Following Rockafellar (1970), the property of being polyhedral reflects the notion that a given set is the intersection of finitely many closed half-spaces, *i.e.*, it is the solution set to some finite system of linear inequalities. (Note that we obtain what is termed a ***polyhedral convex cone*** if the said system is homogeneous,*i.e.*, all bounding hyperplanes pass through the origin.) In this regard, the quality of being polyhedral imposes a "finiteness" condition on the outer or ***external representation*** of a convex set (an n-dimensional closed convex set is the intersection of its set of supporting or tangent closed half-spaces). Dually, a finiteness condition can also be placed on the ***internal representation*** of a convex set (here a polyhedral convex set can be represented as the convex hull of its set of extreme points plus the conical combination of its set of extreme directions). We may note further that if the polyhedral convex sets under discussion are bounded, then they are called ***convex polytopes.*** Relative to the discussion of polyhedral faces mentioned above, the topics covered are: degenerate and adjacent extreme points; the dimension, minimual representation, and affine hull of a convex polyhedron; and proper face structures.

The preceding bit of material on convex polyhedra, along with some of the developments in chapter 7, now sets the stage for determining the location of extreme points. In particular, we posit a necessary and sufficient characterization of an extreme point by demonstrating that there exists a one-to-one correspondence between basic feasible solutions to the equalities defining a convex polyhedron and the extreme points of the polyhedron. This is then followed by an existence theorem (for extreme points) and a representation theorem which states that every convex polytope is the convex hull of its set of extreme points and conversely.

The definition of a recession direction and a recession cone appears next. It is then shown that, under certain conditions, a recession direction for a convex polyhedron is also an extreme direction. After examining a set of unboundedness criteria for convex polyhedra, the discussion turns to the development of an existence theorem for extreme directions.

At this point we are now able to offer a combined extreme point and extreme direction representation theorem for polyhedral convex sets. This is followed by an analysis of the resolution or decomposition of convex polyhedra. In particular, after stating a finite basis theorem for polyhedra, it is demonstrated that every polyhedral convex set is resolvable into the sum of a bounded convex polyhedron (or polytope) and a polyhedral convex cone. This is then followed by a finite bases theorem for polytopes.

The next section appearing in chapter 8 involves the separation of convex polyhedra. Here we extend some of the fundamental separation results for convex sets developed in chapter 3 to the case where at least one of the sets being separated is a convex polyhedron.

In chapter 9 we develop the notion of a k-dimensional simplex along with the definition of a standard n-simplex and unit simplex. Simplicial faces, facets, and carrier faces are next introduced along with concepts such as a simplicial complex and a simplicial decomposition (triangulation). Additional definitions such as a subdivision, and in particular the barycentric subdivision of a simplex, a simplicial mapping, and an (integer) labeling function set the stage for the development of Sperner's lemma, the Knaster-Kuratowski-Mazurkiewicz (K-K-M) theorem, Brouwer's (fixed point) theorem (along with a modification by Schauder), and Kakutani's (fixed point) theorem. Specifically, Sperner's lemma informs us that a properly labeled simplex contains an odd number of completely labeled subsimplexes. the K-K-M theorem provides us with a set of conditions which guarantee that the intersection of a collection of closed sets on a simplex is nonempty. And Brouwer's theorem demonstrates that a continuous point-to-point mapping of a simplex into itself admits at least one fixed point, *i.e.*, a point which is transformed into itself under the mapping. It is further shown that these three theorems are mathematically equivalent. Finally, Kakatani's theorem, which is a generalization of Brouwer's theorem to multivalued functions, states that an upper hemicontinuous point-to-set mapping of a compact convex set into itself has a fixed point.

0.2. A Note on the Method of Mathematical Induction

Quite often in mathematical analysis there are theorems which can be formulated in terms of "n" in that they assert a certain equation or proposition holds where n is any positive integer. For theorems such as these, an appropriate method of proof is ***mathematical induction.*** This procedure consists of the following two steps:

(1) Verify that the theorem/proposition holds for $n = 1$ (usually);

(2) Assume that the theorem/proposition holds for $n = p$ (the ***induction hypothesis***) and then prove that it holds for $n = p+1$.

Clearly the process of mathematical induction involves a type of "domino effect," *i.e.* once the proposition is proved for a particular integer, then the proposition will automatically follow for the next integer, and the next one, and the next, and so on ***ad infinitum.*** Given that steps 1, 2 have been executed, the "chain reaction" inherent in the process is set in motion and subsequently applies for any $n > 0$.

Suppose steps 1, 2 have been carried out for some theorem which is to be proved. How can we be sure that this procedure actually proves the theorem? To answer this let us assume, to the contrary, that the theorem under consideration is false. In this instance there exist positive integers for which the theorem is false and thus there must be some smallest integer, say $M+1$, for which the theorem is false. Since Step 1 precludes the integer $M+1$ from equaling 1, there is an integer M preceding $M+1$ and, for $n = M$, the theorem is true. Then under step 2, the theorem follows for $n = M+1$ and thus a contradiction occurs. But then this means that the assumption "the theorem is false" was incorrect and thus the theorem must be true.

0.3. Vector Notation

In the material which follows we shall deal with n-dimensional vector spaces taken over a field $\mathbf{R}$ of real scalars. Since the elements in $\mathbf{R}$ are "ordered," concepts such as "positive, semipositive, or nonnegative" can be defined. Specifically, the relations " $>$, $\geq$, and $\geqq$ " constitute a partial ordering on the vectors in $\mathbf{R}^n$, *i.e.*, for

$$\mathbf{x} = \begin{bmatrix} x_1 \\ x_2 \\ \vdots \\ x_n \end{bmatrix} \in \mathbf{R}^n \quad \text{or} \quad \mathbf{x} = (x_i) \in \mathbf{R}^n,\ i = 1, \ldots, n,$$

(1) $\mathbf{x}$ is a ***strictly positive vector*** (written $\mathbf{x} > \mathbf{0}$) if $x_i > 0$ for all i;

(2) $\mathbf{x}$ is a ***nonnegative vector*** (denoted $\mathbf{x} \geqq \mathbf{0}$) if $x_i \geqq 0$ for all i, and

(3) $\mathbf{x}$ is a ***semipositive vector*** (written $\mathbf{x} \geq \mathbf{0}$) if $\mathbf{x} \geqq \mathbf{0}$ but $\mathbf{x} \neq \mathbf{0}$, *i.e.*, $\mathbf{x}$ has at least one positive component.

Moreover, for vectors $\mathbf{x}_1, \mathbf{x}_2 \in \mathbf{R}^n$, we may write $\mathbf{x}_1 > \mathbf{x}_2$, $\mathbf{x}_1 \geqq \mathbf{x}_2$, or $\mathbf{x}_1 \geq \mathbf{x}_2$ according to whether $\mathbf{x}_1 - \mathbf{x}_2$ is strictly positive, nonnegative, or semipositive.

0.4 Acknowledgements

I am deeply indebted to a host of outstanding researchers whose accomplishments have served as an important source of inspiration and who, through their insight and genius, made the path traversed herein highly accessible and a joy to follow. In particular, I have borrowed quite liberally from the writings of Claude Berge, Werner Fenchel, David Gale, A.J. Goldman, Samuel Karlin, H.W. Kuhn, Olvi Mangasarian, R. Tyrell Rockafellar, and A.W. Tucker, among others. Citations to their works are interspersed throughout the text.

Most of this project was completed during my recent sabbatical year stay as a visiting fellow at Yale University. The kindness of the Department of Economics, Graduate School, and Cowles Library staff is sincerely appreciated. An additional note of thanks is extended to the Coffin Grant Committee and the Barney School of Business and Public Administration at the University of Hartford for providing financial support.

Finally, a sincere thank you goes to Marilyn Baleshiski for her extreme accuracy and dedication in typing a highly complex and involved manuscript.

CHAPTER 1
PRELIMINARY MATHEMATICS

1.1 Vector Spaces and Subspaces

A ***vector*** is an ordered n-tuple of elements expressed as a column $(n \times 1)$ or row $(1 \times n)$ matrix

$$\mathbf{x} = \begin{bmatrix} x_1 \\ \cdot \\ \cdot \\ \cdot \\ x_n \end{bmatrix}, \quad \mathbf{x}' = (x_1, \ldots, x_n)$$

respectively. The n elements of a vector are called its ***components.***

In all of what follows our discussion will take place in an "n-dimensional real vector space" (denoted $\mathbf{R}^n$). To fully develop the notion of such a space, let us begin by defining the concept of an "abstract vector space." Specifically, let $\mathcal{F}$ be a given field[1] of scalars. A ***vector (linear) space over*** $\mathcal{F}$ is a set $\mathcal{V}$ of vectors which is closed under the operations of addition and scalar multiplication. That is, the vectors must satisfy the following axioms:

A. If $\mathbf{x}, \mathbf{y} \in \mathcal{V}$, then $\mathbf{x}+\mathbf{y} \in \mathcal{V}$ and

1. $\mathbf{x}+\mathbf{y} = \mathbf{y}+\mathbf{x}$ (commutative law).
2. $\mathbf{x}+(\mathbf{y}+\mathbf{z}) = (\mathbf{x}+\mathbf{y})+\mathbf{z}$ (associative law).
3. there is a unique element $\mathbf{0} \in \mathcal{V}$ such that $\mathbf{x}+\mathbf{0} = \mathbf{x}$ for each $\mathbf{x} \in \mathcal{V}$.

---

[1] A set $\mathcal{F}$ of elements together with the operations of addition and multiplication on $\mathcal{F}$ constitutes a ***field*** if: both addition and multiplication are associative and commutative; additive and multicative inverses and identity elements exist; and multiplication distributes over addition.

4. for each $\mathbf{x} \in \mathcal{V}$ there is a unique element $-\mathbf{x} \in \mathcal{V}$ such that $\mathbf{x} + (-\mathbf{x}) = \mathbf{0}$.

B. For $c \in \mathcal{F}$ and $\mathbf{x} \in \mathcal{V}$, it follows that $c\mathbf{x} \in \mathcal{V}$ and

1. $c(\mathbf{x}+\mathbf{y}) = c\mathbf{x} + c\mathbf{y}$
2. $(c_1+c_2)\mathbf{x} = c_1\mathbf{x} + c_2\mathbf{x}$

} (distributive laws).

3. $(c_1 c_2)\mathbf{x} = c_1(c_2\mathbf{x})$ (associative law).
4. $1\mathbf{x} = \mathbf{x}$ for every $\mathbf{x} \in \mathcal{V}$.

Next, let $\mathcal{V}_n(\mathcal{F})$ be the "set of all ordered n-tuples" $(x_1, \ldots, x_n) = \mathbf{x}'$, where the components $\mathbf{x}_i$, i=1,..., n, are elements of $\mathcal{F}$. Let the vectors of $\mathcal{V}_n(\mathcal{F})$ satisfy the following conditions:

C. If $\mathbf{x}, \mathbf{y}$ represent arbitrary vectors in $\mathcal{V}_n(\mathcal{F})$, then

1. $\mathbf{x}=\mathbf{y}$ if and only if $x_i = y_i$.
2. $\mathbf{x}' + \mathbf{y}' = (x_1+y_1, \ldots, x_n+y_n)$.
3. $\mathbf{0}' = (0, \ldots, 0)$.
4. $c\mathbf{x}' = (cx_1, \ldots, cx_n)$.

Clearly the vectors of $\mathcal{V}_n(\mathcal{F})$ satisfying (C) must also satisfy axiom sets A, B above so that $\mathcal{V}_n(\mathcal{F})$ can be termed an ***n-dimensional "coordinate" vector space over*** $\mathcal{F}$. And since every n-dimensional vector space over $\mathcal{F}$ is isomorphic to $\mathcal{V}_n(\mathcal{F})$, we can identify $\mathcal{V}$ with $\mathcal{V}_n(\mathcal{F})$. Specifically, when $\mathcal{F}$ is taken to be the set of real scalars $\mathbf{R}$, then $\mathcal{V}_n(\mathbf{R})$ is an ***n-dimensional "real" vector space*** and is denoted simply as $\mathbf{R}^n$.

The ***Cartesian product*** (or ***product set***) $\mathcal{A} \times \mathcal{B}$ of two sets $\mathcal{A}, \mathcal{B}$ in $\mathbf{R}^n$ is the set of ordered pairs (a, b), where $\mathrm{a} \in \mathcal{A}$, $\mathrm{b} \in \mathcal{B}$. The product set $R \times R \times \cdots \times R = R^n = \{\mathrm{x}' = (x_1, x_2, \ldots, x_n) \mid x_1, x_2, \ldots, x_n \in \mathbf{R}\}$.

To establish a ***metric*** or concept of distance between two points $\mathbf{x}$, $\mathbf{y} \in \mathbf{R}^n$, let us first state that the ***scalar product*** of $\mathbf{x}$, $\mathbf{y}$ is the scalar

$$\mathbf{x} \cdot \mathbf{y} = \mathbf{x}'\mathbf{y} = \sum_{i=1}^{n} x_i y_i, \tag{1.1.1}$$

where

(a) $\mathbf{x}\cdot\mathbf{x} \geq 0$ and $\mathbf{x}\cdot\mathbf{x} = 0$ if and only if $\mathbf{x} = \mathbf{0}$;

(b) $\mathbf{x}\cdot\mathbf{y} = \mathbf{y}\cdot\mathbf{x}$ (commutative law);

(c) $(\mathbf{x}+\mathbf{y})\cdot\mathbf{z} = \mathbf{x}\cdot\mathbf{z}+\mathbf{y}\cdot\mathbf{z}$ (distributive law); and

(d) $(c\mathbf{x})\cdot\mathbf{y} = c(\mathbf{x}\cdot\mathbf{y})$, $c\in\mathfrak{F}$.

Next, a ***norm*** on $\mathbf{R}^n$ is a function which assigns to each $\mathbf{x}\in\mathbf{R}^n$ some number $\|\mathbf{x}\|$ such that:

(a) $\|\mathbf{x}\| \geq 0$ and $\|\mathbf{x}\| = 0$ if and only if $\mathbf{x} = \mathbf{0}$;

(b) $\|\mathbf{x}+\mathbf{y}\| \leq \|\mathbf{x}\| + \|\mathbf{y}\|$ (the triangle inequality);

(c) $\|c\mathbf{x}\| = |c|\,\|\mathbf{x}\|$ (homogeneity); and

(d) $|\mathbf{x}\cdot\mathbf{y}| \leq \|\mathbf{x}\|\,\|\mathbf{y}\|$ (the Cauchy-Schwarz inequality).

In this regard, the ***distance between points*** $\mathbf{x}$, $\mathbf{y}\in\mathbf{R}^n$ induced by the norm "$\|\cdot\|$" on $\mathbf{R}^n$ is

$$\|\mathbf{x}-\mathbf{y}\| = \left[(\mathbf{x}-\mathbf{y})'(\mathbf{x}-\mathbf{y})\right]^{1/2} = \left[\sum_{i=1}^{n}|x_i-y_i|^2\right]^{1/2}. \tag{1.1.2}$$

An important special case of this definition occurs when we consider the distance between any $\mathbf{x}\in\mathbf{R}^n$ and the origin $\mathbf{0}$, *i.e.*, the ***length (magnitude)*** of $\mathbf{x}$ or norm of $\mathbf{x}$ is

$$\|\mathbf{x}\| = (\mathbf{x}'\mathbf{x})^{1/2} = \left[\sum_{i=1}^{n}|x_i|^2\right]^{1/2}. \tag{1.1.3}$$

The ***direction of*** $\mathbf{x}$ ($\neq \mathbf{0}$) is a unit vector $\mathbf{x}/\|\mathbf{x}\|$ which "points in the direction of $\mathbf{x}$."

Given an arbitrary set $\mathfrak{X}$ and a metric d defined on $\mathfrak{X}\times\mathfrak{X}$, the pair $(\mathfrak{X}, d)$ is called a ***metric space.*** If $\mathfrak{X} = \mathbf{R}^n$ and d is given by (1.1.2), then $\mathbf{R}^n$ can always be considered as a metric space with metric (1.1.2).

Let $\mathbf{x}$, $\mathbf{y}\in\mathbf{R}^n$ with $\mathbf{x}, \mathbf{y} \neq \mathbf{0}$. The ***cosine of the angle*** $\boldsymbol{\theta}$ between $\mathbf{x}$, $\mathbf{y}$ is

$$\cos\theta = (\mathbf{x}\cdot\mathbf{y})/\|\mathbf{x}\|\,\|\mathbf{y}\|, \quad 0 \leq \theta \leq \pi.$$

The Cauchy-Schwarz inequality implies that $-1 \leq \cos\theta \leq 1$. Also, the angle between $\mathbf{x}, \mathbf{y}$ is acute $(<\pi/2)$, right $(=\pi/2)$, or obtuse $(>\pi/2)$ according to whether the scalar product $\mathbf{x}\cdot\mathbf{y}$ is positive, zero, or negative. Conversely, the vectors $\mathbf{x}, \mathbf{y}\in\mathbf{R}^n$ with $\mathbf{x}, \mathbf{y} \neq \mathbf{0}$ are said to be ***orthogonal*** (*i.e.*, mutually perpendicular) if $\mathbf{x}\cdot\mathbf{y}=0$. Since $\mathbf{x}\cdot\mathbf{y} = \|\mathbf{x}\|\,\|\mathbf{y}\|\cos\theta$, $\mathbf{x}\cdot\mathbf{y}=0$ if and only if $\theta=\pi/2$ and thus $\cos\theta=0$.

A vector $\mathbf{x}\in\mathbf{R}^n$ is a ***linear combination*** of the vectors $\mathbf{x}_j\in\mathbf{R}^n$, $j=1,\ldots, m$, if there exist scalars λ_j, $j=1,\ldots, m$, such that

$$\mathbf{x} = \sum_{j=1}^{m} \lambda_j \mathbf{x}_j. \tag{1.1.4}$$

In this light we can state that a set of vectors $\{\mathbf{x}_j\in\mathbf{R}^n,\ j=1,\ldots, m\}$ is ***linearly dependent*** if there exist scalars λ_j, $j=1,\ldots, m$, not all zero such that

$$\sum_{j=1}^{m} \lambda_j \mathbf{x}_j = \mathbf{0}, \tag{1.1.5}$$

i.e., the null vector is a linear combination of the vectors $\mathbf{x}_j$. If the only set of scalars λ_j for which (1.1.5) holds is $\lambda_j=0$, $j=1,\ldots, m$, then the vectors $\mathbf{x}_j$ are ***linearly independent***, *i.e.*, the trivial combination $0\mathbf{x}_1+\cdots+0\mathbf{x}_m$ is the only linear combination of the $\mathbf{x}_j$ which equals the null vector.

The essential features of a set of linearly dependent (independent) vectors $\{\mathbf{x}_j\in\mathbf{R}^n,\ j=1,\ldots, m\}$ are:

(a) if the set $\{\mathbf{x}_j\in\mathbf{R}^n,\ j=1,\ldots, m\}$ is linearly dependent, one of the vectors $\mathbf{x}_j$ is a linear combination of the others; if no vector in the set can be written as a linear combination of the others, the set is linearly independent;

(b) the set $\{\mathbf{x}_j\in\mathbf{R}^n,\ j=1,\ldots, m\}$ is linearly dependent if at least one of the vectors $\mathbf{x}_j$ is the null vector. Conversely, if $\{\mathbf{x}_j\in\mathbf{R}^n,$

$j=1,\ldots,m\}$ is a linearly independent set, it cannot contain the null vector;

(c) the set $\{\mathbf{x}\in R^n\}$ containing a single vector $\mathbf{x}$ is linearly independent if and only if $\mathbf{x}\neq\mathbf{0}$;

(d) if the set $\{\mathbf{x}_j\in R^n,\ j=1,\ldots,m\}$ is linearly independent, then any nonempty subset of this set is linearly independent;

(e) if the set $\{\mathbf{x}_j\in R^n,\ j=1,\ldots,m\}$ is linearly dependent, then the set containing $m+1$ vectors $\{\mathbf{x},\mathbf{x}_j\in R^n,\ j=1,\ldots,m\}$ is also linearly dependent and conversely;

(f) if $k<m$ is the maximum number of linearly independent vectors in the set $\{\mathbf{x}_j\in R^n,\ j=1,\ldots,m\}$, every other vector in the set is linearly dependent upon (*i.e.*, can be expressed as a linear combination of) the k linearly independent vectors.

Moreover,

1.1.1. THEOREM [Steinitz, 1913]. Let $\mathcal{S}_1,\mathcal{S}_2$ be linearly independent sets of vectors in R^n with $\mathcal{S}_1$ containing fewer elements than $\mathcal{S}_2$. Then there exists at least one element $\mathbf{x}\in\mathcal{S}_2$ such that the augmented set $\mathcal{S}_1\cup\{\mathbf{x}\}$ is also linearly independent.

A set of vectors $\{\mathbf{x}_j\in R^n,\ j=1,\ldots,m\}$ is a ***spanning set*** for R^n if every vector $\mathbf{x}\in R^n$ can be written as a linear combination of the vectors $\mathbf{x}_j$, *i.e.*, if $\mathbf{x}\in R^n$, then

$$\mathbf{x}=\sum_{j=1}^{m}\lambda_j\mathbf{x}_j.$$

Here the set $\{\mathbf{x}_j\in R^n,\ j=1,\ldots,m\}$ is said to ***span*** or ***generate*** R^n since every vector in R^n is (uniquely) linearly dependent on the spanning set. It is important to note that:

(a) the vectors which span R^n need not be linearly independent; but

(b) any set of vectors spanning $\mathbf{R}^n$ which contains the smallest possible number of vectors is linearly independent.

A ***basis*** for $\mathbf{R}^n$ is a linearly independent subset of vectors from $\mathbf{R}^n$ which spans $\mathbf{R}^n$. Moreover:

(a) a basis for $\mathbf{R}^n$ is not unique; but the vectors in the basis are unique;

(b) every basis for $\mathbf{R}^n$ contains the same number of basis vectors; and there are precisely n vectors in every basis for $\mathbf{R}^n$;

(c) any set of n linearly independent vectors from $\mathbf{R}^n$ forms a basis for $\mathbf{R}^n$; and

(d) any set of $n+1$ vectors from $\mathbf{R}^n$ is linearly dependent.

The ***dimension of a vector space*** $\mathcal{V}$, $dim(\mathcal{V})$, is the maximum number of linearly independent vectors that span the space, *i.e.*, it is the number of vectors in a basis for $\mathcal{V}$. Clearly $dim(\mathcal{V}_n(\mathcal{F})) = dim(\mathbf{R}^n) = n$.

A nonempty subset $\mathcal{M}$ of $\mathbf{R}^n$ is termed a ***subspace*** or ***linear manifold*** if it is closed under the operations of vector additional and scalar multiplication, *i.e.*, given vectors $\mathbf{x}_1, \mathbf{x}_2 \in \mathcal{M}$, their linear combination $\lambda_1\mathbf{x}_1+\lambda_2\mathbf{x}_2$ is also a member of $\mathcal{M}$ for every pair of scalars $\lambda_1, \lambda_2 \in \mathbf{R}$. Both the origin $\{\mathbf{0}\}$ and $\mathbf{R}^n$ are themselves (trivially) subspaces. In fact, since $\mathbf{0} = 0\mathbf{x}$ for any $\mathbf{x} \in \mathbf{R}^n$, it follows that $\mathbf{0}$ is (trivially) a member of every subspace. Additionally, if $\mathcal{M}$ is a subspace of $\mathbf{R}^n$, then $\mathcal{M}$ must contain "all" linear combinations of its elements. In fact, the set $\ell(\mathcal{M})$ of all finite linear combination of points of $\mathcal{M}$ is called the ***linear hull*** of $\mathcal{M}$ and constitutes a subspace of $\mathbf{R}^n$. Here

$$\ell(\mathcal{M}) = \Big\{\mathbf{x} \,|\, \mathbf{x} = \sum_{i=1}^{k} \lambda_i\mathbf{x}_i, \mathbf{x}_i \in \mathcal{M},\ \lambda_i \in \mathbf{R} \text{ for all } i, \\ \text{and } k \text{ is an arbitrary positive integer}\Big\}.$$

Moreover, $\ell(\mathcal{M})$ is the smallest subspace containing $\mathcal{M}$. Hence $\mathcal{M}$ is a subspace of $\mathbf{R}^n$ if and only if $\mathcal{M} = \ell(\mathcal{M})$. By convention, $\ell(\phi) = \{\mathbf{0}\}$. A set of vectors $\mathcal{S}$ is a ***spanning set*** of $\mathcal{M}$ in $\mathbf{R}^n$ if $\mathcal{M} = \ell(\mathcal{S})$.

A linearly independent set $\mathcal{S}$ of vectors contained in a linear subspace $\mathcal{M}$ of $\mathbf{R}^n$ is termed ***maximal*** in $\mathcal{M}$ if $\mathcal{S}$ is not a proper subset of any other linearly independent collection of vectors in $\mathcal{M}$. A spanning set $\mathcal{S}$ of $\mathcal{M}$ is ***minimal*** for $\mathcal{M}$ if no proper subset of $\mathcal{S}$ spans $\mathcal{M}$. In this regard:

(a) a spanning set $\mathcal{S}$ of $\mathcal{M}$ is minimal for $\mathcal{M}$ if and only if it is linearly independent: and

(b) a linear subset $\mathcal{S}$ of $\mathcal{M}$ is maximal in $\mathcal{M}$ if and only if it spans $\mathcal{M}$.

Thus each maximal linearly independent subset is a ***minimal spanning set*** and conversely. Clearly a minimal spanning set constitutes a ***basis*** for subspace $\mathcal{M}$. All bases for $\mathcal{M}$ possess a finite set of linearly independent vectors.

Given that $\mathcal{M}$ is a subspace of $\mathbf{R}^n$, the ***dimension of*** $\mathcal{M}$, $dim(\mathcal{M})$, is $\leq n$ and equals the maximum number of linearly independent vectors that constitute a basis for $\mathcal{M}$. So if $dim(\mathcal{M}) = m < n$, a basis for $\mathcal{M}$ is a subset of m linearly independent vectors selected from a set of n linearly independent vectors in $\mathbf{R}^n$.

If $\mathcal{M}_1$, $\mathcal{M}_2$ are subspaces of $\mathbf{R}^n$, $\mathcal{M}_1 \cap \mathcal{M}_2=\{\mathbf{0}\}$, and, taken together, $\mathcal{M}_1$ and $\mathcal{M}_2$ **span** $\mathbf{R}^n$ in the sense that each vector in $\mathbf{R}^n$ is uniquely expressible as $\mathbf{z} = \mathbf{x}+\mathbf{y}$, where $\mathbf{x} \in \mathcal{M}_1$ and $\mathbf{y} \in \mathcal{M}_2$, then $\mathbf{R}^n$ is expressible as the ***direct sum of*** $\mathcal{M}_1$, $\mathcal{M}_2$ and written as $\mathbf{R}^n = \mathcal{M}_1 \oplus \mathcal{M}_2 = \{\mathbf{x}+\mathbf{y} | \mathbf{x} \in \mathcal{M}_1, \mathbf{y} \in \mathcal{M}_2\}$. Moreover, if $\{\mathbf{x}_1, \ldots, \mathbf{x}_r\}$ is a basis for $\mathcal{M}_1$ and $\{\mathbf{y}_1, \ldots, \mathbf{y}_s\}$ is a basis for $\mathcal{M}_2$, then the combined set $\{\mathbf{x}_1, \ldots, \mathbf{x}_r, \mathbf{y}_1, \ldots, \mathbf{y}_s\}$ is a basis for $\mathbf{R}^n$. In this regard, $dim(\mathbf{R}^n) = n = dim(\mathcal{M}_1 \oplus \mathcal{M}_2) = dim(\mathcal{M}_1)+dim(\mathcal{M}_2) = r+s$. If a subspace $\mathcal{M}_1$ is contained in a subspace $\mathcal{M}_2$, then $dim(\mathcal{M}_1) \leq dim(\mathcal{M}_2)$.

For $\mathcal{M}$ a subspace of $\mathbf{R}^n$, the ***orthogonal complement*** or ***dual subspace of*** $\mathcal{M}$, $\mathcal{M}^\perp$, is also a subspace of $\mathbf{R}^n$ and consists of all vectors that are orthogonal to those in $\mathcal{M}$. Hence $\mathcal{M}^{\perp} = \{\mathbf{y} | \mathbf{y}'\mathbf{x} = 0, \ \mathbf{x} \in \mathcal{M}\}$. Additionally, $(\mathcal{M}^{\perp})^{\perp} = \mathcal{M}$.

If $\mathbf{x}$ is an arbitrary point in $\mathbf{R}^n$, the ***orthogonal projection*** of $\mathbf{x}$ into $\mathcal{M}$ is the point $\bar{\mathbf{x}} \in \mathcal{M}$ for which $\|\mathbf{x} - \bar{\mathbf{x}}\|^2 = (\mathbf{x} - \bar{\mathbf{x}})'(\mathbf{x} - \bar{\mathbf{x}})$ assumes a minimum. There exists at most one such projection of a point $\mathbf{x}$ into a subspace $\mathcal{M}$. And if $\bar{\mathbf{x}}$ is the projection of $\mathbf{x}$ into $\mathcal{M}$, then $\mathbf{x} - \bar{\mathbf{x}}$ is the projection of $\mathbf{x}$ into $\mathcal{M}^{\perp}$. $\mathcal{M}$ and $\mathcal{M}^{\perp}$ span $\mathbf{R}^n$ (any vector $\mathbf{z} \in \mathbf{R}^n$ can be uniquely expressed as $\mathbf{z} = \mathbf{x} + \mathbf{y}$ for $\mathbf{x} \in \mathcal{M}$, $\mathbf{y} \in \mathcal{M}^{\perp}$) so that $\mathbf{R}^n = \mathcal{M} \oplus \mathcal{M}^{\perp}$. In this regard, $\mathbf{x}$ and $\mathbf{y}$ are termed ***orthogonal projections*** of $\mathbf{z}$ onto the subspaces $\mathcal{M}$, $\mathcal{M}^{\perp}$ respectively. Clearly $dim(M^{\perp}) = n - dim(\mathcal{M})$.

If the set of vectors $\{\mathbf{x}_1, \ldots, \mathbf{x}_m\}$ is a basis for a subspace $\mathcal{M}$ of $\mathbf{R}^n$, then $\mathcal{M}^{\perp} = \{\mathbf{y} | \mathbf{y}'\mathbf{x}_i = 0, \ i=1, \ldots, m, \ \mathbf{x}_i \in \mathcal{M}\}$.

1.2. The Solution Set of a System of Simultaneous Linear Equations

To set the stage for the developments of this section we first state that the ***rank*** of an $(m \times n)$ matrix $\mathbf{A}$, denoted $\rho(\mathbf{A})$, is the order of the largest non-singular submatrix of $\mathbf{A}$, *i.e.*, it is the order of the largest nonvanishing determinant in $\mathbf{A}$; moreover, if $\rho(\mathbf{A}) = k$, all submatrices of order $k+1$ are singular. Also, an nth order matrix $\mathbf{B}$ is said to be of ***full rank*** if $\rho(\mathbf{B}) = n$; and this occurs if and only if $\mathbf{B}$ is nonsingular (*i.e.*, $|\mathbf{B}| \neq 0$). Hence, for an nth order matrix $\mathbf{B}$, $\rho(\mathbf{B}) < n$ if and only if $\mathbf{B}$ is singular.

A system of m linear equations in n unknowns may be written in matrix form as $\mathbf{A}\mathbf{x} = \mathbf{C}$, where $\mathbf{A}$ is an $(m \times n)$ matrix of (constant) coefficients a_{ij}, $\mathbf{x}$ is an $(n \times 1)$ matrix of unknown variables x_i, and C is an $(m \times 1)$ matrix of constants c_i, $i=1, \ldots, m$; $j=1, \ldots, n$. Let us now look to the structure of the solution set for $\mathbf{A}\mathbf{x} = \mathbf{C}$. Here we shall be interested in the existence and uniqueness of a solution to this simultaneous linear system. To facilitate this discussion we note that a system of equations is ***consistent*** if it has at least one solution; it is ***inconsistent*** if it does not possess a solution. In this regard, the "structure of the solution set for $\mathbf{Ax} = \mathbf{C}$," which involves three mutually exclusive and collectively exhaustive cases, appears as:

1. No solution exists ($\mathbf{Ax} = \mathbf{C}$ is inconsistent);
2. Exactly one solution exists; } ($\mathbf{Ax} = \mathbf{C}$ is consistent).
3. An infinity of solutions exists }

We may easily relate the above concept of rank to the consistency notion by stating

1.2.1. THEOREM.[2] Given the system $\mathbf{Ax} = \mathbf{C}$, where $\mathbf{A}$ is of order $(m \times n)$ if:

1. $\rho[\mathbf{A}, \mathbf{C}] > \rho(\mathbf{A})$, the system is inconsistent;
2. $\rho[\mathbf{A}, \mathbf{C}] = \rho(\mathbf{A}) =$ number of unknowns n, the system is consistent and possess a unique solution;
3. $\rho[\mathbf{A}, \mathbf{C}] = \rho(\mathbf{A}) = k <$ number of unknowns n, the system is consistent and possesses an infinity of solutions, where arbitrary values may be assigned to $n-k$ of the variables.

As a particularization of case (2), if $\mathbf{A}$ is $(n \times n)$ and nonsingular, then $\mathbf{x} = \mathbf{A}^{-1}\mathbf{C}$.

If $\mathbf{C} = \mathbf{0}$, then $\mathbf{Ax} = \mathbf{0}$ is termed a ***homogeneous*** linear equation system. Clearly this sort of system can never be inconsistent, *i.e.*, $\mathbf{Ax} = \mathbf{0}$ always has a solution since $\rho(\mathbf{A}) = \rho[\mathbf{A}, \mathbf{0}]$. Moreover, $\mathbf{x} = \mathbf{0}$ is always a (trivial) solution. What about the existence of nontrivial solutions? The answer is provided by a special case of Theorem 1.2.1, namely

1.2.2. THEOREM.[3] Given the system $\mathbf{Ax} = \mathbf{0}$, where $\mathbf{A}$ is of order $(m \times n)$, if:

[2]For a proof see Noble (1969), p. 91.

[3]This theorem is also proven by Noble (1969), p. 91.

1. $\rho(\mathbf{A}) =$ number of unknowns n, the system has a unique (trivial) solution $\mathbf{x} = \mathbf{0}$;
2. $\rho(\mathbf{A}) = k <$ number of unknowns n, the system has an infinity of nontrivial solutions, where arbitrary values may be assigned to $n-k$ of the variables.

If $\mathbf{A}$ is $(m \times \mathrm{n})$ and $m < \mathrm{n}$, then $\mathbf{Ax} = \mathbf{0}$ always possesses an infinity of nontrivial solutions. In general, for $\mathbf{A}$ of order $(m \times n)$ and $\rho(\mathbf{A}) = k$, the solution set $\mathscr{S}$ of the homogeneous linear system $\mathbf{Ax} = \mathbf{0}$, $\mathbf{x} \in \mathbf{R}^n$, is an $(n-k)$ - dimensional subspace of $\mathbf{R}^n$, *i.e.*, a basis for $\mathscr{S}$ consists on $n-k$ linearly independent vectors.

To further elaborate on the structure of the solution set of $\mathbf{Ax} = \mathbf{C}$, let us concentrate on case (3) above -- an infinite number of particular solutions exists. As we shall soon see, a detailed study of this possibility will enable us to further scrutinize cases (1), (2). To this end let us assume that $\mathbf{x}_1, \mathbf{x}_2$ are both solutions to $\mathbf{Ax} = \mathbf{C}$. Then

$$\mathbf{Ax}_1 - \mathbf{Ax}_2 = \mathbf{A}(\mathbf{x}_1 - \mathbf{x}_2) = \mathbf{Ah} = \mathbf{C} - \mathbf{C} = \mathbf{0},$$

i.e., $\mathbf{h} = \mathbf{x}_1 - \mathbf{x}_2$ solves the "homogeneous system $\mathbf{Ah} = \mathbf{0}$ associated with the original system $\mathbf{Ax} = \mathbf{C}$." Conversely, if $\mathbf{h}$ is any solution to $\mathbf{Ah} = \mathbf{0}$ and $\mathbf{x}_0$ is a particular solution to $\mathbf{Ax} = \mathbf{C}$, then $\mathbf{x}_0 + \mathbf{h}$ is another specific solution to $\mathbf{Ax} = \mathbf{C}$ since $\mathbf{A}(\mathbf{x}_0 + \mathbf{h}) = \mathbf{Ax}_0 + \mathbf{Ah} = \mathbf{C} + \mathbf{0} = \mathbf{C}$. These observations are summarized by

1.2.3. THEOREM. Let $\mathbf{x}_0$ represent a particular solution to $\mathbf{Ax} = \mathbf{C}$, where $\mathbf{A}$ is of order $(m \times \mathrm{n})$. Then the set of all solutions $\mathbf{x}$ to $\mathbf{Ax} = \mathbf{C}$ is equivalent to the set of all solutions $\mathbf{x}_0 + \mathbf{h}$, where $\mathbf{h}$ is a member of the set of all solutions of the homogeneous system $\mathbf{Ah} = \mathbf{0}$ associated with $\mathbf{Ax} = \mathbf{C}$.

In light of this theorem it should be clear that a consistent system $\mathbf{Ax} = \mathbf{C}$ has a unique solution $\mathbf{x}_0$ if and only if $\mathbf{Ah} = \mathbf{0}$ has only the trivial solution $\mathbf{h} = \mathbf{0}$.

This set of results may be strengthened somewhat if **A** is taken to be an nth order matrix of full rank n. Then it must always be true that $\mathbf{Ax} = \mathbf{C}$ has a unique solution for any **C** if and only if the associated homogeneous system $\mathbf{Ah} = \mathbf{0}$ has the unique solution $\mathbf{h} = \mathbf{0}$. To operationalize this result let us consider the following "theorem of the alternative"[4] for equalities

1.2.4. THEOREM. Given the system $\mathbf{Ax} = \mathbf{C}$, where **A** is of order $(n \times n)$, either:

1. $\mathbf{Ax} = \mathbf{C}$ has exactly one solution for each **C** or
2. $\mathbf{Ah} = \mathbf{0}$ has a nontrivial solution,

but never both.

So to determine whether or not the nth order system $\mathbf{Ax} = \mathbf{C}$ has a unique solution for "every" **C**, we need only determine whether it has a unique (trivial) solution for "only one" particular **C**, namely $\mathbf{C} = \mathbf{0}$.

Let us now examine in greater detail the conditions under which the simultaneous linear system $\mathbf{Ax} = \mathbf{C}$ has no solutions. Based upon our earlier observations it is easily seen that $\mathbf{Ax} = \mathbf{C}$ is inconsistent if $\rho[\mathbf{A}, \mathbf{C}] = 1+\rho(\mathbf{A})$.

Another approach to this problem involves an application of

1.2.5. FARKAS' THEOREM OF THE ALTERNATIVE [Farkas, 1902]. For any $(m \times n)$ matrix **A** and a vector $\mathbf{b} \in \mathbf{R}^m$, either

(I) $\mathbf{Ax} = \mathbf{C}$ has a solution $\mathbf{x} \geqq \mathbf{0}$, $\mathbf{x} \in \mathbf{R}^n$, or

(II) $\mathbf{A'y} \geqq \mathbf{0}$, $\mathbf{C'y} < 0$ has a solution $\mathbf{y} \in \mathbf{R}^m$,

but never both.

(A variety of proofs of this theorem as well as its geometric interpretation will be offered in subsequent chapters.) In this regard, if the assertion "$\mathbf{Ax} = \mathbf{C}$ is consistent" is to serve as Farkas' (I), then what should be the appropriate form of Farkas' (II)? To answer this question and to fully mirror (I), let $\mathbf{x} = \mathbf{u} - \mathbf{v}$ with $\mathbf{u}, \mathbf{v} \geqq \mathbf{0}$. Then

---

[4] A detailed discussion of such theorems is presented in Chapter 6.

$$\mathbf{A}(\mathbf{u}-\mathbf{v}) = \mathbf{C} \text{ has a solution } \mathbf{u},\ \mathbf{v} \geqq \mathbf{0} \text{ or}$$

$$(\mathrm{I}')\quad [\mathbf{A}, -\mathbf{A}]\begin{bmatrix}\mathbf{u}\\ \mathbf{v}\end{bmatrix} = \mathbf{C} \text{ has a nonnegative solution}$$

has the form of (I). From (II), the alternative to (I′) is

$$\begin{bmatrix}\mathbf{A}'\\ -\mathbf{A}'\end{bmatrix}\mathbf{y} \geqq \mathbf{0},\ \mathbf{C}'\mathbf{y} < 0 \text{ or}$$

$$(\mathrm{II}')\quad \mathbf{A}'\mathbf{y} = \mathbf{0},\ \mathbf{C}'\mathbf{y} < 0 \text{ has a solution } \mathbf{y}.$$

Actually, the sign of $\mathbf{C}'\mathbf{y}$ is irrelevant in the sense that if $\mathbf{y}$ is replaced by $-\mathbf{y}$, then (II′) becomes

$$(\mathrm{II}'')\quad \mathbf{A}'\mathbf{y} = 1,\ \mathbf{C}'\mathbf{y} \neq 0 \text{ has a solution } \mathbf{y}.$$

In sum, either $\mathbf{A}\mathbf{x} = \mathbf{C}$ is consistent of there exists a vector $\mathbf{y}$ which is orthogonal to the columns of $\mathbf{A}$ but not orthogonal to $\mathbf{C}$. These observations are consolidated in the following

1.2.6. THEOREM. The simultaneous linear system $\mathbf{A}\mathbf{x} = \mathbf{C}$, where $\mathbf{A}$ is of order $(m \times n)$ and $\mathbf{x} \in \mathbf{R}^n$, is inconsistent if and only if the linear system $\mathbf{A}'\mathbf{u} = \mathbf{0}$, $\mathbf{C}'\mathbf{u} = \alpha$, $0 \neq \alpha \in \mathbf{R}$, has a solution $\mathbf{u} \in \mathbf{R}^m$.

To close this discussion, let us relate the results obtained in theorem 1.2.2 to the concept of linear independence (dependence) presented in Section 1.1 above. Specifically, if we redefine the ***rank*** of an $(m \times \mathrm{n})$ matrix $\mathbf{A}$ as the maximum number of linearly independent vectors $\mathbf{x}_j \in \mathbf{R}^m$, $j=1, \ldots, n$, which span the columns of $\mathbf{A}$ (*i.e.*, it is at most the number of vectors in a basis for $\mathbf{R}^m$), then a test for the linear independence of a set of vectors of the form $\{\mathbf{x}_j \in \mathbf{R}^m,\ j=1, \ldots, n\}$ may be executed by considering the vector $\mathbf{x}_j$ as the jth column of an $(m \times \mathrm{n})$ matrix $\mathbf{A} = [\mathbf{x}_1, \ldots, \mathbf{x}_n]$. Then:

1. if $m \geq n$ and $\rho(\mathbf{A}) = n$, the set of vectors $\{\mathbf{x}_j \in \mathbf{R}^m,\ j=1, \ldots, n\}$ is linearly independent (by part 1 of theorem 1.2.2); but
2. if $m \geq n$ and $\rho(\mathbf{A}) < n$, or if $m < \mathrm{n}$, the set of vectors $\{\mathbf{x}_j \in \mathbf{R}^m,\ j=1, \ldots, n\}$ is linearly dependent (by part 2 of theorem 1.2.2).

An alternative way of expressing (1) is provided by

1.2.7 THEOREM. Let $\mathbf{A}' = [\mathbf{a}_1,\ldots,\mathbf{a}_m]$ be of order $(n\times m)$. If the vectors $\mathbf{a}_1,\ldots,\mathbf{a}_m$ are linearly independent in $\mathbf{R}^n$, then there exists a vector $\mathbf{y}\in\mathbf{R}^n$ such that $\mathbf{Ay}=\mathbf{b}$ for any $\mathbf{b}\in\mathbf{R}^m$.

1.3 Point-Set Theory: Topological Properties of $\mathbf{R}^n$

Let δ be any positive scalar. A ***δ-neighborhood of a point*** $\mathbf{x}_0\in\mathbf{R}^n$ (or an ***open ball or sphere of radius δ about*** $\mathbf{x}_0$) is the set $B(\mathbf{x}_0,\delta)$ consisting of all points $\mathbf{x}$ such that $\|\mathbf{x}-\mathbf{x}_0\|<\delta$, *i.e.*, $B(\mathbf{x}_0,\delta)=\{\mathbf{x}\,|\,\|\mathbf{x}-\mathbf{x}_0\|<\delta,\ \delta>0\}$.

A point $\bar{\mathbf{x}}$ is an ***interior point*** of a set $\mathcal{S}$ in $\mathbf{R}^n$ if there exists an open ball of radius δ about $\bar{\mathbf{x}}$ which contains only points of $\mathcal{S}$. The ***interior*** of a set $\mathcal{S}$ in $\mathbf{R}^n$, denoted $\mathcal{S}^o$, is the collection of all its interior points.

A set $\mathcal{S}$ in $\mathbf{R}^n$ is said to be ***open*** if, given any point $\mathbf{x}_0\in\mathcal{S}$, there exists a positive scalar δ such that $B(\mathbf{x}_0,\delta)\subseteq\mathcal{S}$. Hence $\mathcal{S}$ is open if it contains only interior points. Moreover:

(a) ϕ, $B(\mathbf{x}_0,\delta)$, and $\mathbf{R}^n$ are all open sets.

(b) A subset $\mathcal{S}$ of $\mathbf{R}^n$ is open if and only if it is the union of open spheres. The interior of $\mathcal{S}$ thus equals the union of all open subsets of $\mathcal{S}$.

(c) Any union of open sets if $\mathbf{R}^n$ is open; and any finite intersection of open sets in $\mathbf{R}^n$ is open.

(d) The interior of $\mathcal{S}$ is the largest open subset of $\mathcal{S}$.

(e) $\mathcal{S}$ is an open set if and only if $\mathcal{S}=\mathcal{S}^o$.

(f) For $\mathcal{A}$, $\mathcal{B}$ subsets of $\mathbf{R}^n$, $\mathcal{A}^o\cup\mathcal{B}^o\subseteq(\mathcal{A}\cup\mathcal{B})^o$ while $\mathcal{A}^o\cap\mathcal{B}^o=(\mathcal{A}\cap\mathcal{B})^o$.

Let $\mathcal{S}$ be a set in $\mathbf{R}^n$. Then the ***complementary set of*** $\mathcal{S}$, $\mathcal{S}^c$, is the collection of all points of $\mathbf{R}^n$ exclusive (lying outside) of $\mathcal{S}$. A point $\bar{\mathbf{x}}\in\mathcal{S}^c$ is an ***exterior point*** of $\mathcal{S}$ in $\mathbf{R}^n$ if there exists an open ball of radius δ about $\bar{\mathbf{x}}$ which contains only points of $\mathcal{S}^c$. The ***exterior*** of a set $\mathcal{S}$ in $\mathbf{R}^n$, denoted $\mathcal{S}^e$, is the collection of all its exterior points.

A point $\bar{\mathbf{x}}$ is a ***boundary point*** of a set $\mathcal{S}$ in $\mathbf{R}^n$ if every open ball of radius δ about $\bar{\mathbf{x}}$ encompasses points in $\mathcal{S}$ and in $\mathcal{S}^c$, *i.e.*, it is a point which is neither an interior nor an exterior point of $\mathcal{S}$. The ***boundary*** of a set $\mathcal{S}$ in $\mathbf{R}^n$, $\partial(\mathcal{S})$, is the collection of all its boundary points.

A set $\mathcal{S}$ in $\mathbf{R}^n$ is ***bounded*** if there exists a scalar $M \epsilon \mathbf{R}$ such that $\|\mathbf{x}\| < M$ for all $\mathbf{x} \epsilon \mathcal{S}$. Equivalently, $\mathcal{S}$ in $\mathbf{R}^n$ is bounded if it has a finite ***diameter*** $d(\mathcal{S})$, *i.e.*, $d(\mathcal{S}) = sup\{\|\mathbf{x}-\mathbf{y}\| \mid \mathbf{x}, \ \mathbf{y} \epsilon \mathcal{S}\} < +\infty$. Two additional concepts which utilize the notion of the distance between points in $\mathbf{R}^n$ will at times prove useful. First, for $\mathbf{x}$ a vector in $\mathbf{R}^n$ and $\mathcal{S}$ a subset of $\mathbf{R}^n$, the ***distance between a vector* x *and a set* $\mathcal{S}$** is $d(\mathbf{x}, \mathcal{S}) = \ inf\{\|\mathbf{x}-\mathbf{y}\| \mid \mathbf{y} \epsilon \mathcal{S}\}$. Next, for sets $\mathcal{S}$, $\mathcal{T}$ in $\mathbf{R}^n$, the ***distance between sets* $\mathcal{S}$, $\mathcal{T}$** is $d(\mathcal{S}, \mathcal{T}) = inf\ \{\|\mathbf{x}-\mathbf{y}\| \mid \mathbf{x} \epsilon \mathcal{S}, \ \mathbf{y} \epsilon \mathcal{T}\}$. If $\mathcal{S} \cap \mathcal{T} \neq \phi$, then $d(\mathcal{S}, \mathcal{T}) = 0$.

It is said that a set $\mathcal{S}$ in $\mathbf{R}^n$ has an ***open cover*** if there exists a collection $\{\mathcal{G}_i\}$ of open subsets from $\mathbf{R}^n$ such that $\mathcal{S} \subseteq \cup_i \mathcal{G}_i$. The open cover $\{\mathcal{G}_i\}$ of $\mathcal{S}$ in $\mathbf{R}^n$ is said to contain a ***finite subcover*** if there are finitely many indices $i_1, \ldots, i_m$ for which $\mathcal{S} \subseteq \cup_{j=1}^{m} \mathcal{G}_{i_j}$.

A point $\bar{\mathbf{x}}$ is termed a ***point of closure*** or ***adherent point*** of a set $\mathcal{S}$ in $\mathbf{R}^n$ if every open ball of radius δ centered on $\bar{\mathbf{x}}$ contains at least one point of $\mathcal{S}$, *i.e.*, $B(\bar{\mathbf{x}}, \ \delta) \cap \mathcal{S} \ \neq \phi$. Hence every open ball of radius δ about $\bar{\mathbf{x}}$ meets $\mathcal{S}$. It is important to note that a point of closure of $\mathcal{S}$ need not be a member of $\mathcal{S}$; however, every element within $\mathcal{S}$ is also a point of closure of $\mathcal{S}$. In this regard, all that is required is that there exist points in $\mathcal{S}$ that are "arbitrarily close" to $\bar{\mathbf{x}}$. A subset $\mathcal{S}$ of $\mathbf{R}^n$ is said to be ***closed*** if every point of closure of $\mathcal{S}$ is contained in $\mathcal{S}$, *i.e.*, every point that is arbitrarily close to $\mathcal{S}$ is a member of $\mathcal{S}$. The ***closure*** of a set $\mathcal{S}$ in $\mathbf{R}^n$, denoted $\bar{\mathcal{S}}$, is the set of points of closure of $\mathcal{S}$. Clearly a set $\mathcal{S}$ in $\mathbf{R}^n$ is closed if and only if $\mathcal{S} = \bar{\mathcal{S}}$. A set $\mathcal{S}$ in $\mathbf{R}^n$ has a ***closed cover*** if there exists a collection $\{\mathcal{G}_i\}$ of closed subsets from $\mathbf{R}^n$ such that $\mathcal{S} \subseteq \cup_i \mathcal{G}_i$.

Closely related to the concept of a point of closure of $\mathcal{S}$ is the notion of a ***limit point of a set*** $\mathcal{S}$ in $\mathbf{R}^n$ (also termed a ***cluster point*** or ***point of accumulation***). Specifically, $\bar{\mathbf{x}}$ is a limit point of a set $\mathcal{S}$ if each open ball of radius δ about $\bar{\mathbf{x}}$ contains at least one point of $\mathcal{S}$ different from $\bar{\mathbf{x}}$. Hence $B(\bar{\mathbf{x}}, \delta)$ meets $\mathcal{S} - \{\bar{\mathbf{x}}\}$; points of $\mathcal{S}$ different from $\bar{\mathbf{x}}$ "pile up" at $\bar{\mathbf{x}}$. In this regard, if $\bar{\mathbf{x}}$ is a limit point of a set $\mathcal{S}$ in $\mathbf{R}^n$, then for every $\delta > 0$, the set $\mathcal{S} \cap B(\bar{\mathbf{x}}, \delta)$ is an infinite set, *i.e.*, every neighborhood of $\bar{\mathbf{x}}$ contains infinitely many points of $\mathcal{S}$. So if $\mathcal{S}$ is a finite set in $\mathbf{R}^n$, then it has no limit point. Here also a limit point of $\mathcal{S}$ need not be an element of $\mathcal{S}$. The collection of all limit points of $\mathcal{S}$ in $\mathbf{R}^n$ is called the ***derived set*** of $\mathcal{S}$ (denoted $\mathcal{S}'$). It is made up of all points of $\mathbf{R}^n$ which are arbitrarily close to $\mathcal{S}$. Finally, a point $\mathbf{x}_0 \in \mathcal{S}$ which is not a limit point of $\mathcal{S}$ is termed an ***isolated point*** of $\mathcal{S}$.

On the basis of the preceding discussion we can alternatively conclude that a set $\mathcal{S}$ in $\mathbf{R}^n$ is ***closed*** if it contains each of its limit points or if $\mathcal{S}' \subseteq \mathcal{S}$. Additionally, we can equivalently state that the closure of a set $\mathcal{S}$ in $\mathbf{R}^n$ is $\mathcal{S}$ itself together with its set of limit points, *i.e.*, $\bar{\mathcal{S}} = \mathcal{S} \cup \mathcal{S}'$. Furthermore:

(a) ϕ, a single vector or point, and $\mathbf{R}^n$ are all closed sets.

(b) Any finite union of closed sets in $\mathbf{R}^n$ is closed; any intersection of closed sets in $\mathbf{R}^n$ is closed.

(c) The closure of any set $\mathcal{S}$ in $\mathbf{R}^n$ is the smallest closed set containing $\mathcal{S}$, *i.e.*, $\mathcal{S} \subseteq \bar{\mathcal{S}}$.

(d) A subset $\mathcal{S}$ of $\mathbf{R}^n$ is closed if and only if its complementary set $\mathcal{S}^c$ is open.

(e) For $\mathcal{S}$ a set in $\mathbf{R}^n$, the boundary of $\mathcal{S}$, $\partial(\mathcal{S})$, is the closed set $\bar{\mathcal{S}} \cap \bar{\mathcal{S}}^c$. Thus it must be true that $\bar{\mathcal{S}} = \mathcal{S} \cup \partial(\mathcal{S})$.

(f) A subset $\mathcal{S}$ in $\mathbf{R}^n$ is closed if and only if $\mathcal{S}$ contains its boundary.

(g) The closure of a set $\mathcal{S}$ in $\mathbf{R}^n$ is the intersection of all closed subsets of $\mathbf{R}^n$ which contain $\mathcal{S}$.

Let $\mathcal{S}$ be a subset of $\mathbf{R}^n$. A ***sequence of points or vectors*** in $\mathcal{S}$ is a function whose domain is the set of all positive integers I and whose range appears in $\mathcal{S}$. If the value of the function at $k \epsilon I$ is $\mathbf{x}_k \epsilon \mathcal{S}$, then the range of the sequence will be denoted by $\{\mathbf{x}_k\}$ and simply interpreted as "the sequence of points $\mathbf{x}_1, \mathbf{x}_2, \ldots$ in $\mathcal{S}$." It is important to note that the sequence of points $\{\mathbf{x}_k\}$ mapped into $\mathcal{S}$ is not a subset of $\mathcal{S}$. By deleting certain elements of the sequence $\{\mathbf{x}_k\}$, we obtain the ***subsequence*** $\{\mathbf{x}_k\}_{k \epsilon \mathcal{K}}$, where $\mathcal{K}$ is a subset of positive integers. A subsequence can also be represented as $\{\mathbf{x}_{k_j}\}$, $j=1, 2, \ldots$.

A sequence $\{\mathbf{x}_k\}$ in $\mathbf{R}^n$ ***converges (in norm) to a limit*** $\bar{\mathbf{x}}$ if and only if $\lim_{k \to \infty} \|\mathbf{x}_k - \bar{\mathbf{x}}\| = 0$. (This is alternatively expressed as $\lim_{k \to \infty} \mathbf{x}_k = \bar{\mathbf{x}}$ or $\mathbf{x}_k \to \bar{\mathbf{x}}$ as $k \to \infty$.) That is, $\bar{\mathbf{x}}$ is the limit of $\{\mathbf{x}_k\}$ if for each $\epsilon > 0$ there exists an index value $\bar{k}_\epsilon$ such that $k \geq \bar{k}_\epsilon$ implies $\|\mathbf{x}_k - \bar{\mathbf{x}}\| < \epsilon$. Looked at in another fashion, $\{\mathbf{x}_k\}$ converges to $\bar{\mathbf{x}}$ if for each open ball of radius $\epsilon > 0$ centered on $\bar{\mathbf{x}}$, $B(\bar{\mathbf{x}}, \epsilon)$, there exists a $\bar{k}_\epsilon$ such that $\mathbf{x}_k$ is in $B(\bar{\mathbf{x}}, \epsilon)$ for all $k \geq \bar{k}_\epsilon$. Hence $B(\bar{\mathbf{x}}, \epsilon)$ contains all points of $\{\mathbf{x}_k\}$ from $\mathbf{x}_{\bar{k}_\epsilon}$ on.

A point $\hat{\mathbf{x}}$ in $\mathbf{R}^n$ is a ***limit point (cluster point*** or ***point of accumulation)*** of an infinite sequence $\{\mathbf{x}_k\}$ if and only if there exists an infinite subsequence $\{\mathbf{x}_k\}_{k \epsilon \mathcal{K}}$ of $\{\mathbf{x}_k\}$ which converges to $\hat{\mathbf{x}}$, *i.e.*, there exists an infinite subsequence $\{\mathbf{x}_k\}_{k \epsilon \mathcal{K}})$ such that $\lim_{j \to \infty} \left\|\mathbf{x}_{k_j} - \hat{\mathbf{x}}\right\| = 0$ or $\mathbf{x}_{k_j} \to \hat{\mathbf{x}}$ as $j \to \infty$. More formally, $\hat{\mathbf{x}}$ is a limit point of $\{\mathbf{x}_k)$ if, given $\epsilon > 0$ and given an index value $\bar{k}$, there exists some $k \geq \bar{k}$ such that $\|\mathbf{x}_k - \hat{\mathbf{x}}\| < \epsilon$, *i.e.*, $B(\hat{\mathbf{x}}, \epsilon)$ contains infinitely many terms of $\{\mathbf{x}_k\}$, though not necessarily consecutive ones.

To solidify the distinction between the limit of a sequence and a limit point of a sequence, let us note that:

(a) $\bar{\mathbf{x}}$ is the limit of a sequence $\{\mathbf{x}_k\}$ in $\mathbf{R}^n$ if, given $\epsilon > 0$, all but a finite number of terms of the sequence are within ϵ of $\bar{\mathbf{x}}$.

(b) $\hat{\mathbf{x}}$ is a limit point of $\{\mathbf{x}_k\}$ in $\mathbf{R}^n$ if, given $\epsilon > 0$ and given $\bar{k}$, infinitely many terms of the sequence are within ϵ of $\hat{\mathbf{x}}$.

(c) A sequence $\{\mathbf{x}_k\}$ in $\mathbf{R}^n$ may have a limit but no limit point. However, if a convergent sequence $\{\mathbf{x}_k\}$ in $\mathbf{R}^n$ has infinitely many distinct points, then its limit is a limit point of $\{\mathbf{x}_k\}$. Likewise, $\{\mathbf{x}_k\}$ may posses a limit point but no limit. In fact, if the sequence $\{\mathbf{x}_k\}$ in $\mathbf{R}^n$ has a limit point $\hat{\mathbf{x}}$, then there is a subsequence $\{\mathbf{x}_k\}_{k \epsilon \mathcal{K}}$ of $\{\mathbf{x}_k\}$ which has $\hat{\mathbf{x}}$ as a limit; but this does not necessarily mean that the entire sequence $\{\mathbf{x}_k\}$ converges to $\hat{\mathbf{x}}$. For example:

(c.1) If $\mathbf{x}_k = n =$ constant for all k, then $\{\mathbf{x}_k\}$ converges to the limit n. But since the range of this sequence is but a single point, it follows that the sequence has no limit point.

(c.2) If $\mathbf{x}_k = 1/k$, then clearly the sequence $\{\mathbf{x}_k\}$ converges to a limit of zero, which is also a limit point.

(c.3) If $\mathbf{x}_k = (-1)^k$, then the sequence $\{\mathbf{x}_k\}$ has limit points at ± 1, but has no limit.

(d) A sufficient condition that at least one limit point of an infinite sequence $\{\mathbf{x}_k\}$ in $\mathbf{R}^n$ exists is that $\{\mathbf{x}_k\}$ is ***bounded***, ***i.e.***, there exists a scalar $M \epsilon \mathrm{R}$ such that $\| \mathbf{x}_k \| \leq M$ for all k. In this regard, if an infinite sequence of points or vectors $\{\mathbf{x}_k\}$ in $\mathbf{R}^n$ is bounded and if it has only one limit point, then the sequence converges and has as its limit that single limit point.

The above definition of the convergence of a sequence $\{\mathbf{x}_k\}$ in $\mathbf{R}^n$ explicitly incorporated its actual limit $\bar{\mathbf{x}}$. Clearly this definition is of little value in proving convergence unless one knows what $\bar{\mathbf{x}}$ is. In fact, one frequently has no information about the value of $\bar{\mathbf{x}}$. The theorem which follows, called ***Cauchy's convergence criterion***, enables us to prove that a sequence is convergent even if its limit is unknown. To this end let us state first that a sequence $\{\mathbf{x}_k\}$ in $\mathbf{R}^n$ is a ***Cauchy sequence*** if for each $\epsilon > 0$ there exists an index value $N_{\epsilon/2}$ such that $m,\ n \geq N_{\epsilon/2}$ implies $\|\mathbf{x}_m - \mathbf{x}_n\| < \epsilon$.[5] Secondly, $\mathbf{R}^n$ is ***complete*** in that to every Cauchy sequence $\{\mathbf{x}_k\}$ defined on $\mathbf{R}^n$ there corresponds a point $\bar{\mathbf{x}}$ such that $\lim_{k \to \infty} \mathbf{x}_k = \bar{\mathbf{x}}$. In view of these concepts we may new state the

> **CAUCHY CONVERGENCE CRITERION.** Given that $\mathbf{R}^n$ is complete, a sequence $\{\mathbf{x}_k\}$ in $\mathbf{R}^n$ converges to a limit $\bar{\mathbf{x}}$ if and only if it is a Cauchy sequence, *i.e.*, a necessary and sufficient condition for $\{\mathbf{x}_k\}$ to be convergent in $\mathbf{R}^n$ is that $\|\mathbf{x}_m - \mathbf{x}_n\| \to 0$ as $m,\ n \to \infty$.

Hence every convergent sequence on $\mathbf{R}^n$ is a Cauchy sequence and conversely. The implication of this statement is that if the terms of a sequence approach a limit, then, beyond some point, the distance between pairs of terms diminishes.

To relate the preceding notions of a limit and a limit point of a sequence in $\mathbf{R}^n$ to some of the above topological properties of sets in the same, let us take another look at the point of closure concept. Specifically, a limit point (as well as a limit) of a sequence $\{\mathbf{x}_k\}$ in $\mathbf{R}^n$ is a ***point of closure*** of a set $\mathcal{S}$ in $\mathbf{R}^n$ if $\mathcal{S}$ contains $\{\mathbf{x}_k\}$. Conversely, if $\bar{\mathbf{x}}$ is a point of closure of a set $\mathcal{S}$ in $\mathbf{R}^n$, then there exists a sequence $\{\mathbf{x}_k\}$ in $\mathcal{S}$ (and hence also a subsequence $\{\mathbf{x}_k\}_{k \in \mathcal{K}}$ in $\mathcal{S}$) such that $\bar{\mathbf{x}}$ is a limit point of $\{\mathbf{x}_k\}$ (and thus a limit of $\{\mathbf{x}_k\}_{k \in \mathcal{K}}$). Hence the ***closure*** of a set $\mathcal{S}$, $\bar{\mathcal{S}}$, consists of all limits of convergent sequences $\{\mathbf{x}_k\}$ from $\mathcal{S}$.

---

[5] That is, for $\epsilon > 0$ there exists a positive integer $N_{\epsilon/2}$ such that: $m \geq N_{\epsilon/2}$ implies $\|\mathbf{x}_m - \bar{\mathbf{x}}\| < \epsilon/2$; and $n \geq N_{\epsilon/2}$ implies $\|\mathbf{x}_n - \bar{\mathbf{x}}\| < \epsilon/2$. Hence both $m, n > N_{\epsilon/2}$ imply, via the triangle inequality, that

$$\|\mathbf{x}_m - \mathbf{x}_n\| \leq \|\mathbf{x}_m - \bar{\mathbf{x}}\| + \|\mathbf{x}_n - \bar{\mathbf{x}}\| < \epsilon/2 + \epsilon/2 = \epsilon.$$

In a similar vein we state that a subset $\mathcal{S}$ of $\mathbf{R}^n$ is ***closed*** if and only if every convergent sequence of points $\{\mathbf{x}_k\}$ from $\mathcal{S}$ has a limit in $\mathcal{S}$, *i.e.*, $\mathcal{S}$ is closed if for $\{\mathbf{x}_k\}$ in $\mathcal{S}$, $\lim_{k\to\infty} \mathbf{x}_k = \bar{\mathbf{x}} \in \mathcal{S}$.

A set $\mathcal{S}$ in $\mathbf{R}^n$ is ***bounded*** if every sequence of points $\{\mathbf{x}_k\}$ formed from $\mathcal{S}$ is bounded.

We previously defined the concepts of a closed set and a bounded set. If a set $\mathcal{S}$ in $\mathbf{R}^n$ is both closed and bounded, then it is termed ***compact***. While this is one way to view compactness, there are some useful alternative characterizations. To this end let us first note that a set $\mathcal{S}$ in $\mathbf{R}^n$ has the ***Bolzano-Weierstrass property*** if every infinite subset of $\mathcal{S}$ has at least one limit point in $\mathcal{S}$ (or if every infinite sequence of points $\{\mathbf{x}_k\}$ in $\mathcal{S}$ has at least one limit point in $\mathcal{S}$). Next, a set $\mathcal{S}$ in $\mathbf{R}^n$ possesses the ***Heine-Borel property*** if every open covering of $\mathcal{S}$ has a finite subcovering. Finally, let $\{\mathcal{F}_i\}$ be a class of closed subsets of a nonempty set $\mathcal{S}$ in $\mathbf{R}^n$. Then $\{\mathcal{F}_i\}$ has the ***finite intersection property*** if every finite subclass has a nonempty intersection.[6] In this light, a set $\mathcal{S}$ in $\mathbf{R}^n$ has the finite intersection property if every class $\{\mathcal{F}_i\}$ of closed subsets in $\mathcal{S}$ with the finite intersection property has a nonempty intersection, *i.e.*, $\cap_{j=1}^{m} \mathcal{F}_{i_j} \neq \phi$ for all subsets of indices $i_1, \ldots, i_m$ implies $\cap_i \mathcal{F}_i \neq \phi$.

Based upon the preceding set of definitions, we can now state that a set $\mathcal{S}$ in $\mathbf{R}^n$ is compact if it satisfies any of the following equivalent conditions:

1. $\mathcal{S}$ is closed and bounded.
2. $\mathcal{S}$ has the Bolzano-Weierstrass property.
3. $\mathcal{S}$ has the Heine-Borel property.
4. $\mathcal{S}$ has the finite intersection property.

---

[6] In general, any class of sets has the finite intersection property if every finite subclass has a nonvacuous intersection.

The notion of compactness now enables us to offer the following collection of theorems. We state first

1.3.1. LEBESGUE'S THEOREM. Let $\{\mathcal{F}_1, \mathcal{F}_2, \ldots, \mathcal{F}_m\}$ be a finite closed covering of a compact metric space $\mathcal{S}$ in $\mathbf{R}^n$. Then there exists a number ϵ (called a ***Lebesgue number***) such that for every set $\mathcal{A}$ with $d(\mathcal{A}) \leq \epsilon$, the intersection of the sets $\mathcal{F}_i$ meeting $\mathcal{A}$ is not empty.

Given that a ***decreasing sequence*** of sets $\mathcal{F}_1, \mathcal{F}_2, \ldots$ in $\mathbf{R}^n$ is a sequence such that, for every n, $\mathcal{F}_{n+1} \subseteq \mathcal{F}_n$, we next have

1.3.2. THEOREM. A decreasing sequence of non-null compact sets $\mathcal{F}_1, \mathcal{F}_2, \ldots$ in $\mathbf{R}^n$ has a non-null intersection.

Finally,

1.3.3. THEOREM. Let $\mathcal{F}_1, \mathcal{F}_2, \cdots$ be a decreasing sequence of compact sets in $\mathbf{R}^n$ with $\mathcal{F}(\neq \phi)$ their intersection. Then for each $\in > 0$ there exists an integer n_0 such that $n \geq n_0$ implies $\mathcal{F}_n \subset \bigcup_{\mathbf{x} \in \mathcal{F}} B(\mathbf{x}, \in)$, *i.e.*, there is an n such that every point of $\mathcal{F}_n$ is within $\in$ of $\mathcal{F}$.

We note briefly that:

(a) A closed subset of a compact set $\mathcal{S}$ in $\mathbf{R}^n$ is compact.

(b) The union of a finite number of compact sets in $\mathbf{R}^n$ is compact; the intersection of any number of compact sets in $\mathbf{R}^n$ is compact.

(c) A set $\mathcal{S}$ in $\mathbf{R}^n$ is compact if and only if it is complete and bounded.

(d) Any finite set $\{\mathbf{x}_1, \ldots, \mathbf{x}_k\}$ in $\mathbf{R}^n$ is compact.

(e) If $\mathcal{S}$ in $\mathbf{R}^n$ is the set consisting of a convergent sequence $\{\mathbf{x}_k\}$ and its limit $\bar{\mathbf{x}} = \lim_{k \to \infty} \mathbf{x}_k$, then $\mathcal{S}$ is compact. Conversely, if $\mathcal{S}$ in $\mathbf{R}^n$ is compact, every sequence $\{\mathbf{x}_k\}$ has a convergent subsequence $\{\mathbf{x}_k\}_{k \in \mathcal{K}}$ whose limit belongs to $\mathcal{S}$. (That is, if $\mathcal{S}$

is bounded, then $\{\mathbf{x}_k\}$ is bounded and consequently admits a convergent subsequence. If $\mathcal{S}$ is also closed, then the limit of the subsequence belongs to $\mathcal{S}$.)

(f) If $\mathcal{S}_1$, $\mathcal{S}_2$ are respectively compact and closed subsets of $\mathbf{R}^n$, then $\mathcal{S} = \mathcal{S}_1 + \mathcal{S}_2 = \{\mathbf{x} | \mathbf{x} = \mathbf{x}_1 + \mathbf{x}_2,\ \mathbf{x}_1 \in \mathcal{S}_1,\ \mathbf{x}_2 \in \mathcal{S}_2\}$ is closed. To see this choose $\{\mathbf{x}_k\}$ in $\mathcal{S}$ so that $\lim_{k \to \infty} \mathbf{x}_k = \bar{\mathbf{x}}$. By definition of $\mathcal{S}$, $\mathbf{x}_k = \mathbf{x}_{1k} + \mathbf{x}_{2k}$ for $\mathbf{x}_{1k} \in \mathcal{S}_1$, $\mathbf{x}_{2k} \in \mathcal{S}_2$. Since $\mathcal{S}_1$ is compact, there exists a subsequence $\{\mathbf{x}_{1k}\}_{k \in \mathcal{K}}$ with limit $\bar{\mathbf{x}}_1 \in \mathcal{S}_1$. And since $\mathbf{x}_{1k} + \mathbf{x}_{2k} \to \bar{\mathbf{x}}$ and $\mathbf{x}_{1k} \to \bar{\mathbf{x}}_1$ for $k \in \mathcal{K}$, it follows that $\mathbf{x}_{2k} \to \bar{\mathbf{x}}_2 \in \mathcal{S}_2$ given that $\mathcal{S}_2$ is closed. Hence $\bar{\mathbf{x}} = \bar{\mathbf{x}}_1 + \bar{\mathbf{x}}_2$ with $\bar{\mathbf{x}}_1 \in \mathcal{S}_1$, $\bar{\mathbf{x}}_2 \in \mathcal{S}_2$. Thus $\bar{\mathbf{x}} \in \mathcal{S}$ and thus $\mathcal{S}$ is closed.

A set $\mathcal{S}$ in $\mathbf{R}^n$ is said to be ***locally compact*** if each of its points has a neighborhood with compact closure, *i.e.*, for each $\mathbf{x} \in \mathcal{S}$ there is an open sphere of radius $\delta > 0$ centered on $\mathbf{x}$, $B(\mathbf{x}, \delta)$, such that $\overline{B(\mathbf{x}, \delta)}$ is compact. It should be evident that any compact space is locally compact but not conversely, *e.g.*, $\mathbf{R}^n$ is locally compact but not compact. Conversely, if $\mathcal{S}$ in $\mathbf{R}^n$ is compact, every sequence $\{\mathbf{x}_k\}$ has a convergent subsequence $\{\mathbf{x}_k\}_{k \in K}$ whose limit beolngs to $\mathcal{S}$. (That is, if $\mathcal{S}$ is bounded, then $\{\mathbf{x}_k\}$ is bounded and consequently admits a convergent subsequence. If $\mathcal{S}$ is also closed, then the limit of the subsequence belongs to $\mathcal{S}$.)

1.4 Hyperplanes and Half-Planes (-Spaces)

A set $\mathcal{L}$ in $\mathbf{R}^n$ is termed a ***linear variety*** or ***affine set*** if for vectors $\mathbf{x}_1, \mathbf{x}_2 \in \mathcal{L}$, their ***affine combination***

$$\lambda_1 \mathbf{x}_1 + \lambda_2 \mathbf{x}_2,\ \lambda_1 + \lambda_2 = 1, \text{ and } \lambda_1, \lambda_2 \in \mathbf{R}$$

is also a member of $\mathcal{L}$. Geometrically, this definition implies that the line passing through any two distinct points of $\mathcal{L}$ lies entirely in $\mathcal{L}$. Alternatively, a set $\mathcal{L}$ in $\mathbf{R}^n$ is a linear variety or affine set if and only if $\mathcal{L} = \mathcal{P} + \mathbf{x}_0$, where $\mathcal{P}$ is a subspace of $\mathbf{R}^n$ and $\mathbf{x}_0$ is an arbitrary point of $\mathbf{R}^n$. Here $\mathcal{P} + \mathbf{x}_0 = \{\mathbf{x}+\mathbf{x}_0 | \mathbf{x} \epsilon \mathcal{P} \subset \mathbf{R}^n, \ \mathbf{x}_0 \epsilon \mathbf{R}^n\}$ is termed the ***translation of* P *by* $\mathbf{x}_0$**. In this regard, for $\mathcal{P}$ a subspace of $\mathbf{R}^n$ and $\mathbf{x}_0$ a vector in $\mathbf{R}^n$, choose $\lambda_1, \lambda_2 \epsilon \mathbf{R}$ so that $\lambda_1 + \lambda_2 = 1$. If $\mathbf{x}_1, \ \mathbf{x}_2 \epsilon \mathcal{P} + \mathbf{x}_0$, then both $\mathbf{x}_1 - \mathbf{x}_0$ and $\mathbf{x}_2 - \mathbf{x}_0$ are members of $\mathcal{P}$. Since $\lambda_1(\mathbf{x}_1 - \mathbf{x}_0) + \lambda_2(\mathbf{x}_2 - \mathbf{x}_0) = \ (\lambda_1\mathbf{x}_1 + \lambda_2\mathbf{x}_2) - \mathbf{x}_0 \epsilon \mathcal{P}$, it follows that $\lambda_1\mathbf{x}_1 + \lambda_2\mathbf{x}_2$ is a member of $\mathcal{P} + \mathbf{x}_0$ so that $\mathcal{P} + \mathbf{x}_0$ is a linear variety. Conversely, it can be shown that every linear variety is of the form $\mathcal{P} + \mathbf{x}_0$, *i.e.*, if $\mathcal{L}$ is a linear variety, then there exists a subspace $\mathcal{P}$ and a point $\mathbf{x}_0 \epsilon \mathbf{R}^n$ such that $\mathcal{L} = \mathcal{P} + \mathbf{x}_0$. Based upon this discussion, it is evident that an affine set $\mathcal{L}$ in $\mathbf{R}^n$ is simply the translate (by a vector $\mathbf{x}_0$) of a subspace $\mathcal{L}$ in $\mathbf{R}^n$. An affine set $\mathcal{L}$ is said to be ***parallel*** to an affine set $\mathcal{P}$ if $\mathcal{L} = \mathcal{P} + \mathbf{x}$.

The ***dimension of an affine set*** $\mathcal{L}$ in $\mathbf{R}^n$ is the dimension of the subspace $\mathcal{P}$ of $\mathbf{R}^n$ obtained by translating $\mathcal{L}$ so that it contains the origin $\{\mathbf{0}\}$. That is, for some $\mathbf{x}_0 \epsilon \mathcal{L}$, a translation by $-\mathbf{x}_0$ renders $\mathcal{P} = \mathcal{L} - \mathbf{x}_0 = \{\mathbf{x} - \mathbf{x}_0 \,|\, \mathbf{x} \epsilon \mathcal{L}\}$, a unique linear subspace which runs parallel to $\mathcal{L}$ through the origin of $\mathbf{R}^n$. Then $dim(\mathcal{L}) = dim(\mathcal{P})$, *i.e.*, the dimension of $\mathcal{L}$ is the dimension of the subspace $\mathcal{P}$ parallel to it; it is the dimension of the subspace of which it is a translate. If $\mathcal{L} = \phi$, then $dim(\mathcal{L}) = -1$. Additionally, affine sets of dimension 0, 1, and 2 are points, lines, and planes respectively.

As we shall now demonstrate, for a vector $\mathbf{C}(\neq \mathbf{0}) \epsilon \mathbf{R}^n$ and a scalar $\alpha \epsilon \mathbf{R}$, the ***hyperplane***

$$\mathcal{H} = \{\mathbf{x} | \mathbf{C}'\mathbf{x} = \alpha, \ \mathbf{x} \epsilon \mathbf{R}^n\}$$

is an $(n-1)$-dimensional linear variety or affine subset of $\mathbf{R}^n$. To see this let $\mathbf{x}_1 \epsilon \mathcal{H}$. Translating by $-\mathbf{x}_1$ yields the linear subspace $\mathcal{M} = \mathcal{H} - \mathbf{x}_1$ (Figure 1.1).

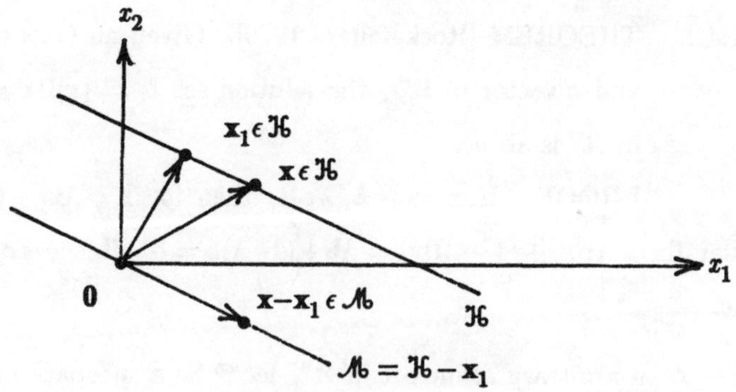

Figure 1.1

What about the structure of $\mathcal{M}$? Since $\mathbf{C}'\mathbf{x}_1 = \alpha$, it follows that $\mathbf{C}'\mathbf{x} - \mathbf{C}'\mathbf{x}_1 = \mathbf{C}'(\mathbf{x}-\mathbf{x}_1) = \mathbf{C}'\hat{\mathbf{x}} = \alpha - \alpha = 0$. Hence $\hat{\mathbf{x}}(=\mathbf{x}-\mathbf{x}_1) \epsilon \mathcal{M}$ so that $\mathcal{M}$ must be the set of all vectors orthogonal to $\mathbf{C}$. Likewise, $\mathbf{C}$ must be (mutually) orthogonal to the subspace $\mathcal{M}$ so that $\mathbf{C}$ amounts to the one-dimensional subspace $\mathcal{M}^\perp = \{\mathbf{C}\}$. And since $\mathcal{M}$ and $\{\mathbf{C}\}$ taken together span $\mathbf{R}^n$, it follows that $dim(\mathcal{M}) = dim(\mathcal{H}) = n - dim(\{\mathbf{C}\}) = n-1$.

Any hyperplane $\mathcal{H} = \{\mathbf{x} | \mathbf{C}'\mathbf{x} = \alpha,\ \mathbf{x} \epsilon \mathbf{R}^n\}$ divides $\mathbf{R}^n$ into the two ***closed half-planes (−spaces)***

$$[\mathcal{H}^+] = \{\mathbf{x} | \mathbf{C}'\mathbf{x} \geq \alpha,\ \mathbf{x} \epsilon \mathbf{R}^n\},$$
$$[\mathcal{H}^-] = \{\mathbf{x} | \mathbf{C}'\mathbf{x} \leq \alpha,\ \mathbf{x} \epsilon \mathbf{R}^n\}.$$

If the boundary line $\mathbf{C}'\mathbf{x} = \alpha$ is excluded from the preceding two sets, then we have the two ***open half-planes (−spaces)***

$$(\mathcal{H}^+) = \{\mathbf{x} | \mathbf{C}'\mathbf{x} > \alpha,\ \mathbf{x} \epsilon \mathbf{R}^n\},$$
$$(\mathcal{H}^-) = \{\mathbf{x} | \mathbf{C}'\mathbf{x} < \alpha,\ \mathbf{x} \epsilon \mathbf{R}^n\}.$$

Clearly $[\mathcal{H}^+]$, $[\mathcal{H}^-]$ are closed sets while $(\mathcal{H}^+)$, $(\mathcal{H}^-)$ are open sets.

The following theorem informs us that every affine subset $\mathcal{L}$ of $\mathbf{R}^n$ corresponds to the solution set of a system of m simultaneous linear equations in n unknowns. Specifically, we state

1.4.1. THEOREM [Rockafellar, 1970]. Given an $(m \times n)$ matrix $\mathbf{B}$ and a vector $\mathbf{b} \in \mathbf{R}^m$, the solution set $\mathcal{L} = \{\mathbf{x} | \mathbf{B}\mathbf{x} = \mathbf{b}, \ \mathbf{x} \in \mathbf{R}^n\}$ in $\mathbf{R}^n$ is affine.

PROOF. If $\mathbf{x}_1, \mathbf{x}_2 \in \mathcal{L}$, $\lambda \in \mathbf{R}$, then for $\mathbf{x} = \lambda \mathbf{x}_1 + (1-\lambda)\mathbf{x}_2$, it follows that $\mathbf{B}\mathbf{x} = \lambda \mathbf{B}\mathbf{x}_1 + (1-\lambda)\mathbf{B}\mathbf{x}_2 = \lambda \mathbf{b} + (1-\lambda)\mathbf{b} = \mathbf{b}$. Hence $\mathbf{x} \in \mathcal{L}$ and thus $\mathcal{L}$ is affine.

For $\mathcal{L}$ an arbitrary affine set in $\mathbf{R}^n$, let $\mathcal{P}$ be a subspace parallel to $\mathcal{L}$ and let $\beta_1, \ldots, \beta_m$ represent a basis for $\mathcal{P}^\perp$. Then $\mathcal{P} = (\mathcal{P}^\perp)^\perp = \{\mathbf{v} \mid \beta_i \mathbf{v} = 0, \ i=1, \ldots, m, \ \mathbf{v} \in \mathbf{R}^n\} = \{\mathbf{v} \mid \mathbf{B}\mathbf{v} = \mathbf{0}, \ \mathbf{v} \in \mathbf{R}^n\}$, where β_i is the ith row of $\mathbf{B}$. Since $\mathcal{L}$ is parallel to $\mathcal{P}$, let us write, for some $\mathbf{y} \in \mathbf{R}^n$, $\mathcal{L} = \mathcal{P} + \mathbf{y} = \{\mathbf{v} \mid \mathbf{B}\mathbf{v} = \mathbf{0}, \ \mathbf{v} \in \mathbf{R}^n\} + \mathbf{y} = \{\mathbf{v} + \mathbf{y} \mid \mathbf{B}\mathbf{v} = \mathbf{0}, \mathbf{v} \in \mathbf{R}^n\} = \{\mathbf{x} \mid \mathbf{B}(\mathbf{x}-\mathbf{y}) = \mathbf{0}, \mathbf{x} \in \mathbf{R}^n\}$ $= \{\mathbf{x} \mid \mathbf{B}\mathbf{x} = \mathbf{b}, \ \mathbf{x} \in \mathbf{R}^n\}$, where $\mathbf{b} = \mathbf{B}\mathbf{y}$. Q.E.D.

If in theorem 1.4.1 the affine set $\mathcal{L}$ is rewritten as $\mathcal{L} = \{\mathbf{x} \mid \beta_i \mathbf{x} = b_i, \ i=1, \ldots, m, \ \mathbf{x} \in \mathbf{R}^n\} = \cap_{i=1}^m \mathcal{H}_i$, where β_i is the ith row of $\mathbf{B}$, b_i is the ith component of $\mathbf{b}$, and $\mathcal{H}_i = \{\mathbf{x} \mid \beta_i \mathbf{x} = b_i, \ \mathbf{x} \in \mathbf{R}^m\}, \ i = 1, \ldots, m,$ is a hyperplane, then it is evident that

1.4.2. COROLLARY. Every affine subset in $\mathbf{R}^n$ is expressible as the intersection of a finite collection of hyperplanes from $\mathbf{R}^n$.

Next, given any set $\mathcal{S}$ in $\mathbf{R}^n$, the ***affine hull*** (or the ***affine subspace spanned by $\mathcal{S}$***) of $\mathcal{S}$, denoted $aff(\mathcal{S})$, is the set of all affine combinations of points from $\mathcal{S}$ or

$$aff(\mathcal{S}) = \left\{ \mathbf{x} \,\middle|\, \mathbf{x} = \sum_{i=1}^{m} \lambda_i \mathbf{x}_i, \ \mathbf{x}_i \in \mathcal{S}, \ \sum_{i=1}^{m} \lambda_i = 1, \ \lambda_i \in R \text{ for all } i \right\}.$$

Here $aff(\mathcal{S})$ is unique and can be thought of as the intersection of all affine sets $\mathcal{A}_i$ in $\mathbf{R}^n$ which contain $\mathcal{S}$, *i.e.*, $aff(\mathcal{S}) = \cap_i \mathcal{A}_i$, where $\mathcal{S} \subset \mathcal{A}_i$ for all i. In this regard, an affine set $\mathcal{T}$ in $\mathbf{R}^n$ is the smallest affine set containing $\mathcal{S}$ in $\mathbf{R}^n$ if and only if $\mathcal{T} = aff(\mathcal{S})$. Then $dim(\mathcal{T}) = dim(aff(\mathcal{S}))$.

Let $\{\mathbf{x}_0, \mathbf{x}_1, \ldots, \mathbf{x}_k\}$ represent a finite set of $k+1$ vectors in $\mathbf{R}^n$. The affine hull of $\{\mathbf{x}_0, \mathbf{x}_1, \ldots, \mathbf{x}_k\}$, denoted $aff\{\mathbf{x}_0, \mathbf{x}_1, \ldots, \mathbf{x}_k\}$, is the set of all affine combinations of its members. The set $\{\mathbf{x}_0, \mathbf{x}_1, \ldots, \mathbf{x}_k\}$ is ***affinely independent*** if $aff\{\mathbf{x}_0, \mathbf{x}_1, \ldots, \mathbf{x}_k\} = aff\{\mathbf{0}, \mathbf{x}_1-\mathbf{x}_0, \ldots, \mathbf{x}_k-\mathbf{x}_0\}+\mathbf{x}_0 = \mathcal{L}+\mathbf{x}_0$ has dimension k. Since $\mathcal{L}$ is the smallest subspace spanned by the ***translated set*** $\{\mathbf{x}_1-\mathbf{x}_0, \ldots, \mathbf{x}_k-\mathbf{x}_0\}$, $dim(aff\{\mathbf{x}_0, \mathbf{x}_1, \ldots, \mathbf{x}_k\}) = dim(\mathcal{L}) = k$ if the translated set of vectors is linearly independent. Hence a set of $k+1$ vectors $\{\mathbf{x}_0, \mathbf{x}_1, \ldots, \mathbf{x}_k\}$ in $\mathbf{R}^n$ is said to be affinely independent if and only if the translated set $\{\mathbf{x}_1-\mathbf{x}_0, \ldots, \mathbf{x}_k-\mathbf{x}_0\}$ is linearly independent. Here the set of vectors $\{\mathbf{x}_0, \mathbf{x}_1, \ldots, \mathbf{x}_k\}$ in $\mathbf{R}^n$ is sometimes called a ***maximal finite subset*** of $\mathcal{L}$ if the translated set $\{\mathbf{x}_1-\mathbf{x}_0, \ldots, \mathbf{x}_k-\mathbf{x}_0\}$ is linearly independent. Note: "maximal" means that the augmented translated set $\{\mathbf{x}_1-\mathbf{x}_0, \ldots, \mathbf{x}_{k+1}-\mathbf{x}_0\}$ becomes linearly dependent when $\mathbf{x}_{k+1}$ is included in the initial set $\{\mathbf{x}_0, \mathbf{x}_1, \ldots, \mathbf{x}_k\}$.

If $\mathcal{P} = aff\{\mathbf{x}_0, \mathbf{x}_1, \ldots, \mathbf{x}_k\}$, the vectors in the subspace $\mathcal{L}$ parallel to $\mathcal{P}$ are expressible as linear combinations of the translated set of vectors, *i.e.*, if $\mathbf{x} \in \mathcal{L}$, then

$$\begin{aligned}\mathbf{x} &= \sum_{i=1}^{k} \lambda_i(\mathbf{x}_i-\mathbf{x}_0) + \mathbf{x}_0 \\ &= \lambda_0\mathbf{x}_0 + \sum_{i=1}^{k} \lambda_i\mathbf{x}_i, \quad \lambda_0 + \sum_{i=1}^{k} \lambda_i = 1.\end{aligned}$$

If the set of vectors $\{\mathbf{x}_0, \mathbf{x}_1, \ldots, \mathbf{x}_k\}$ is affinely independent, then λ_0 and the λ_i's, $i=1, \ldots, k$, are uniquely determined.

Additional features of affine sets or linear varieties are:

(a) The translation of an affine set is affine.

(b) ϕ, $\mathbf{R}^n$, and the singleton $\{\mathbf{x}\}$ are all affine sets.

(c) The intersection of an arbitrary collection of affine sets is affine.

(d) Each nonempty affine set $\mathcal{P}$ is parallel to a unique affine subspace $\mathcal{L}$.

(e) If $\mathbf{x}_1, \mathbf{x}_2 \in \mathcal{P} = \mathcal{L} + \mathbf{x}$, then $\mathbf{x}_1 - \mathbf{x}$, $\mathbf{x}_2 - \mathbf{x}$ are elements of $\mathcal{L}$; and if $\mathbf{x}_1, \mathbf{x}_2 \in \mathcal{L} - \mathbf{x}$, then $\mathbf{x}_1 + \mathbf{x}$, $\mathbf{x}_2 + \mathbf{x}$ are members of $\mathcal{L}$.

(f) A subset $\mathcal{L}$ of $\mathbf{R}^n$ is affine if and only if $\mathcal{L} - \mathbf{x}_0$ is a subspace of $\mathbf{R}^n$ for every vector $\mathbf{x}_0 \in \mathcal{L}$.

(g) The dimension of any subset $\mathcal{S}$ of $\mathbf{R}^n$ equals the dimension of its affine hull $aff(\mathcal{S})$.

(h) The set of vectors $\{\mathbf{x}_0, \mathbf{x}_1, \ldots, \mathbf{x}_k\}$ in $\mathbf{R}^n$ is affinely independent if the unique solution of $\Sigma_{i=0}^k \lambda_i \mathbf{x}_i = \mathbf{0}$, $\Sigma_{i=0}^k \lambda_i = 0$, $\lambda_i \in \mathbf{R}$, is $\lambda_i = 0$ for all i.

(i) Linear independence implies affine independence but not conversely.

(j) The maximum number of affinely independent vectors in $\mathbf{R}^n$ is $n+1$ (consisting of n linearly independent vectors and the null vector).

(k) For $\mathbf{B}$ of order $(m \times n)$, if $\mathbf{Bx} = \mathbf{b}$ is consistent, the maximum number of affinely independent solutions of $\mathbf{Bx} = \mathbf{b}$ is $n+1-\rho(\mathbf{B})$.

(l) The affine hull of a finite set of points $\{\mathbf{x}_1, \ldots, \mathbf{x}_k\}$ in $\mathbf{R}^n$ is a subset of its linear hull, *i.e.*, $aff\{\mathbf{x}_1, \ldots, \mathbf{x}_k\} \subseteq \ell\{\mathbf{x}_1, \ldots, \mathbf{x}_k\}$.

(m) The ***translate*** of a set $\mathcal{S}$ in $\mathbf{R}^n$ to a point $\bar{\mathbf{x}}$ is the set $\{\mathbf{y} \mid \mathbf{y} = \mathbf{x} + \bar{\mathbf{x}},\ \mathbf{x} \in \mathcal{S}\}$. In this light, for $\{\mathbf{x}_1, \ldots, \mathbf{x}_k\}$ a finite set of points in $\mathbf{R}^n$, let

$$\begin{aligned}\mathbf{x} &= \Sigma_{i=1}^k \lambda_i \mathbf{x}_i = \lambda_1 \mathbf{x}_1 + \Sigma_{i=2}^k \lambda_i \mathbf{x}_i \\ &= \left(1 - \Sigma_{i=2}^k \lambda_i\right)\mathbf{x}_1 + \Sigma_{i=2}^k \lambda_i \mathbf{x}_i = \mathbf{x}_1 + \Sigma_{i=2}^k \lambda_i (\mathbf{x}_i - \mathbf{x}_1).\end{aligned}$$

Hence we may write $aff\{\mathbf{x}_1, \ldots, \mathbf{x}_k\} = \mathbf{x}_1 + \ell\{\mathbf{x}_2 - \mathbf{x}_1, \ldots, \mathbf{x}_k - \mathbf{x}_1\}$, *i.e.*, the affine hull of $\{\mathbf{x}_1, \ldots, \mathbf{x}_k\}$ is the translate of the linear hull of $\{\mathbf{x}_2 - \mathbf{x}_1, \ldots, \mathbf{x}_k - \mathbf{x}_1\}$ to $\mathbf{x}_1$.

(n) Given the simultaneous linear system $\mathbf{Ax} = \mathbf{C}$, $\mathbf{x} \in \mathbf{R}^n$, let $\rho(\mathbf{A}) = \rho[\mathbf{A}, \mathbf{C}] = k$. Then the solution set of $\mathbf{Ax} = \mathbf{C}$ is an $(n-k)$ - dimensional affine set in $\mathbf{R}^n$

1.5. Exercises

1. For $\mathbf{x}' = (1,-1,3)$ and $\mathbf{y}' = (4,1,1)$, find: $\mathbf{x}\cdot\mathbf{y}$, $\|\mathbf{x}\|$, $\|\mathbf{y}\|$, $\|\mathbf{x}-\mathbf{y}\|$, the direction of $\mathbf{x}$, and the cosine of the angle θ between $\mathbf{x}, \mathbf{y}$. Are $\mathbf{x}, \mathbf{y}$ orthogonal?

2. Express $\mathbf{x}' = (1,4)$ as a linear combination of the vectors $\mathbf{x}_1' = (2,3)$, $\mathbf{x}_2' = (1,-1)$. Are $\mathbf{x}_1, \mathbf{x}_2$ linearly dependent? Do $\mathbf{x}_1, \mathbf{x}_2$ span $\mathbf{R}^2$? Do they form a basis for $\mathbf{R}^2$?

3. Are the vectors $\mathbf{x}_1' = (1,3,1), \mathbf{x}_2' = (-1,1,0)$, and $\mathbf{x}_3' = (0,1,4)$ linearly independent?

4. Demonstrate that the vectors $\mathbf{x}_1' = (1,5)$, $\mathbf{x}_2' - (3,1)$, and $\mathbf{x}_3' = (1,-1)$ are linearly dependent.

5. Do the vectors $\mathbf{x}_1' = (3,6)$, $\mathbf{x}_2' = \left(\frac{1}{2}, 2\right)$ provide a basis for $\mathbf{R}^2$?

6. Find the value of α which renders the vectors $\mathbf{x}_1' = (2,4)$, $\mathbf{x}_2' = (\alpha, 1)$ linearly dependent. Is it unique?

7. Determine the rank of the matrix

$$\mathbf{A} = \begin{bmatrix} 1 & 0 & 3 \\ -3 & 1 & 4 \\ 6 & 0 & 18 \end{bmatrix}.$$

How may we characterize the columns of $\mathbf{A}$?

8. Determine if the following systems are consistent. Which one (if any) has a unique solution?

$$\begin{aligned} \tfrac{2}{3}x_1 + \tfrac{1}{3}x_2 &= 3 \\ x_1 + \tfrac{1}{2}x_2 &= 1 \end{aligned} \qquad \begin{aligned} \tfrac{2}{3}x_1 + x_2 &= 2 \\ \tfrac{1}{3}x_1 + \tfrac{1}{2}x_2 &= 1 \end{aligned} \qquad \begin{aligned} \tfrac{2}{3}x_1 + x_2 &= 2 \\ x_1 - x_2 &= 1 \end{aligned}$$

9. Do the following homogeneous systems possess trivial solutions?

$$\begin{aligned} x_1 - 4x_2 &= 0 \\ 3x_1 + 6x_2 &= 0 \end{aligned} \qquad \begin{aligned} 2x_1 - x_2 &= 0 \\ \tfrac{1}{2}x_1 - \tfrac{1}{4}x_2 &= 0 \end{aligned}$$

10. Does the system

$$\begin{aligned} x_1 + 6x_2 &= c_1 \\ \tfrac{1}{2}x_1 - x_2 &= c_2 \end{aligned}$$

have a unique solution for each set of c_1, c_2 values?

(Hint: apply theorem 1.2.4.)

11. Verify that the system

$$\begin{aligned} \tfrac{2}{3}x_1 + \tfrac{1}{3}x_2 &= 3 \\ x_1 + \tfrac{1}{2}x_2 &= 1 \end{aligned}$$

is inconsistent using theorem 1.2.6.

12. Verify that $\mathbf{0} \in \mathbf{R}^3$ as well as the plane $\mathcal{S} = \{\mathbf{u}' = (u_1, u_2, u_3) \mid u_2 = 0\} \in \mathbf{R}^3$ passing through the origin represent linear subspaces.

13. Show that the set of all linear combinations of the columns of the identity matrix $\mathbf{I}_n$ is a linear subspace of $\mathbf{R}^n$. Does this set equal $\mathbf{R}^n$?

14. Verify that $\mathcal{S} = \{\mathbf{x}' = (x_1, x_2, x_3) \mid x_1 - x_2 = 0\} \in \mathbf{R}^n$ is a linear subspace. Do the vectors $\mathbf{x}_1' = (1,1,0)$, $\mathbf{x}_2' = (0,1,0)$ span $\mathcal{S}$?

(Hint: let $x_1 = x_2$.)

15. For $\mathcal{S} = \{\mathbf{x}_1' = (1,0,0),\ \mathbf{x}_2' = (0,1,0)\}$ in $\mathbf{R}^3$, demonstrate that the linear hull of $\mathcal{S}$ is the x_1, x_2-plane. In addition, verify that the affine hull of $\mathcal{S}$ is a straight line passing through both $\mathbf{x}_1, \mathbf{x}_2$.

16. For

$$\mathbf{A} = \begin{bmatrix} -1 & 1 & 0 \\ 2 & 2 & 1 \end{bmatrix}, \quad \mathbf{C} = \begin{bmatrix} 1 \\ 0 \\ 2 \end{bmatrix},$$

does the system $\mathbf{Ax} \leqq \mathbf{0}$, $\mathbf{C}'\mathbf{x} > 0$ have a solution $\mathbf{x} \epsilon \mathbf{R}^3$?

17. The ***unit column vector*** $\mathbf{e}_i$ has a 1 in its ith position and 0's elsewhere. It is said that the unit column vectors $\mathbf{e}_i, i=1,\ldots,n$, form a "natural basis" for $\mathbf{R}^n$. Verify.

18. Prove property (f) at the end of section 1.4.

19. Are the following two sets in $\mathbf{R}^2$ affine?

$$\mathcal{S}_1 = \{\mathbf{x}' = (x_1, x_2) \mid x_1 + x_2 = 1\};$$
$$\mathcal{S}_2 = \{\mathbf{x}' = (x_1, x_2) \mid x_2 \geq 0\}.$$

20. Let $\mathcal{S} = \{\mathbf{x}' = (x_1, x_2) \mid 3x_1 + x_2 = 2;\ x_1, x_2 \geq 0\} \epsilon \mathbf{R}^2$. Find $aff(\mathcal{S})$. Does $\mathcal{S} \subseteq aff(\mathcal{S})$? For $\mathbf{x}_0' = (\frac{1}{3}, 1)$ find the subspace $\mathcal{S} - \mathbf{x}_0$.
(Hint: if $\mathbf{x} \epsilon \mathcal{S} - \mathbf{x}_0$, then $\mathbf{x} + \mathbf{x}_0 \epsilon \mathcal{S}$.) Also, find $aff(\mathcal{S}) - \mathbf{x}_0$.

21. Verify that the vectors $\mathbf{x}_0' = (1,1), \mathbf{x}_1' = (3,-1)$, and $\mathbf{x}_2' = (0,1)$ are affinely independent.

22. Are the following sets open or closed? Verify.

$$\mathcal{S}_1 = \mathbf{R}^n;$$
$$\mathcal{S}_2 = (a, b] \in \mathbf{R};$$
$$\mathcal{S}_3 = \{\mathbf{x} \mid \|\mathbf{x}\|^2 \leq 1 \text{ for } x_2 \geq 0;\ \|\mathbf{x}\|^2 < 1 \text{ for } x_2 < 0\} \in \mathbf{R}^2;$$
$$\mathcal{S}_4 = \{\mathbf{x} \mid 3x_1 - 4x_2 - 6 = 0\} \in \mathbf{R}^2.$$

23. Demonstrate that $\mathcal{S} = [\alpha, \beta] \in \mathbf{R}$ is closed.

24. Do the following sets in **R** possess limit or accumulation points?

$$\mathcal{S}_1 = \{1, 2, 3\};$$
$$\mathcal{S}_2 = \{x \mid 1 < x < 2\};$$
$$\mathcal{S}_3 = \{x \mid 1 \leq x \leq 2\};$$
$$\mathcal{S}_4 = \left\{1, \tfrac{1}{2}, \tfrac{1}{3}, \ldots, \tfrac{1}{n}, \ldots\right\}.$$

What are the derived sets associated with $\mathcal{S}_1, \ldots, \mathcal{S}_4$?

What is the closure of each of the above sets?

25. Let $\mathcal{S}$ be a set of points in $\mathbf{R}^n$. Prove that $\mathcal{S} \subset \overline{\mathcal{S}}$.

What if $\mathcal{S}$ were termed a finite set?

26. Are the following sets bounded?

$$\mathcal{S}_1 = \{x \mid x \geq 0\} \in \mathbf{R};$$
$$\mathcal{S}_2 = \{\mathbf{x} \mid \|\mathbf{x}\|^2 < 1\} \in \mathbf{R}^n;$$
$$\mathcal{S}_3 = \{x \mid x = \tfrac{1}{n},\ n=1, 2, 3, \ldots\} \in \mathbf{R};$$
$$\mathcal{S}_4 = \{x \mid a \leq x \leq b\} \in \mathbf{R}.$$

Are $\mathcal{S}_1, \ldots, \mathcal{S}_4$ compact?

27. Demonstrate that if $\mathcal{A}, \mathcal{B}$ are compact sets in $\mathbf{R}^n$, then there are points $a \in \mathcal{A}$, $b \in \mathcal{B}$ such $\|a - b\| = d(\mathcal{A}, \mathcal{B})$.

28. Comment on the statement: "If a set is not open, then it must be closed." Is the set $\mathscr{S} = \{x \mid 0 < x \leq 1\} \in \mathbf{R}$ open? Closed? Verify that any finite set is closed. Show that if a set $\mathscr{S} \in \mathbf{R}$ is open, then each of its points is a point of accumulation of $\mathscr{S}$.

29. Show that $\mathscr{S}^o$ for $\mathscr{S}$ in $\mathbf{R}^n$ is open.

30. Determine all the limit or accumulation points for the following sets in $\mathbf{R}^2$:

$$\mathscr{S}_1 = \{\mathbf{x}' = (x_1, x_2) \mid x_1^2 - x_2^2 < 1\};$$
$$\mathscr{S}_2 = \{\mathbf{x}' = (x_1, x_2) \mid x_1 > 0\};$$
$$\mathscr{S}_3 = \{\mathbf{x}' = (x_1, x_2) \mid x_2 \geq 0\}.$$

31. Can a sequence have two different limit points? An infinite number of limit points? No limit points? What if the sequence is bounded?

32. Show that a bounded sequence with exactly one limit point is convergent. Verify that if $\{x_k\}$ converges to x_0, then every subsequence of $\{x_k\}$ converges to x_0.

33. For each positive integer k, let $\{x_k\} = \begin{cases} 1 & \text{if} \quad k \text{ is odd;} \\ 0 & \text{if} \quad k \text{ is even.} \end{cases}$

Is this a bounded sequence? Is it convergent? Does this sequence contain a convergent subsequence? (Hint: while boundedness alone is not sufficient to guarantee convergence of a sequence, it nevertheless enables us to state the ***Bolzano-Weierstrass theorem for sequences***: every bounded sequence $\{x_k\}$ in $\mathbf{R}$ has a convergent subsequence.)

34. Prove that if two subsequences of a given sequence converges to distinct limits, then the sequence does not converge.

35. Let $\{x_k\}$ be a convergent sequence with $lim\ x_k = A$. For $c \in \mathbf{R}$, let $y_k = x_k + c$ for each $k > 0$. Verify that $lim\ y_k = A + c$.

CHAPTER 2
CONVEX SETS IN R^n

2.1. Convex Sets

Let $\mathbf{x}_1$, $\mathbf{x}_2$ be points or vectors in $\mathbf{R}^n$ and λ a scalar in $\mathbf{R}$. The ***closed line segment*** joining $\mathbf{x}_1$ and $\mathbf{x}_2$ is the set $[\mathbf{x}_1, \mathbf{x}_2] = \{\mathbf{x} | \mathbf{x} = \lambda \mathbf{x}_1 + (1-\lambda)\mathbf{x}_2, 0 \leq \lambda \leq 1\}$.

A set $\mathcal{S}$ in $\mathbf{R}^n$ is said to be ***convex*** if for points $\mathbf{x}_1, \mathbf{x}_2 \in \mathcal{S}$, their ***convex combination*** (or ***internal average***)

$$\mathbf{x} = \lambda_1 \mathbf{x}_1 + \lambda_2 \mathbf{x}_2,\ \lambda_1 + \lambda_2 = 1, \text{and } 0 \leq \lambda_1, \lambda_2 \in \mathbf{R}$$

is also a member of $\mathcal{S}$. Geometrically, this definition implies that the closed line segment joining $\mathbf{x}_1$ and $\mathbf{x}_2$ lies entirely in $\mathcal{S}$, *i.e.*, $[\mathbf{x}_1, \mathbf{x}_2] \in \mathcal{S}$ (see Figure 2.1). (A convex set differs somewhat from an affine set in that, for the latter, the entire line passing through $\mathbf{x}_1$ and $\mathbf{x}_2$ lies wholly within the set.) A set $\mathcal{S}$ in $\mathbf{R}^n$ containing only a single element is taken to be (trivially) convex.

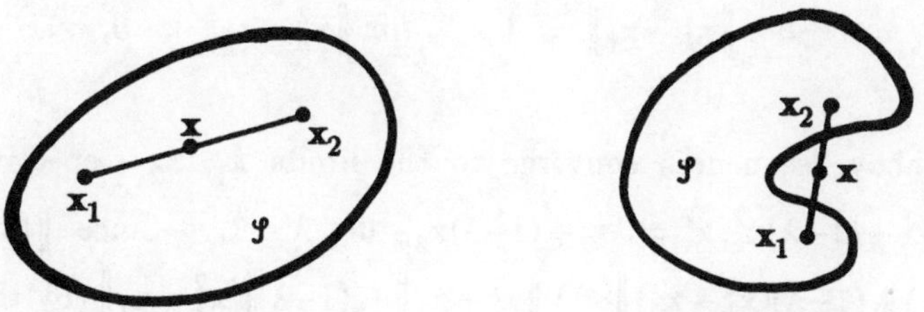

$\mathcal{S}$ convex
a.

$\mathcal{S}$ not convex
b.

Figure 2.1

An important property of a given collection of convex sets is provided by

2.1.1. THEOREM. Let $\{\mathcal{S}_i\}$ be a family of convex sets in $\mathbf{R}^n$. Then their intersection $\cap_i \mathcal{S}_i$ is a convex set.

PROOF. Let $\mathbf{x}_1, \mathbf{x}_2 \in \cap_i \mathcal{S}_i$. Then $\mathbf{x}_1, \mathbf{x}_2 \in \mathcal{S}_i$ for all i. Since each $\mathcal{S}_i$ is convex, $[\mathbf{x}_1, \mathbf{x}_2] \in \mathcal{S}_i$ for all i. Thus $[\mathbf{x}_1, \mathbf{x}_2] \in \cap_i \mathcal{S}_i$ so that the intersection is convex. Q.E.D.

Also of particular interest is

2.1.2. THEOREM. Let $\mathcal{S}$ be a convex set in $\mathbf{R}^n$. Then the closure $\overline{\mathcal{S}}$ of $\mathcal{S}$ is also a convex set.

PROOF. Let $\{\mathcal{S}_i\}$ be a family of closed convex subsets of $\mathbf{R}^n$ which contain $\mathcal{S}$. Then by theorem 2.1.1 and the intersection property of closed sets, $\cap_i \mathcal{S}_i$ is closed and convex. Since $\overline{\mathcal{S}}$ is the intersection of all closed (and here also convex) subsets of $\mathbf{R}^n$ which contain $\mathcal{S}$, it follows that $\overline{\mathcal{S}}$ must also be convex. Q.E.D.

It is instructive to examine the convexity of $\overline{\mathcal{S}}$ from a slightly different perspective, namely, one which is based upon the notion that $\overline{\mathcal{S}}$ contains all limits of convergent sequences from $\mathcal{S}$. Let us choose sequences $\{\mathbf{x}_k^1\}$, $\{\mathbf{x}_k^2\}$ in $\mathcal{S}$ so that

$$\lim_{k\to\infty} \left\| \mathbf{x}_k^1 - \mathbf{x}_1 \right\| = 0, \quad \lim_{k\to\infty} \left\| \mathbf{x}_k^2 - \mathbf{x}_2 \right\| = 0,$$

i.e., the above sequences converge to the limits $\mathbf{x}_1$, $\mathbf{x}_2$ respectively. Let $\mathbf{x}_k^c = \lambda \mathbf{x}_k^1 + (1-\lambda)\mathbf{x}_k^2, \mathrm{x}^c = \lambda \mathbf{x}_1 + (1-\lambda)\mathbf{x}_2$, $0 \leq \lambda \leq 1$. Since $\left\| (\mathbf{x}_k^c - \mathbf{x}^c \right\| = \left\| \lambda(\mathbf{x}_k^1 - \mathbf{x}_1) + (1-\lambda)(\mathbf{x}_k^2 - \mathbf{x}_2) \right\| \leq \lambda \left\| \mathbf{x}_k^1 - \mathbf{x}_1 \right\| + (1-\lambda) \left\| \mathbf{x}_k^2 - \mathbf{x}_2 \right\|$ (by the triangle inequality), it follows that

$$\lim_{k\to\infty} \left\| \mathbf{x}_k^c - \mathbf{x}^c \right\| = \lambda \cdot 0 + (1-\lambda) \cdot 0 = 0,$$

i.e., the convex combination of elements of the two sequences converges to the convex combination of their individual limits. We have thus shown that for $\mathbf{x}_1, \mathbf{x}_2 \in \overline{\mathcal{S}}$, their convex combination $\mathbf{x}^c \in \overline{\mathcal{S}}$ so that $\overline{\mathcal{S}}$ is convex.

Some additional properties of convex sets are:

(a) Trivially, ϕ and $\mathbf{R}^n$ are each convex sets.

(b) If $\mathcal{S}$ in $\mathbf{R}^n$ is a convex set, then $\mathcal{S}^o$ (possible empty) is convex.

(c) In general, the union of a collection of convex sets in $\mathbf{R}^n$ is not convex.

(d) If $\mathcal{S}$ is a convex set in $\mathbf{R}^n$ and α is a scalar in $\mathbf{R}$, then $\alpha\mathcal{S} = \{\alpha\mathbf{x} \,|\, \mathbf{x}\in\mathcal{S}\}$ is convex.

(e) If $\mathcal{S}$ is a convex set in $\mathbf{R}^n$, then $\partial(\overline{\mathcal{S}}) = \partial(\mathcal{S})$.

(f) If $\mathcal{S}$ is a convex set in $\mathbf{R}^n$, with $\mathcal{S}^o \neq \phi$, then $\overline{\mathcal{S}^o} = \overline{\mathcal{S}}$ and $(\overline{\mathcal{S}})^o = \mathcal{S}^o$.

(g) For $\mathcal{S}_1, \mathcal{S}_2$ convex subsets of $\mathbf{R}^n$, the ***linear sum*** $\mathcal{S}_1 + \mathcal{S}_2 = \{\mathbf{x} \,|\, \mathbf{x} = \mathbf{x}_1 + \mathbf{x}_2,\ \mathbf{x}_1 \in \mathcal{S}_1,\ \mathbf{x}_2 \in \mathcal{S}_2\}$ is a convex set. Similarly, $\mathcal{S}_1 - \mathcal{S}_2 = \{\mathbf{x} \,|\, \mathbf{x} = \mathbf{x}_1 - \mathbf{x}_2,\ \mathbf{x}_1 \in \mathcal{S}_1,\ \mathbf{x}_2 \in \mathcal{S}_2\}$ is convex. In addition, the set $\mathcal{S}_1 + \{\mathbf{x}_0\} = \{\mathbf{x} \,|\, \mathbf{x}_1 + \mathbf{x}_0,\ \mathbf{x}_1 \in \mathcal{S}_1\}$ is convex.

(h) A hyperplane along with both open and closed half-spaces in $\mathbf{R}^n$ are all convex sets. (For instance, if $\mathbf{x}_1, \mathbf{x}_2 \in \mathcal{H}$, then $\mathbf{x} = \lambda\mathbf{x}_1 + (1-\lambda)\mathbf{x}_2 \in \mathcal{H}$ for $0 \leq \lambda \leq 1$ since $\mathrm{C}'\mathbf{x} = \mathrm{C}'[\lambda\mathbf{x}_1 + (1-\lambda)\mathbf{x}_2] = \lambda\mathrm{C}'\mathbf{x}_1 + (1-\lambda)\mathrm{C}'\mathbf{x}_2 = \lambda\alpha + (1-\lambda)\alpha = \alpha$. Similarly, if $\mathbf{x}_1, \mathbf{x}_2 \in [\mathcal{H}^+]$ and $\mathbf{x} = \lambda\mathbf{x}_1 + (1-\lambda)\mathbf{x}_2$ $(0 \leq \lambda \leq 1)$, then $\mathrm{C}'\mathbf{x} = \mathrm{C}'[\lambda\mathbf{x}_1 + (1-\lambda)\mathbf{x}_2] = \lambda\mathrm{C}'\mathbf{x}_1 + (1-\lambda)\,\mathrm{C}'\mathbf{x}_2 \leq \lambda\,\alpha + (1-\lambda)\alpha = \alpha$.)

(i) Let $\mathcal{S}$ be a convex set in $\mathbf{R}^n$ with a nonempty interior $\mathcal{S}^o$. Additionally, let $\mathbf{x}_1 \in \mathcal{S}^o$ and $\mathbf{x}_2 \in \overline{\mathcal{S}}$. Then $\lambda\mathbf{x}_1 + (1-\lambda)\mathbf{x}_2 \in \mathcal{S}^o$ for all $0 < \lambda \leq 1$. (This result is sometimes referred to as the "accessibility lemma.")

(j) If $\mathcal{S}_1$, $\mathcal{S}_2$ are each convex sets in $\mathbf{R}^n$, then their sum $\mathcal{S}_1 + \mathcal{S}_2 = \{\mathbf{x}_1 + \mathbf{x}_2 \,|\, \mathbf{x}_1 \in \mathcal{S}_1,\ \mathbf{x}_2 \in \mathcal{S}_2\}$ is also convex.

(k) If $\mathcal{S}_1, \ldots, \mathcal{S}_m$ are each convex sets in $\mathbf{R}^n$, then their linear combination $\mathcal{S} = \Sigma_{i=1}^m \lambda_i \mathcal{S}_i$ is also convex for $\lambda_i \in \mathbf{R}$ for all i.

(l) If $\mathcal{S}$ is a convex set in $\mathbf{R}^n$ and $0 \leq \lambda_i \in \mathbf{R}$, $i = 1, 2$, then $(\lambda_1 + \lambda_2)\mathcal{S} = \lambda_1\mathcal{S} + \lambda_2\mathcal{S}$ is convex.

(m) A convex set $\mathcal{S}$ in $\mathbf{R}^n$ is termed ***symmetric*** if $\mathcal{S} = -\mathcal{S}$; $\mathcal{S}$ must contain $\mathbf{x}$, $-\mathbf{x}$, $\mathbf{0}$, and the closed line segment $[\mathbf{x}, -\mathbf{x}]$.

(n) Given convex sets $\mathcal{S}_1, \mathcal{S}_2$ in $\mathbf{R}^n$, the unique largest convex set included in both $\mathcal{S}_1, \mathcal{S}_2$ is $\mathcal{S}_1 \cap \mathcal{S}_2$.

(o) Let $\mathcal{X}$, $\mathcal{Y}$ be convex sets in $\mathbf{R}^m, \mathbf{R}^n$ respectively. The ***direct sum of*** $\mathcal{X}$ ***and*** $\mathcal{Y}$, $\mathcal{X} \oplus \mathcal{Y} = \{\mathbf{v} = (\mathbf{x}, \mathbf{y}) | \mathbf{x} \epsilon \mathcal{X}, \mathbf{y} \epsilon \mathcal{Y}\}$, is a convex set.

(p) Every affine set is convex, but not conversely.

To set the stage for our next concept let us first consider the following.

2.1.3. THEOREM. Given an $(m \times n)$ matrix $\mathbf{B}$ and a vector $\mathbf{b} \epsilon \mathbf{R}^m$, the solution set $\mathcal{S} = \{\mathbf{x} | \mathbf{B}\mathbf{x} \leqq \mathbf{b}, \mathbf{x} \epsilon \mathbf{R}^n\}$ in $\mathbf{R}^n$ is convex.

PROOF. For β_i, $i=1,\ldots,m$, the ith row of $\mathbf{B}$, let $[\mathcal{H}_i^-] = \{\mathbf{x} | \beta_i \mathbf{x} \leq b_i, \mathbf{x} \epsilon \mathbf{R}^n\}$, $i=1,\ldots,m$, where b_i is the ith component of $\mathbf{b}$. Clearly $[\mathcal{H}_i^-]$ is a closed half-space. And since each of these closed half-spaces is convex, it follows from theorem 2.1.1 that $\mathcal{S} = \cap_{i=1}^m [\mathcal{H}_i^-]$ is convex. Q.E.D.

We can now assert that a set $\mathcal{S}$ in $\mathbf{R}^n$ which can be expressed as the intersection of finitely many closed half-spaces in $\mathbf{R}^n$ is a ***polyhedral convex set*** (this particular convex structure will be studied in considerable detail in chapter 8).

A set of theorems (offered without proof) which address certain salient features of a collection of convex sets in $\mathbf{R}^n$ now follows.

2.1.4. HELLY'S THEOREM [Helly, 1921]. Let $\mathcal{S}_1,\ldots,\mathcal{S}_r$, $r \geq n+1$, be a finite collection of convex sets in $\mathbf{R}^n$. If the intersection of any $n+1$ (the ***Helly number***) of the sets $\mathcal{S}_i$, i=1,…, r, is nonempty, then the intersection of all r sets is nonempty.

If we further stipulate that the convex sets $\mathcal{S}_1,\ldots,\mathcal{S}_r$ are also compact, then the extension of Helly's theorem to infinite collections of convex sets appears as

2.1.4.1. THEOREM. Let $\mathcal{S}$ be a collection of at least $n+1$ compact convex sets in $\mathbf{R}^n$. If the intersection of any $n+1$ members of $\mathcal{S}$ is nonempty, then the intersection of all members of $\mathcal{S}$ is nonempty.

A theorem similar to the one developed by Helly is

2.1.5. BERGE'S THEOREM [Berge, 1959, 1963]. Let $\mathcal{S}_1,\ldots,\mathcal{S}_r$, $r \geq 2$, be closed convex sets in $\mathbf{R}^n$ such that $\cup_{i=1}^r \mathcal{S}_i$ is also convex. If all intersections of any $r-1$ of these sets is nonempty, then the intersection of all r sets is nonempty.

Two results which follow from Berge's theorem are:

2.1.6. COROLLARY. Let $\mathcal{S}_1,\ldots,\mathcal{S}_r$, $r \geq 2$, be closed convex sets in $\mathbf{R}^n$. If the intersection of $r-1$ of these sets is nonempty but the set $\cap_{i=1}^r \mathcal{S}_i = \phi$, then $\cap_{i=1}^r \mathcal{S}_i$ cannot be convex.

2.1.7. COROLLARY. Let $\mathcal{S}_1,\ldots,\mathcal{S}_r$, $r \geq 1$, be closed convex sets in $\mathbf{R}^n$. If there exists a convex set $\mathcal{S}$ such that

$$\mathcal{S} \cap \left(\cap_{\substack{i=1 \\ i \neq j}}^r \mathcal{S}_i \right) \neq \phi, \; j=1, \ldots r,$$

but

$$\mathcal{S} \cap \left(\cap_{i=1}^r \mathcal{S}_i \right) = \phi,$$

then $\mathcal{S}$ is not a subset of $\cup_{i=1}^r \mathcal{S}_i$.

2.2 Convex Combination

A ***convex combination*** of a finite number of points $\mathbf{x}_i$, $i=1,\ldots, m$, in $\mathbf{R}^n$ is the point

$$\mathbf{x} = \sum_{i=1}^m \lambda_i \mathbf{x}_i, \; \sum_{i=1}^m \lambda_i = 1, \quad 0 \leq \lambda_i \in \mathbf{R} \text{ for all } i.$$

The following two theorems relate this construct back to the notion of a convex set presented earlier.

2.2.1. THEOREM. A set $\mathcal{S}$ in $\mathbf{R}^n$ is convex if and only if every convex combination of finitely many points of $\mathcal{S}$ is an element of $\mathcal{S}$.

PROOF. (necessity) Given that $\mathscr{S}$ is convex, we desire to demonstrate that each convex combination of points from $\mathscr{S}$ belongs of $\mathscr{S}$. The proof proceeds by induction on the number m of points from $\mathscr{S}$ used to form the convex combination.

For $m=1$, $\mathbf{x}_1 \in \mathscr{S}$ trivially.

For $m=2$, $[\mathbf{x}_1, \mathbf{x}_2] \in \mathscr{S}$ by definition.

Next, assume that the assertation "every convex combination of $m-1$ points from $\mathscr{S}$ lies in $\mathscr{S}$" is true. Hence one must construct a linear combination of m points from $\mathscr{S}$, *i.e.*, for $\mathbf{x}_1, \ldots, \mathbf{x}_m$ all in $\mathscr{S}$, form

$$\mathbf{x} = \sum_{i=1}^{m} \lambda_i \mathbf{x}_i, \quad \sum_{i=1}^{m} \lambda_i = 1, \; \lambda_i \geq 0 \text{ for all } i.$$

If $\lambda_m = 1$, then $\mathbf{x} = \mathbf{x}_m \in \mathscr{S}$. If $\lambda_m < 1$, then $\Sigma_{i=1}^{m-1} \lambda_i = 1 - \lambda_m > 0$ so that

$$\mathbf{x} = \sum_{i=1}^{m-1} \lambda_i \mathbf{x}_i + \lambda_m \mathbf{x}_m = (1-\lambda_m) \sum_{i=1}^{m-1} \left(\frac{\lambda_i}{1-\lambda_m}\right) \mathbf{x}_i + \lambda_m \mathbf{x}_m$$

$$= (1-\lambda_m)\bar{\mathbf{x}} + \lambda_m \mathbf{x}_m.$$

By the above assertion (the induction hypothesis), $\bar{\mathbf{x}} \in \mathscr{S}$. Since $\mathbf{x}$ is a convex combination of $\bar{\mathbf{x}}$ and $\mathbf{x}_m$, it follows that $\mathbf{x} \in \mathscr{S}$.

(sufficiency) That every convex combination of points from $\mathscr{S}$ implies that $\mathscr{S}$ is convex follows from the definition of convexity (simply take $m=2$). Q.E.D.

The next theorem considers a key property of "the set of all convex combinations of a finite number of points." Specifically,

2.2.2. THEOREM. Let $\{\mathbf{x}_1, \ldots, \mathbf{x}_m\}$ be a finite collection of points in $\mathbf{R}^n$. Then the set of all convex combinations of the $\mathbf{x}_i, \mathrm{i}=1, \ldots, m$,

$$\mathfrak{G} = \left\{\mathbf{x} \mid \mathbf{x} = \sum_{i=1}^{m} \lambda_i \mathbf{x}_i, \; \sum_{i=1}^{m} \lambda_i = 1, \; 0 \leq \lambda_i \in \mathbf{R} \text{ for all } i\right\}$$

is convex.

PROOF. Let $\mathbf{u}$, $\mathbf{v}$ be any vectors expressible as convex combinations of the $\mathbf{x}_i$, $i=1,\ldots,m$, so that

$$\mathbf{u} = \sum_{i=1}^{m} \lambda_i' \,\mathbf{x}_i, \quad \sum_{i=1}^{m} \lambda_i' = 1, \quad \lambda_i' \geq 0 \text{ for all } i;$$

$$\mathbf{v} = \sum_{i=1}^{m} \lambda_i'' \,\mathbf{x}_i, \quad \sum_{i=1}^{m} \lambda_i'' = 1, \ \lambda_i'' \geq 0 \text{ for all } i.$$

Let us form

$$\theta\mathbf{u} + (1-\theta)\mathbf{v} = \sum_{i=1}^{m} \left[\theta\lambda_i' + (1-\theta)\,\lambda_i''\right]\mathbf{x}_i,$$

$$\sum_{i=1}^{m} \left[\theta\lambda_i' + (1-\theta)\,\lambda_i''\right] = \theta \sum_{i=1}^{m} \lambda_i' + (1-\theta) \sum_{i=1}^{m} \lambda_i'' = 1,$$

$$\left[\theta\,\lambda_i' + (1-\theta)\lambda_i''\right] \geq 0 \text{ for all } i.$$

Since $\theta\mathbf{u} + (1-\theta)\,\mathbf{v}$ is clearly a convex combination of the $\mathbf{x}_i$ for $0 \leq \theta \leq 1$, it follows that $\theta\mathbf{u} + (1-\theta)\mathbf{v} = [\mathbf{u}, \mathbf{v}] \in \mathfrak{S}$ and thus $\mathfrak{S}$ is a convex set. Q.E.D.

We note briefly that if $\mathcal{S}_1, \ldots, \mathcal{S}_m$ are each convex sets in $\mathbf{R}^n$, then their convex combination $\mathcal{S} = \Sigma_{i=1}^{m} \lambda_i \mathcal{S}_i$, $\Sigma_{i=1}^{m} \lambda_i = 1$, $0 \leq \lambda_i \in \mathbf{R}$ is also convex.

2.3 Convex Hull

The ***convex hull*** (or ***convex closure***) $co(\mathcal{S})$ of a set $\mathcal{S}$ in $\mathbf{R}^n$ is the collection of all convex combinations of points from $\mathcal{S}$ or

$$co(\mathcal{S}) = \left\{ \mathbf{x} \,|\, \mathbf{x} = \sum_{i=1}^{m} \lambda_i \mathbf{x}_i, \ \mathbf{x}_i \in \mathcal{S}, \ \sum_{i=1}^{m} \lambda_i = 1, \ 0 \leq \lambda_i \in \mathbf{R} \text{ for all } i, \right.$$

$$\left. \text{and } m \text{ is an arbitrary positive integer} \right\}.$$

2.3.1. THEOREM. A set $\mathcal{T}$ in $\mathbf{R}^n$ is the smallest convex set containing an arbitrary set $\mathcal{S}$ in $\mathbf{R}^n$ if and only if $\mathcal{T} = co(\mathcal{S})$.

PROOF. (sufficiency) We first demonstrate that if $\mathcal{T} = co(\mathcal{S})$, then $co(\mathcal{S})$ is the smallest convex set containing $\mathcal{S}$. By theorem 2.2.2, $co(\mathcal{S})$ is convex. Moreover, for every convex set $\mathcal{K}$ in $\mathbf{R}^n$ containing $\mathcal{S}$, $co(\mathcal{S}) \subseteq \mathcal{K}$. This inclusion statement must be true since $\mathcal{K}$ is convex and, given $\mathcal{S} \subseteq \mathcal{K}$, it follows from theorem 2.3.1 that every convex combination of elements of $\mathcal{S}$ belongs to $\mathcal{K}$.

(necessity) We next demonstrate that if $\mathcal{T}$ is the smallest convex set containing $\mathcal{S}$, then $\mathcal{T} = co(\mathcal{S})$. Now, we know that $\mathcal{S} \subseteq \mathcal{T}$ and $\mathcal{T} \subseteq \mathcal{K}$ for every convex set $\mathcal{K}$ containing $\mathcal{S}$. Since $co(\mathcal{S})$ is a convex set containing $\mathcal{S}$, it follows that $\mathcal{T} \subseteq co(\mathcal{S})$. However, from the first part of the proof and from $\mathcal{S} \subseteq \mathcal{T}$, we have $co(\mathcal{S}) \subseteq \mathcal{T}$ and thus $\mathcal{T} = co(\mathcal{S})$. Q.E.D.

Figure 2.2 illustrates the convex hull of a set $\mathcal{S}$ and of a finite collection of points in $\mathbf{R}^2$. Relative to this latter case, a set $\mathcal{S}$ in $\mathbf{R}^n$ which is the convex hull of finitely many points is said to be "spanned or generated" by those points and is called a ***convex polytope*** (see chapter 8).

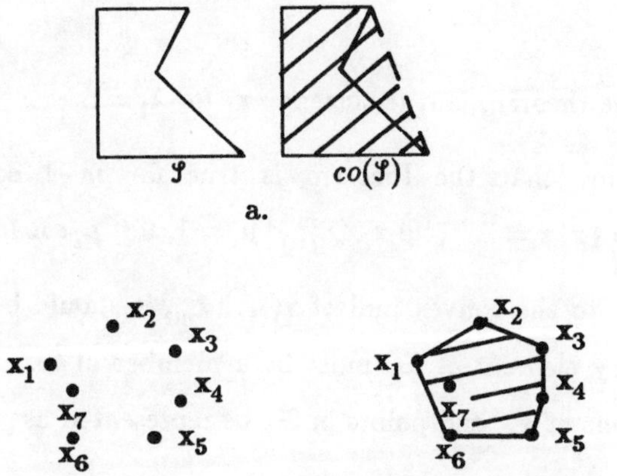

$\{\mathbf{x}_1,\ldots,\mathbf{x}_7\}$ $\quad$ $co\{\mathbf{x}_1,\ldots,\mathbf{x}_7\} = \{\mathbf{x}\,|\,\mathbf{x} = \Sigma_{i=1}^{7}\lambda_i\mathbf{x}_i,\ \Sigma_{i=1}^{7}\lambda_i=1,\ \lambda_i \geq 0 \text{ for all } i\}$

b.

Figure 2.2

How should the word "smallest" in the preceding theorem be interpreted? The answer is provided by

2.3.2. THEOREM. The convex hull of a set $\mathscr{S}$ in $\mathbf{R}^n$ is the intersection of all convex sets $\mathscr{A}_i$ in $\mathbf{R}^n$ which contains $\mathscr{S}$, *i.e.*, $co(\mathscr{S}) = \cap_i A_i$, where $\mathscr{S} \subset \mathscr{A}_i$ for all i.

PROOF. Clearly $\cap_i\mathscr{A}_i$ must contain $\mathscr{S}$ and, by theorem 2.1.1, is convex. In addition, for any convex set $\mathscr{A}_i$ containing $\mathscr{S}$, $\cap_i\mathscr{A}_i$ is the smallest convex set which contains $\mathscr{S}$. Hence, by theorem 2.3.1, $\cap_i\mathscr{A}_i = co(\mathscr{S})$. Q.E.D.

We next have

2.3.3. THEOREM. Let $\{\mathbf{x}_1,\ldots,\mathbf{x}_m\}$ be a finite collection of points in $\mathbf{R}^n$. Then the convex hull of the $\mathbf{x}_i$, i=1,…, m, is the set of all their convex combinations, *i.e.*, $\mathscr{K} = co\{\mathbf{x}_1,\ldots,\mathbf{x}_m\} = \{\mathbf{x}\,|\,\mathbf{x} = \Sigma_{i=1}^{m}\lambda_i\mathbf{x}_i,\ \Sigma_{i=1}^{m}\lambda_i = 1,\ 0 \leq \lambda_i \in \mathbf{R} \text{ for all } i\}$.

PROOF. We proceed by induction on the number of points m. Our objective is to demonstrate that every convex set containing $\mathbf{x}_1,\ldots,\mathbf{x}_m$ also contains $co\{\mathbf{x}_1,\ldots,\mathbf{x}_m\}$.

For $m=1$, the theorem is true since $\mathbf{x}=\mathbf{x}_1$ for $\lambda_1=1$.

Let us assume that the theorem is true for $m-1$ so that $\mathcal{S}_1 = co\{\mathbf{x}_1,\ldots,\mathbf{x}_{m-1}\} = \{\mathbf{x}\,|\,\mathbf{x}=\Sigma_{i=1}^{m-1}\mu_i\mathbf{x}_i,\ \Sigma_{i=1}^{m-1}\mu_i=1,\ 0\leq\mu_i\,\epsilon\,\mathbf{R}$ for all $i\}$.

Looking next to the convex hull of $\mathbf{x}_1,\ldots,\mathbf{x}_m$, it should be evident that $\mathbf{x}_m$ as well as every element of $\mathcal{S}_1$ must be a member of $\mathcal{S}$. Let the set of convex combinations of $\mathbf{x}_m$ and points in $\mathcal{S}_1$ be represented as

$$\mathcal{S}_2 = \{\mathbf{x}\,|\,\mathbf{x}=\lambda\mathbf{x}_m+(1-\lambda)\,\Sigma_{i=1}^{m-1}\mu_i\mathbf{x}_i,\ 0\leq\lambda\leq 1\}.$$

If $\lambda_i=(1-\lambda)\mu_i$, $i=1,\ldots,m-1$, and $\lambda_m=\lambda$, then $\Sigma_{i=1}^{m}\lambda_i=\Sigma_{i=1}^{m-1}(1-\lambda)\mu_i+\lambda=1$, $\lambda_i\geq 0$ for all i. Hence $\mathcal{S}_2$ must be the set of all convex combinations of $\mathbf{x}_1,\ldots,\mathbf{x}_m$ and consequently represents the convex hull of $\mathbf{x}_1,\ldots,\mathbf{x}_m$ given that $\mathcal{S}_1$ is the convex hull of $\mathbf{x}_1,\ldots,\mathbf{x}_{m-1}$; it is the smallest convex set containing $\mathbf{x}_m$ and $\mathcal{S}_1$. So by induction, the convex hull of $\mathbf{x}_1,\ldots,\mathbf{x}_m$ is the set of all convex combinations of these points. Q.E.D.

In forming the convex hull of a set $\mathcal{I}$ in $\mathbf{R}^n$, we shall now see that there exists an upper limit to the number of points from $\mathcal{I}$ needed to generate the collection of all convex combinations. That is,

2.3.4. CARATHÉODORY'S THEOREM [Carathéodory, 1907]. Let $\mathbf{x}\,\epsilon\,co(\mathcal{I})$ for $\mathcal{I}$ a set in $\mathbf{R}^n$. Then $\mathbf{x}$ is a convex combination of, at most, $n+1$ points from $\mathcal{I}$.

PROOF. Suppose, to the contrary, that at least $r > n+1$ vectors are required to form

$$\mathbf{x} = \sum_{i=1}^{r}\lambda_i\mathbf{x}_i,\ \sum_{i=1}^{r}\lambda_i = 1,\ 0\leq\lambda_i\,\epsilon\,\mathbf{R} \text{ for all } i.$$

Since a basis for $\mathbf{R}^n$ contains exactly (and thus no more than) n linearly independent vectors, it follows that the set of vectors $\{\mathbf{x}_i-\mathbf{x}_r,\ i=1,\ldots,r-1\}$ must be linearly dependent. Hence there exist scalars α_i not all zero such that $\Sigma_{i=1}^{r-1}\alpha_i(\mathbf{x}_i-\mathbf{x}_r)=\mathbf{0}$. Let $\alpha_r = -\Sigma_{i=1}^{r-1}\alpha_i$. Then

$$\sum_{i=1}^{r-1} \alpha_i(\mathbf{x}_i-\mathbf{x}_r) = \sum_{i=1}^{r-1} \alpha_i\mathbf{x}_i+\alpha_r\mathbf{x}_r = \sum_{i=1}^{r} \alpha_i\mathbf{x}_i = 0, \ \sum_{i=1}^{r} \alpha_i = 0.$$

Choose $0 < \beta \epsilon \mathbf{R}$ and small enough so that $\theta_i = \lambda_i - \beta\alpha_i \geq 0$ for all i and, for at least one value of i, let $\theta_i = 0$ (take $i=j$) so that $\theta_j = \lambda_j - \beta\alpha_j = 0$).[1]

Then

$$\sum_{i=1}^{r} \theta_i = \sum_{i=1}^{r} \lambda_i = 1, \ \theta_i \geq 0 \text{ for all } i.$$

Hence

$$\sum_{i=1}^{r} \theta_i\mathbf{x}_i = \sum_{i=1}^{r}(\lambda_i-\beta\alpha_i)\mathbf{x}_i = \sum_{i=1}^{r} \lambda_i\mathbf{x}_i - \beta\sum_{i=1}^{r} \alpha_i\mathbf{x}_i = \sum_{i=1}^{r} \lambda_i\mathbf{x}_i = \mathbf{x}.$$

But since $\theta_j = 0$, this contradicts the assumption that at least r vectors are required to generate $\mathbf{x}$. Q.E.D.

We note briefly that:

(a) If a set $\mathcal{S}$ in $\mathbf{R}^n$ is convex, then $co(\mathcal{S}) = \mathcal{S}$.

(b) For $\mathcal{S}$ a subset of $\mathbf{R}^n$, $co(\mathcal{S})$ is a closed set if $\mathcal{S}$ is a finite set.

(c) If a set $\mathcal{S}$ in $\mathbf{R}^n$ is compact, then $co(\mathcal{S})$ is compact.

(d) If $\mathcal{S}$ is an open set in $\mathbf{R}^n$, then $co(\mathcal{S})$ is open. The convex hull of a closed set in $\mathbf{R}^n$ need not be closed.

(e) All affine sets are convex (hence convex sets are more general than affine sets).

(f) The unique smallest convex set which includes both $\mathcal{S}_1$, $\mathcal{S}_2$ in $\mathbf{R}^n$ is $co(\mathcal{S}_1 \cup \mathcal{S}_2)$.

(g) For $\mathcal{S}$ a subset of $\mathbf{R}^n$, the ***closed convex hull*** of $\mathcal{S}$ is the closure of $co(\mathcal{S})$ or $\overline{co(\mathcal{S})}$.

(h) In general, the convex hull of a closed set need not be closed. If $\mathcal{S}$ is a bounded subset of $\mathbf{R}^n$, then $\overline{co(\mathcal{S})} = co(\overline{\mathcal{S}})$. So if $\mathcal{S}$ is closed and bounded, then $co(\mathcal{S})$ is closed and bounded.

---

[1] Since at least one $\alpha_i > 0$, we can choose $\beta = \min_i \{\lambda_i/\alpha_i \mid \alpha_i > 0\} = \lambda_j/\alpha_j$.

(i) For $\mathcal{S}(\neq\phi)$ an arbitrary set in $\mathbf{R}^n$, $\mathcal{S} \subseteq co(\mathcal{S}) \subseteq aff(\mathcal{S}) \subseteq \ell(\mathcal{S})$.

(j) For a finite set of points $\{\mathbf{x}_1,\ldots,\mathbf{x}_k\}$ in $\mathbf{R}^n$,

$$co\{\mathbf{x}_1,\ldots,\mathbf{x}_k\} \subseteq aff\{\mathbf{x}_1,\ldots,\mathbf{x}_k\} \subseteq \ell\{\mathbf{x}_1,\ldots,\mathbf{x}_k\}$$

(k) For $\mathcal{A}, \mathcal{B}$ nonempty sets in $\mathbf{R}^n$:

(i) if $\mathcal{A} \subset \mathcal{B}$, then $\mathcal{A} \subset co(\mathcal{B})$;

(ii) if $\mathcal{A} \subset \mathcal{B}$ and $\mathcal{B}$ is convex, then $co(\mathcal{A}) \subset \mathcal{B}$;

(iii) if $\mathcal{A} \subset \mathcal{B}$, then $co(\mathcal{A}) \subset co(\mathcal{B})$;

(iv) $(co(\mathcal{A}) \cup co(\mathcal{B})) \subset co(\mathcal{A} \cup \mathcal{B})$;

(v) $co(\mathcal{A} \cap \mathcal{B}) \subset (co(\mathcal{A}) \cap co(\mathcal{B}))$.

In section 3 of Chapter 1 we encountered the concept of the interior of a set in $\mathbf{R}^n$. Interestingly enough, for convex sets, "interior" is subsumed under the notion of "relative interior." Specifically, the ***relative interior*** of a convex set $\mathcal{S}$ in $\mathbf{R}^n$, denoted $ri(\mathcal{S})$, is the interior which results whenever $\mathcal{S}$ is considered a subset of its affine hull $aff(\mathcal{S})$. To put this definition in its proper perspective, let us admit another point $\mathbf{y}$ into our discussion. In this regard, the relative interior of $\mathcal{S}$ consists of those points $\mathbf{x} \in aff(\mathcal{S})$ such that $\mathbf{y} \in \mathcal{S}$ whenever $\mathbf{y}$ is near $\mathbf{x}$ and also a member of $aff(\mathcal{S})$ or $ri(\mathcal{S}) = \{\mathbf{x} \in aff(\mathcal{S}) \,|\, \mathbf{y} \in B(\mathbf{x}, \delta) \cap (aff(\mathcal{S}))$ implies $\mathbf{y} \in \mathcal{S}\}$.

A convex set $\mathcal{S}$ in $\mathbf{R}^n$ is termed ***relatively open*** if $\mathcal{S} = ri(\mathcal{S})$. Furthermore, the ***relative boundary*** of a convex subset $\mathcal{S}$ *of* R^n is $r\partial(\mathcal{S}) = \{\mathbf{x} \,|\, \mathbf{x} \in \overline{\mathcal{S}}$ and $\mathbf{x} \notin ri(\mathcal{S})\}$.

Additional properties or relationships involving the relative interior of a convex set $\mathcal{S}$ in $\mathbf{R}^n$ are:

(a) A nonempty convex set $\mathcal{S}$ in $\mathbf{R}^n$ always has a nonempty relative interior.

(b) For $\mathcal{S}$ a convex subset of $\mathbf{R}^n$, $ri(\mathcal{S}) \subset \mathcal{S} \subset \overline{\mathcal{S}} \subset aff(\mathcal{S})$.

(c) Any affine set $\mathcal{S}$ in $\mathbf{R}^n$ is relatively open (by definition).

(d) For any convex set $\mathcal{S}$ in $\mathbf{R}^n$ and any scalar $\lambda \in \mathbf{R}$, $ri\,(\lambda\mathcal{S}) = \lambda\, ri(\mathcal{S})$.

(e) For $\mathcal{S}_1$, $\mathcal{S}_2$ convex sets in $\mathbf{R}^n$, $ri(\mathcal{S}_1+\mathcal{S}_2) = ri(\mathcal{S}_1)+ri(\mathcal{S}_2)$.

(f) Let $\mathcal{S}$ be a convex set in $\mathbf{R}^n$. In addition, let $\mathbf{x} \in ri(\mathcal{S})$ and $\mathbf{y} \in \overline{\mathcal{S}}$. Then $(1-\lambda)\,\mathbf{x}+\lambda\mathbf{y} \in ri(\mathcal{S})$ for $0 \leq \lambda < 1$, $\lambda \in \mathbf{R}$.

(g) Let $\mathcal{S}$ be a convex set in $\mathbf{R}^n$. Then both $ri(\mathcal{S})$, $\overline{\mathcal{S}}$ are convex sets in $\mathbf{R}^n$. Moreover $aff(\mathrm{ri}(\mathcal{S})) = aff(\overline{\mathcal{S}}) = aff(\mathcal{S})$ and thus $dim\,(ri(\mathcal{S})) = dim(\overline{\mathcal{S}}) = dim(\mathcal{S})$.

(h) For $\mathcal{S}$ a convex set in $\mathbf{R}^n$, $\overline{ri(\mathcal{S})} = \overline{\mathcal{S}}$ and $ri(\overline{\mathcal{S}}) = ri(\mathcal{S})$.

(i) For $\mathcal{S}_1$, $\mathcal{S}_2$ convex sets in $\mathbf{R}^n$, $\overline{\mathcal{S}}_1 = \overline{\mathcal{S}}_2$ if and only if $ri(\mathcal{S}_1) = ri(\mathcal{S}_2)$.

(j) For $\mathcal{S}_1, \mathcal{S}_2$ non-empty convex sets in $\mathbf{R}^n$, let $\mathcal{S} = \mathcal{S}_1 - \mathcal{S}_2$. Then, from (e), $ri(\mathcal{S}) = ri(\mathcal{S}_1) - ri(\mathcal{S}_2)$ and $\mathbf{0} \notin ri(\mathcal{S})$ if and only if $ri(\mathcal{S}_1) \cap ri(\mathcal{S}_2) = \phi$.

Finally, a set $\mathcal{S}$ in $\mathbf{R}^n$ is said to be ***locally convex*** if each point of $\mathcal{S}$ has a neighborhood whose intersection with $\mathcal{S}$ is convex.

2.4. Exercises

1. Are the following sets in $\mathbf{R}^2$ convex? Verify.

$$\mathcal{S}_1 = \{\mathbf{x}' = (x_1, x_2) \mid x_2 \geq |x_1|\};$$
$$\mathcal{S}_2 = \{\mathbf{x}' = (x_1, x_2) \mid x_1 + x_2 \leq 1\};$$
$$\mathcal{S}_3 = \{\mathbf{x}' = (x_1, x_2) \mid \|\mathbf{x}\|^2 \leq 1\};$$
$$\mathcal{S}_4 = \{\mathbf{x}' = (x_1, x_2) \mid x_1 x_2 \geq 1, \mathbf{x} > \mathbf{0}\}.$$

2. Can $\mathbf{x}_0' = (1,2)$ be written as a convex combination of the points $\mathbf{x}_1' = (-1,0)$, $\mathbf{x}_2' = (3,6)$?

3. Express the point $\mathbf{x}_0$ inside a triangle in $\mathbf{R}^2$ as the convex combination of its vertexes $\mathbf{x}_1, \mathbf{x}_2, \mathbf{x}_3$.

4. Prove that the set of solutions to $\mathbf{Ax} = \mathbf{b}$ is a convex set. (Hint: show that if $\mathbf{x}_1, \mathbf{x}_2$ are solutions, then so is their convex combination.)

5. For $\mathcal{S} = \{\mathbf{x}_1' = (1,0,0), \mathbf{x}_2' = (0,1,0)\}$ in $\mathbf{R}^3$, express the convex hull of $\mathcal{S}$ as a straight line segment connecting $\mathbf{x}_1, \mathbf{x}_2$.

6. Prove that if $\mathcal{S}$ is a convex set in $\mathbf{R}^n$, then so is its closure. Let $\mathfrak{X}, \mathfrak{Y}$ in $\mathbf{R}^n$ be respectively closed and compact. Verify that $\mathfrak{X} \cap \mathfrak{Y}$ is compact.

7. Let $\mathcal{S} = \{\mathbf{x}_1' = (1,0,0),\ \mathbf{x}_2' = (1,-1,2)\}$. Find $\ell(\mathcal{S})$, $aff(\mathcal{S})$, and $co(\mathcal{S})$.

8. Prove that, in general, the union of a collection of convex sets in $\mathbf{R}^n$ is not convex.

CHAPTER 3
SEPARATION AND SUPPORT THEOREMS

3.1 Hyperplanes and Half-Planes Revisited

In section 4 of Chapter 1 we encountered the definitions of a hyperplane and a half-plane (–space). Let us now examine these concepts in greater detail. In particular, we shall define a variety of different types of hyperplanes which will be of considerable importance in what follows. To briefly review these definitions, let us note first that for a vector $\mathbf{C}(\neq \mathbf{0}) \in \mathbf{R}^n$ and a scalar $\alpha \in \mathbf{R}$, a ***hyperplane***

$$\mathcal{H} = \{\mathbf{x} \mid \mathbf{C}'\mathbf{x} = \alpha, \mathbf{x} \in \mathbf{R}^n\}$$

is an $(n-1)$ - dimensional linear variety or affine set. Next, any hyperplane $\mathcal{H}$ in $\mathbf{R}^n$ generates the two ***closed half-spaces***

$$[\mathcal{H}^+] = \{\mathbf{x} \mid \mathbf{C}'\mathbf{x} \geq \alpha, \mathbf{x} \in \mathbf{R}^n\},$$

$$[\mathcal{H}^-] = \{\mathbf{x} \mid \mathbf{C}'\mathbf{x} \leq \alpha, \mathbf{x} \in \mathbf{R}^n\}.$$

A point in $\mathbf{R}^n$ lies in $[\mathcal{H}^+]$, $[\mathcal{H}^-]$, or in both of these sets.

If $\mathbf{x}_0$ is taken to be a fixed reference point in $\mathcal{H}$, then $\mathbf{C}'\mathbf{x}_0 = \alpha$ and, for any other $\mathbf{x} \in \mathcal{H}$, $\mathbf{C}'\mathbf{x} = \alpha$ so that $\mathbf{C}'\mathbf{x} - \mathbf{C}'\mathbf{x}_0 = \mathbf{C}'(\mathbf{x}-\mathbf{x}_0) = \alpha - \alpha = 0$. Hence a hyperplane has the alternative representation

$$\mathcal{H} = \{\mathbf{x} \mid \mathbf{C}'(\mathbf{x}-\mathbf{x}_0) = 0, \mathbf{x} = \mathbf{x}_0 \text{ fixed}, \mathbf{x} \in \mathbf{R}^n\}.$$

Clearly $\mathbf{C}$ is orthogonal to $\mathbf{x} - \mathbf{x}_0$ for any $\mathbf{x} \in \mathcal{H}$ and is termed the ***normal*** of the hyperplane (Figure 3.1.a). Additionally, for fixed $\mathbf{x}_0 \in \mathcal{H}$, we now have an alternative representation of

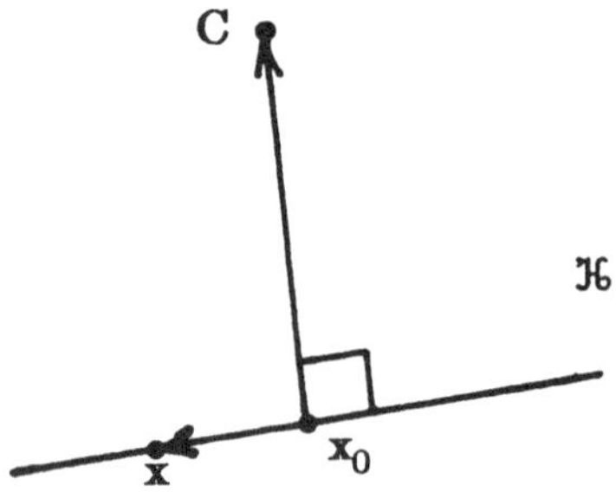

$\mathcal{H} = \{\mathbf{x}\,|\,C'(\mathbf{x}-\mathbf{x}_0) = 0,$
$\mathbf{x} \epsilon \mathbf{R}^n\}$

a.

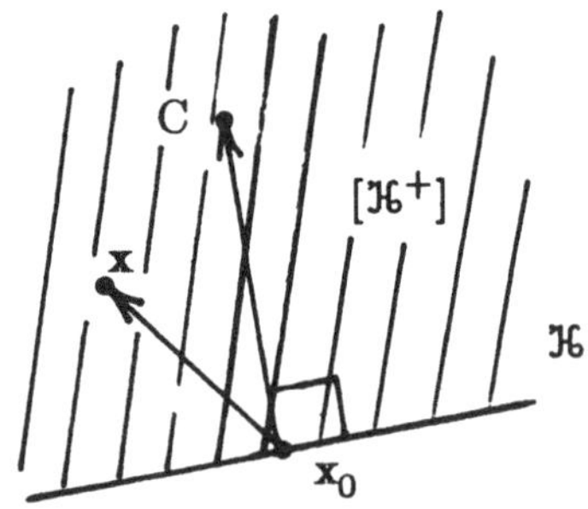

$[\mathcal{H}^+] = \{\mathbf{x}\,|\,C'(\mathbf{x}-\mathbf{x}_0) \geq 0,$
$\mathbf{x} = \mathbf{x}_0 \text{ fixed}, \mathbf{x} \epsilon \mathbf{R}^n\}$

b.

$[\mathcal{H}^-] = \{\mathbf{x}\,|\,C'(\mathbf{x}-\mathbf{x}_0) \leq 0,$
$\mathbf{x} = \mathbf{x}_0 \text{ fixed}, \mathbf{x} \epsilon \mathbf{R}^n\}$

c.

Figure 3.1

the preceding closed half-spaces:

$$[\mathcal{H}^+] = \{\mathbf{x} \mid C'(\mathbf{x}-\mathbf{x}_0) \geq 0,\ \mathbf{x} = \mathbf{x}_0 \text{ fixed}, \mathbf{x} \in R^n\},$$

$$[\mathcal{H}^-] = \{\mathbf{x} \mid C'(\mathbf{x}-\mathbf{x}_0) \leq 0,\ \mathbf{x} = \mathbf{x}_0 \text{ fixed}, \mathbf{x} \in R^n\}$$

(see Figure 3.1.b, c). In what follows either of these equivalent representations of hyperplanes and half-planes will be used depending upon which version is most appropriate for the discussion at hand.

Let $\mathcal{S}_1, \mathcal{S}_2$ be nonempty sets in R^n. Then the hyperplane $\mathcal{H} = \{\mathbf{x} \mid C'\mathbf{x} = \alpha, \mathbf{x} \in R^n\}$ ***weakly separates*** $\mathcal{S}_1, \mathcal{S}_2$ if $C'\mathbf{x} \geq \alpha$ for $\mathbf{x} \in \mathcal{S}_1$ and $C'\mathbf{x} \leq \alpha$ for $\mathbf{x} \in \mathcal{S}_2$. A hyperplane $\mathcal{H}$ ***properly separates*** $\mathcal{S}_1, \mathcal{S}_2$ if $\mathcal{S}_1$ and $\mathcal{S}_2$ are not both contained in $\mathcal{H}$ itself or $\mathcal{S}_1 \cup \mathcal{S}_2 \not\subset \mathcal{H}$ (Figure 3.2.a). Moreover $\mathcal{H}$ ***strictly separates*** $\mathcal{S}_1, \mathcal{S}_2$ if $C'\mathbf{x} > \alpha$ for $\mathbf{x} \in \mathcal{S}_1$ and $C'\mathbf{x} < \alpha$ for $\mathbf{x} \in \mathcal{S}_2$ (Figure 3.2.b depicts the strict separation of open sets $\mathcal{S}_1, \mathcal{S}_2$ by $\mathcal{H}$). Finally, $\mathcal{H}$ is said to ***strongly separate*** $\mathcal{S}_1, \mathcal{S}_2$ if there exists $0 < \in \epsilon R$ such that $C'\mathbf{x} \geq \alpha + \in$ for $\mathbf{x} \in \mathcal{S}_1$ and $C'\mathbf{x} \leq \alpha$ for $\mathbf{x} \in \mathcal{S}_2$ (Figure 3.2.c).

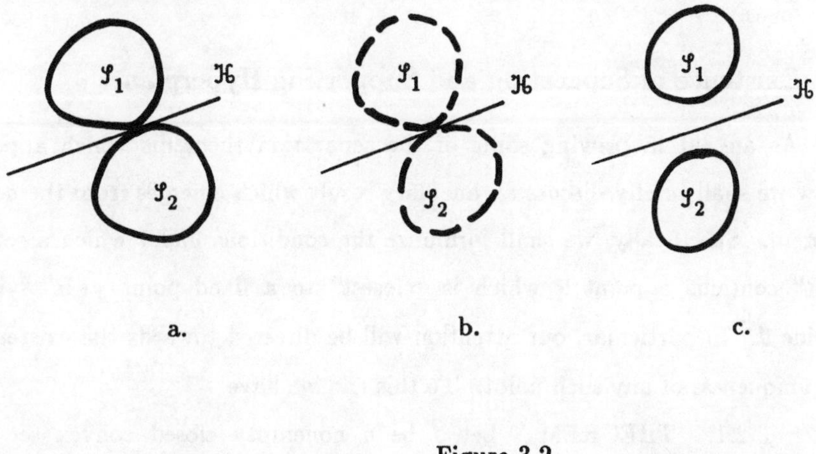

Figure 3.2

As far as the hierarchy of separation characteristics of sets $\mathcal{S}_1$, $\mathcal{S}_2$ in R^n is concerned: strong separation $\rightarrow$ strict separation $\rightarrow$ proper separation $\rightarrow$ weak separation.

For $\mathcal{S}$ a nonempty set in $\mathbf{R}^n$ and $\bar{\mathbf{x}}$ a point in the boundary of $\mathcal{S}$, a hyperplane $\mathcal{H} = \{\mathbf{x} \mid C'(\mathbf{x}-\bar{\mathbf{x}}) = 0, \ \mathbf{x} = \bar{\mathbf{x}} \text{ fixed}, \ \mathbf{x}\in\mathbf{R}^n\}$ is a ***supporting*** (or ***tangent) hyperplane*** of $\mathcal{S}$ at $\bar{\mathbf{x}}$ if either $\mathcal{S} \subset [\mathcal{H}^+]$ (hence $C'(\mathbf{x}-\bar{\mathbf{x}}) \geq 0$ for $\mathbf{x}\in\mathcal{S}$) or $\mathcal{S}\subset[\mathcal{H}^-]$ (so that $C'(\mathbf{x}-\bar{\mathbf{x}}) \leq 0$ for $\mathbf{x}\in\mathcal{S}$). Thus $\mathcal{H}$ is a supporting hyperplane of $\mathcal{S}$ if: (1) $\mathcal{S}$ lies in one of the two closed half-spaces $[\mathcal{H}^+]$, $[\mathcal{H}^-]$ and; (2) $\mathcal{H}$ has a point in common with $\mathcal{S}$. Equivalently, $\mathcal{H}$ is a supporting hyperplane of $\mathcal{S}$ in $\mathbf{R}^n$ at $\bar{\mathbf{x}}\in\partial(\mathcal{S})$ if either $C'\bar{\mathbf{x}} = \inf\{C'\mathbf{x} \mid \mathbf{x}\in\mathcal{S}\}$ or $C'\bar{\mathbf{x}} = \sup\{C'\mathbf{x} \mid \mathbf{x}\in\mathcal{S}\}$. Moreover, $\mathcal{H}$ ***properly supports*** $\mathcal{S}$ in $\mathbf{R}^n$ at $\bar{\mathbf{x}}$ if $\mathcal{S} \not\subset \mathcal{H}$.

If $\mathcal{H}$ is a supporting hyperplane for a convex set $\mathcal{S}$ in $\mathbf{R}^n$ at $\bar{\mathbf{x}}\in\partial(\mathcal{S})$, then the half-spaces

$$[\mathcal{H}_s^+] = \{\mathbf{x} \mid C'\mathbf{x} \geq \alpha = C'\bar{\mathbf{x}}, \ \mathbf{x}\in\mathbf{R}^n\},$$

$$[\mathcal{H}_s^-] = \{\mathbf{x} \mid C'\mathbf{x} \leq \alpha = C'\bar{\mathbf{x}}, \ \mathbf{x}\in\mathbf{R}^n\}$$

are ***supporting closed half-spaces*** for $\mathcal{S}$.

Also, a hyperplane $\mathcal{H}$ is said to ***cut*** a convex set $\mathcal{S}$ in $\mathbf{R}^n$ if and only if: (a) $\mathcal{S} \not\subset \mathcal{H}$; and (b) $\mathcal{H} \cap ri(\mathcal{S}) \neq \phi$.

3.2 Existence of Separating and Supporting Hyperplanes

As an aid in proving some of the separation theorems which appear below we shall briefly discuss an ancillary result which emerges from the next theorem. Specifically, we shall formalize the conditions under which a set $\mathcal{S}$ in $\mathbf{R}^n$ contains a point $\bar{\mathbf{x}}$ which is "closest" to a fixed point $\mathbf{y}\in\mathbf{R}^n$ lying outside $\mathcal{S}$. In particular, our attention will be directed towards the existence and uniqueness of any such point. To this end we have

3.2.1. THEOREM. Let $\mathcal{S}$ be a nonempty closed convex set in $\mathbf{R}^n$ with $\mathbf{y}\in\mathbf{R}^n$ a vector not in $\mathcal{S}$. Then there exists a unique point $\bar{\mathbf{x}}\in\mathcal{S}$ with minimum distance from $\mathbf{y}$ (*i.e.*, $\|\mathbf{y}-\bar{\mathbf{x}}\| \leq \|\mathbf{y}-\mathbf{x}\|$ for all $\mathbf{x}\in\mathcal{S}$) if and only if $(\mathbf{x}-\bar{\mathbf{x}})'(\mathbf{y}-\bar{\mathbf{x}}) \leq 0$ for all $\mathbf{x}\in\mathcal{S}$.

Thus $(\mathbf{x}-\bar{\mathbf{x}})'(\mathbf{y}-\bar{\mathbf{x}}) \leq 0$ for all $\mathbf{x}\in\mathcal{S}$ is a necessary and sufficient condition for $\bar{\mathbf{x}}$

to be the point in $\mathcal{S}$ closest to $\mathbf{y} \notin \mathcal{S}$. As Figure 3.3 reveals, $\bar{\mathbf{x}} \in \mathcal{S}$ is the closest vector to $\mathbf{y} \notin \mathcal{S}$ if and only if $\pi/2 \leq \theta \leq \pi$. But this is equivalent to the requirement that the angle between $\mathbf{y}-\bar{\mathbf{x}}$ and $\mathbf{x}-\bar{\mathbf{x}}$ for any point $\mathbf{x} \in \mathcal{S}$ is $\geq 90°$ or $\cos\theta \leq 0$, *i.e.*,

$$\cos\theta = (\mathbf{x}-\bar{\mathbf{x}})'(\mathbf{y}-\bar{\mathbf{x}})/\|\mathbf{x}-\bar{\mathbf{x}}\|\,\|\mathbf{y}-\bar{\mathbf{x}}\| \leq 0 \quad \text{or}$$

$$(\mathbf{x}-\bar{\mathbf{x}})'(\mathbf{y}-\bar{\mathbf{x}}) = \|\mathbf{x}-\bar{\mathbf{x}}\|\,\|\mathbf{y}-\bar{\mathbf{x}}\| \cos\theta \leq 0.$$

We note briefly that a closest vector $\bar{\mathbf{x}} \in \mathcal{S}$ need not exist for $\mathcal{S}$ an arbitrary convex set in $\mathbf{R}^n$. However, if a convex set $\mathcal{S}$ is also closed, then a closest vector $\bar{\mathbf{x}} \in \mathcal{S}$ does exist. Moreover, it is unique.[1]

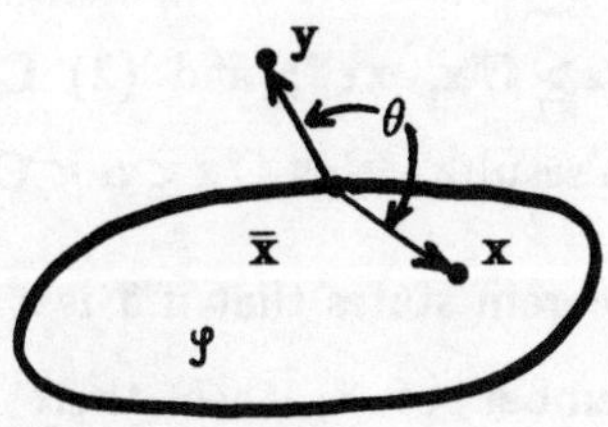

$$(\mathbf{x}-\bar{\mathbf{x}})'(\mathbf{y}-\bar{\mathbf{x}}) \leq 0, \mathbf{x} \in \mathcal{S}$$

Figure 3.3

Based upon this observation, we now have

3.2.2. STRONG SEPARATION THEOREM. Let $\mathcal{S}$ be a nonempty closed convex set in $\mathbf{R}^n$ with $\mathbf{y} \in \mathbf{R}^n$ a vector not in $\mathcal{S}$. Then there exists a vector $\mathbf{C}(\neq \mathbf{0}) \in \mathbf{R}^n$ and a scalar $\alpha \in \mathbf{R}$ such that $\mathbf{C}'\mathbf{x} \leq \alpha < \mathbf{C}'\mathbf{y}$ for all $\mathbf{x} \in \mathcal{S}$.

---

[1] A special case of theorem 3.2.1 emerges if we set $\mathbf{y} = \mathbf{0}$ and consider $\|\mathbf{x}\| = \|\mathbf{x}-\mathbf{0}\|$, $\mathbf{x} \in \mathcal{S}$, as the distance between any point $\mathbf{x}$ in $\mathcal{S}$ and the origin $\mathbf{0}$. We thus have

3.2.1.a. THEOREM. Let $\mathcal{S}$ be a nonempty closed convex set in $\mathbf{R}^n$ with the origin $\mathbf{0}$ not in $\mathcal{S}$. Then there exists a unique point $\bar{\mathbf{x}} \in \mathcal{S}$ with minimum distance from $\mathbf{0}$, *i.e.*,

$$\|\bar{\mathbf{x}}\| = \min\{\|\mathbf{x}\| \mid \mathbf{x} \in \mathcal{S}\}.$$

Hence there exists exactly one point in $\mathcal{S}$, namely $\bar{\mathbf{x}}$, which is closest to the origin in the sense that $\|\bar{\mathbf{x}}\|$ is the minimum distance between any $\mathbf{x} \in \mathcal{S}$ and the origin.

PROOF. Let $\mathcal{S}(\neq\phi)$ be a closed convex set in $\mathbf{R}^n$ with $\mathbf{y}\epsilon\mathbf{R}^n, \mathbf{y}\notin\mathcal{S}$. By theorem 3.2.1, there exists a unique $\bar{\mathbf{x}}\epsilon\mathcal{S}$ such that $(\mathbf{x}-\bar{\mathbf{x}})'(\mathbf{y}-\bar{\mathbf{x}}) \leq 0$ or $-\bar{\mathbf{x}}'(\mathbf{y}-\bar{\mathbf{x}}) \leq -\mathbf{x}'(\mathbf{y}-\bar{\mathbf{x}})$ for all $\mathbf{x}\epsilon\mathcal{S}$. By virtue of this inequality, we have

$$\begin{aligned}\|\mathbf{y}-\bar{\mathbf{x}}\|^2 = (\mathbf{y}-\bar{\mathbf{x}})'(\mathbf{y}-\bar{\mathbf{x}}) &= \mathbf{y}'(\mathbf{y}-\bar{\mathbf{x}}) - \bar{\mathbf{x}}'(\mathbf{y}-\bar{\mathbf{x}}) \\ &\leq \mathbf{y}'(\mathbf{y}-\bar{\mathbf{x}}) - \mathbf{x}'(\mathbf{y}-\bar{\mathbf{x}}) = (\mathbf{y}-\mathbf{x})'(\mathbf{y}-\bar{\mathbf{x}}) \\ &= \mathbf{C}'(\mathbf{y}-\mathbf{x}),\end{aligned}$$

where $\mathbf{C} = \mathbf{y}-\bar{\mathbf{x}}$. Hence

$$\begin{aligned}\mathbf{C}'(\mathbf{y}-\mathbf{x}) &\geq \|\mathbf{y}-\bar{\mathbf{x}}\|^2 = \|\mathbf{C}\|^2 > 0 \quad \text{or} \\ \mathbf{C}'\mathbf{y} &\geq \mathbf{C}'\mathbf{x} + \|\mathbf{C}\|^2 > 0.\end{aligned}$$

If α represents the least upper bound of $\mathbf{C}'\mathbf{x}$ for $\mathbf{x}\epsilon\mathcal{S}$, *i.e.*, $\alpha = sup\{\mathbf{C}'\mathbf{x} \mid \mathbf{x}\epsilon\mathcal{S}\}$, then: (1) clearly $\alpha \geq \mathbf{C}'\mathbf{x}$, $\mathbf{x}\epsilon\mathcal{S}$; and (2) $\mathbf{C}'\mathbf{y} \geq \alpha + \|\mathbf{C}\|^2 > 0$ or $\mathbf{C}'\mathbf{y} > \alpha$. Combining these two results yields $\mathbf{C}'\mathbf{x} \leq \alpha < \mathbf{C}'\mathbf{y}$. Q.E.D.

In sum, this theorem states that if $\mathcal{S}$ is a closed convex set in $\mathbf{R}^n$ and if $\mathbf{y}\epsilon\mathbf{R}^n$ is not a member of $\mathcal{S}$, then there exists a strongly separating hyperplane $\mathcal{H} = \{\mathbf{x} \mid \mathbf{C}'\mathbf{x} = \alpha, \ \mathbf{x}\epsilon\mathbf{R}^n\}$ such that $\mathbf{y}$ is in the open half-space $(\mathcal{H}^+)$ and $\mathcal{S}$ is in the closed half-space $[\mathcal{H}^-]$ determined by $\mathcal{H}$ (Figure 3.4). Note that by implication the hyperplane $\mathcal{H}$ also strictly separates set $\mathcal{S}$ and the vector $\mathbf{y}$.

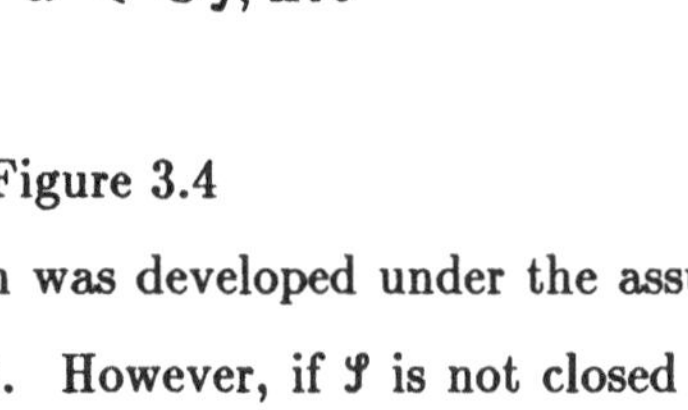

$\mathbf{C}'\mathbf{x} \leq \alpha < \mathbf{C}'\mathbf{y}, \ \mathbf{x}\epsilon\mathcal{S}$

Figure 3.4

The preceding theorem was developed under the assumption that $\mathcal{S}$ is a nonempty closed convex set. However, if $\mathcal{S}$ is not closed and $\mathbf{y}\notin\bar{\mathcal{S}}$, then the basic result of theorem 3.2.2 still holds with $\mathbf{C}'\mathbf{y} > \alpha = sup\{\mathbf{C}'\mathbf{x} \mid \mathbf{x}\epsilon\mathcal{S}\}$. In this instance theorem 3.2.2 may be restated in its less restrictive form as

3.2.2.a. STRONG SEPARATION THEOREM. Let $\mathscr{S}$ be a nonempty convex set in $\mathbf{R}^n$ with $\mathbf{y}\in\mathbf{R}^n$ an exterior point of $\bar{\mathscr{S}}$. Then there exists a vector $\mathbf{C}(\neq\mathbf{0})\in\mathbf{R}^n$ and a scalar $\alpha\in\mathbf{R}$ such that $\mathbf{C}'\mathbf{y} > \alpha = sup\{\mathbf{C}'\mathbf{x}\,|\,\mathbf{x}\in\mathscr{S}\}$.

PROOF. Let $\delta = inf\{\|\mathbf{y}-\mathbf{x}\|\,|\,\mathbf{x}\in\mathscr{S}\} > 0$. Hence there exists a point $\bar{\mathbf{x}}\in\mathbf{R}^n$ on the boundary of $\mathscr{S}$ such that $\|\mathbf{y}-\bar{\mathbf{x}}\| = \delta$.[2] Let $\mathbf{x}\in\mathscr{S}$. For $0<\lambda\leq 1$, $\hat{\mathbf{x}} = \lambda\mathbf{x}+(1-\lambda)\bar{\mathbf{x}} = \bar{\mathbf{x}}+\lambda(\mathbf{x}-\bar{\mathbf{x}})\in\bar{\mathscr{S}}$ and thus $\mathbf{y}-\hat{\mathbf{x}} = (\mathbf{y}-\bar{\mathbf{x}}) -\lambda(\mathbf{x}-\bar{\mathbf{x}})$ (Figure 3.5.a). Then

$$\|\mathbf{y}-\hat{\mathbf{x}}\| = \|(\mathbf{y}-\bar{\mathbf{x}}) - \lambda(\mathbf{x}-\bar{\mathbf{x}})\| \geq \|\mathbf{y}-\bar{\mathbf{x}}\|$$

and thus

$$\|(\mathbf{y}-\bar{\mathbf{x}}) - \lambda(\mathbf{x}-\bar{\mathbf{x}})\|^2 \geq \|\mathbf{y}-\bar{\mathbf{x}}\|^2$$

or, upon expanding

$$(\mathbf{y}-\bar{\mathbf{x}})'(\mathbf{x}-\bar{\mathbf{x}}) - \frac{\lambda}{2}(\mathbf{x}-\bar{\mathbf{x}})'(\mathbf{x}-\bar{\mathbf{x}}) \leq 0.$$

As $\lambda\to 0+$, $(\mathbf{y}-\bar{\mathbf{x}})'(\mathbf{x}-\bar{\mathbf{x}})\leq 0$. Let us rewrite this inequality as $\mathbf{x}'(\mathbf{y}-\bar{\mathbf{x}}) \leq \bar{\mathbf{x}}'(\mathbf{y}-\bar{\mathbf{x}})$. Clearly it must also be true that

$$\mathbf{x}'(\mathbf{y}-\bar{\mathbf{x}}) \leq \bar{\mathbf{x}}'(\mathbf{y}-\bar{\mathbf{x}}) < \mathbf{y}'(\mathbf{x}-\bar{\mathbf{x}}).$$

For $\mathbf{C} = \mathbf{y}-\bar{\mathbf{x}}$, the preceding expression may be modified to read

$$\mathbf{C}'\mathbf{x} \leq \mathbf{C}'\bar{\mathbf{x}} < \mathbf{C}'\mathbf{y}.$$

Then $\alpha = sup\{\mathbf{C}'\mathbf{x}\,|\,\mathbf{x}\in\mathscr{S}\} \leq \mathbf{C}'\bar{\mathbf{x}} < \mathbf{C}'\mathbf{y}$. Q.E.D.

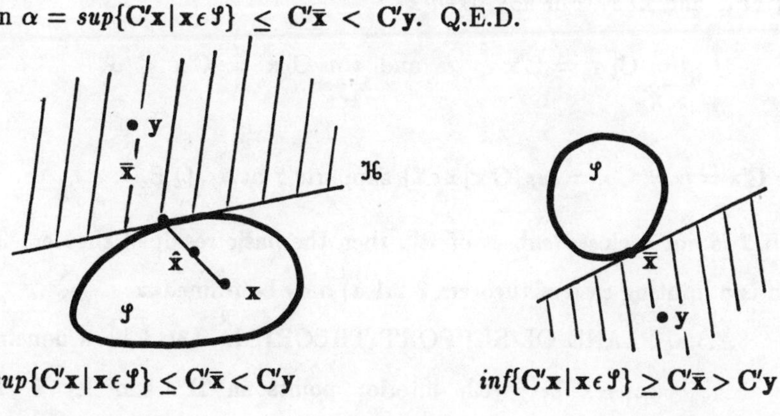

$sup\{\mathbf{C}'\mathbf{x}\,|\,\mathbf{x}\in\mathscr{S}\} \leq \mathbf{C}'\bar{\mathbf{x}} < \mathbf{C}'\mathbf{y}$ $\qquad$ $inf\{\mathbf{C}'\mathbf{x}\,|\,\mathbf{x}\in\mathscr{S}\} \geq \mathbf{C}'\bar{\mathbf{x}} > \mathbf{C}'\mathbf{y}$

a. $\qquad$ b.

Figure 3.5

[2] By Weierstrass' theorem, any continuous function $f(\mathbf{x}) = \|\mathbf{y}-\mathbf{x}\|$ attains a global minimum (as well as a global maximum) over a closed and bounded set in $\mathbf{R}^n$. Since $\bar{\mathscr{S}}\cap B(\mathbf{y},2\delta)(\neq\phi)$ is closed and bounded, there exists an $\mathbf{x} = \bar{\mathbf{x}}\in\bar{\mathscr{S}}\cap B(\mathbf{y},2\delta)$ such that $\delta = \|\mathbf{y}-\bar{\mathbf{x}}\|$.

This result reveals that if $\mathcal{S}$ is a convex set in $\mathbf{R}^n$ and $\mathbf{y} \in \mathbf{R}^n$ is not a member of $\bar{\mathcal{S}}$, then there exists a strongly separating hyperplane $\mathcal{H} = \{\mathbf{x} \mid \mathbf{C}'\mathbf{x} = \alpha,\ \mathbf{x} \in \mathbf{R}^n\}$ passing through $\bar{\mathbf{x}} \in \partial(\mathcal{S})$ and having $\mathcal{S}$ on or beneath it with $\mathbf{y}$ an element of the open half-space $(\mathcal{H}^+)$. Note that if the situation depicted in Figure 3.5.b prevails, then the conclusion of the theorem is that $\mathit{inf}\{\mathbf{C}'\mathbf{x} \mid \mathbf{x} \in \mathcal{S}\} \geq \mathbf{C}'\bar{\mathbf{x}} > \mathbf{C}'\mathbf{y}$.

3.2.3. PLANE OF SUPPORT THEOREM. Let $\mathcal{S}$ be a nonempty closed convex set in $\mathbf{R}^n$ and let $\bar{\mathbf{x}}$ be a boundary point of $\mathcal{S}$. Then there exists a vector $\mathbf{C}(\neq \mathbf{0}) \in \mathbf{R}^n$ and a scalar $\alpha \in \mathbf{R}$ such that $\mathbf{C}'\mathbf{x} \leq \alpha = \mathbf{C}'\bar{\mathbf{x}}$ $(= \mathit{sup}\{\mathbf{C}'\mathbf{x} \mid \mathbf{x} \in \mathcal{S}\})$ for all $\mathbf{x} \in \mathcal{S}$.

PROOF. Choose a sequence of vectors $\{\mathbf{u}_k\}$ external to $\mathcal{S}$ and such that $\lim_{k \to \infty} \mathbf{u}_k = \bar{\mathbf{x}}$. Let $\mathcal{H}_k = \{\mathbf{x} \mid \mathbf{C}_k'\, \mathbf{x} = \alpha_k,\ \mathbf{x} \in \mathbf{R}^n\}$ be a sequence of hyperplanes which strictly separate $\mathbf{U}_k$ from $\bar{\mathbf{x}}$ (that such a sequence of hyperplanes exists follows from theorem 3.2.2). In addition, let $\hat{\mathbf{C}}_k' = (\mathbf{C}_k, \alpha_k) \in \mathbf{R}^{n+1}$. Since $\hat{\mathbf{C}}_k \neq \mathbf{0}$, let us assume that $\left\| \hat{\mathbf{C}}_k \right\| = 1$ for all k so the $\{\hat{\mathbf{C}}_k\}$ is restricted to a compact subset (*i.e.*, a closed ball of radius 1) of $\mathbf{R}^n$. With $\{\hat{\mathbf{C}}_k\}$ bounded, it has a convergent subsequence $\{\hat{\mathbf{C}}_k\}_{k \in K}$ with limit $\hat{\mathbf{C}}' = (\mathbf{C}, \alpha)$ (and with $\|\hat{\mathbf{C}}\| = 1$). For this subsequence $\mathbf{C}_k'\mathbf{u}_k > \alpha_k$ and $\mathbf{C}_k'\mathbf{x} < \alpha_k$ for all $k \in K$ and $\mathbf{x} \in \mathcal{S}$. For any fixed $\mathbf{x} \in \mathcal{S}$,

$$\lim_{\substack{k \to \infty \\ k \in K}} \mathbf{C}_k'\mathbf{u}_k = \mathbf{C}'\bar{\mathbf{x}} \geq \alpha \quad \text{and} \quad \lim_{\substack{k \to \infty \\ k \in K}} \mathbf{C}_k'\mathbf{x} = \mathbf{C}'\mathbf{x} \leq \alpha.$$

Hence $\mathbf{C}'\mathbf{x} = \alpha = \mathbf{C}'\bar{\mathbf{x}} = \mathit{sup}\{\mathbf{C}'\mathbf{x} \mid \mathbf{x} \in \mathcal{S}\}$ supports $\mathcal{S}$ at $\bar{\mathbf{x}}$. Q.E.D.

If $\mathcal{S}$ is not a closed subset of $\mathbf{R}^n$, then the basic result of theorem 3.2.3 (which is a limiting case of theorem 3.2.1.a) may be framed as

3.2.3.a. PLANE OF SUPPORT THEOREM. Let $\mathcal{S}$ be a nonempty convex set with interior points in $\mathbf{R}^n$ and let $\bar{\mathbf{x}}$ be a boundary point of $\mathcal{S}$. Then there exists a vector $\mathbf{C}(\neq \mathbf{0}) \in \mathbf{R}^n$ such that $\mathbf{C}'\mathbf{x} \leq \alpha = \mathbf{C}'\bar{\mathbf{x}}$ $(= \mathit{sup}\{\mathbf{C}'\mathbf{x} \mid \mathbf{x} \in \bar{\mathcal{S}}\})$ for each $\mathbf{x}$ contained within the closure of $\mathcal{S}$, $\bar{\mathcal{S}}$.

PROOF. Let $\bar{\mathbf{x}} \in \partial(\mathcal{S})$. Then there exists a sequence $\{\mathbf{u}_k\}$ not in $\bar{\mathcal{I}}$ such that $\lim_{k\to\infty} \mathbf{u}_k = \bar{\mathbf{x}}$. By theorem 3.2.1, corresponding to each $\mathbf{u}_k$ there exists a vector $\mathbf{C}_k$ ($\neq \mathbf{0}$) such that $\mathbf{C}'_k\mathbf{u}_k > \mathbf{C}'_k\mathbf{x}$ for all $\mathbf{x}\epsilon\bar{\mathcal{I}}$. With $\mathbf{C}_k \neq \mathbf{0}$, we can assume that $\|\mathbf{C}_k\| = 1$ for all k so that $\{\mathbf{C}_k\}$ is contained within a compact subset of $\mathbf{R}^n$. Since $\{\mathbf{C}_k\}$ is bounded, it has a convergent subsequence $\{\mathbf{C}_k\}_{k\epsilon K}$ with limit $\mathbf{C}(\|\mathbf{C}\| = 1)$. For this subsequence $\mathbf{C}'_k\mathbf{u}_k > \mathbf{C}'_k\mathbf{x}$ for all $k\epsilon K$ and $\mathbf{x}\epsilon\bar{\mathcal{I}}$. For fixed $\mathbf{x}\epsilon\bar{\mathcal{I}}$,

$$\lim_{\substack{k\to\infty \\ k\epsilon K}} \mathbf{C}'_k\mathbf{U}_k = \mathbf{C}'\bar{\mathbf{x}} \geq \lim_{\substack{k\to\infty \\ k\epsilon K}} \mathbf{C}'_k\mathbf{x} = \mathbf{C}'\mathbf{x}.$$

Hence $\mathbf{C}'\mathbf{x} = \mathbf{C}'\bar{\mathbf{x}} = sup\{\mathbf{C}'\mathbf{x}\,|\,\mathbf{x}\epsilon\bar{\mathcal{I}}\}$ is a supporting hyperplane for $\mathcal{I}$ at $\bar{\mathbf{x}}$. Q.E.D.

Theorems 3.2.3, 3.2.3.a inform us that a convex subset of $\mathbf{R}^n$ (which may or may not be closed) has a supporting hyperplane at each of its boundary points (Figure 3.6). So given a boundary point of a convex set $\mathcal{I}$ in $\mathbf{R}^n$, there exists a hyperplane that: (1) contains the boundary point, *i.e.*, $\mathcal{H} = \{\mathbf{x}\,|\,\mathbf{C}'\mathbf{x} = \alpha = \mathbf{C}'\bar{\mathbf{x}}, \mathbf{x}\epsilon\mathbf{R}^n\}$; and (2) contains $\mathcal{I}$ in one of its supporting closed half-spaces (as illustrated in Figure 3.6, $\mathcal{I} \subset [\mathcal{H}_s^-] = \{\mathbf{x}\,|\,\mathbf{C}'\mathbf{x} \leq \mathbf{C}'\bar{\mathbf{x}}, \mathbf{x}\epsilon\mathbf{R}^n\}$).

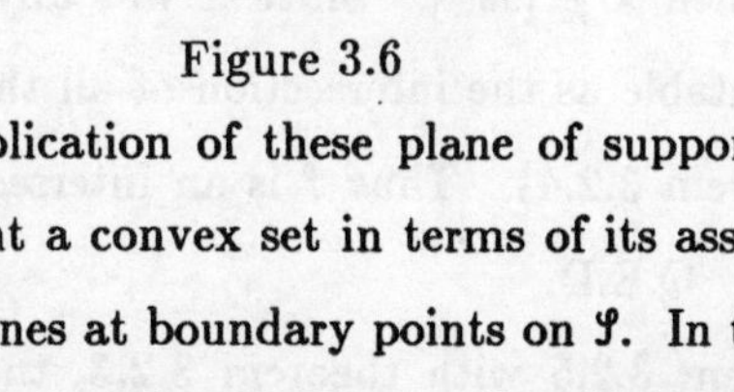

$\mathcal{I} \subset [\mathcal{H}_s^-]$ (or $\bar{\mathcal{I}} \subset [\mathcal{H}_s^-]$)

Figure 3.6

One important implication of these plane of support theorems is that they allow us to represent a convex set in terms of its associated set of supporting or tangent hyperplanes at boundary points on $\mathcal{I}$. In this regard,

3.2.4. REPRESENTATION THEOREM. A closed convex set $\mathcal{I}$ in $\mathbf{R}^n$ is the intersection of all supporting closed half-spaces containing it, *i.e.*, $\mathcal{I} = \cap_i [\mathcal{H}_s^+]_i$ (or $\mathcal{I} = \cap_i [\mathcal{H}_s^-]_i$).

PROOF. We know from our earlier discussions on properties of convex sets that the intersection of an arbitrary number of closed and convex sets in $\mathbf{R}^n$ is itself a closed convex set. Hence the intersection of all supporting closed half-spaces containing $\mathcal{S}$ is the smallest closed convex set containing $\mathcal{S}$ so that $\mathcal{S} = co(\mathcal{S}) = \cap_i[\mathcal{H}_s^+]_i$ (by virtue of theorems 2.3.1, 2.3.2). Q.E.D.

Note that if $\mathcal{S}$ in $\mathbf{R}^n$ is only convex and not also closed, then $\bar{\mathcal{S}} = \cap_i[\mathcal{H}_s^+]_i$ (or $\bar{\mathcal{S}} = \cap_i [\mathcal{H}_s^-]_i$).

A somewhat stronger version of the preceding theorem is

3.2.4.a. REPRESENTATION THEOREM. A closed onvex set $\mathcal{S}$ in $\mathbf{R}^n$ is the intersection of all the closed half-spaces tangent to it.

As will now be demonstrated, an important application of the notion of a supporting hyperplane is that it can be used to characterize a closed convex set with nonempty interior as convex. Specifically,

3.2.5. THEOREM. Let $\mathcal{S}$ be a closed set in $\mathbf{R}^n$ with $\mathcal{S}^o \neq \phi$. If $\mathcal{S}$ has a supporting hyperplane at each of its boundary points, then $\mathcal{S}$ is convex.

PROOF. If $\mathcal{S} = \mathbf{R}^n$, then clearly $\mathcal{S}$ is convex. For $\mathcal{S} \subset \mathbf{R}^n$, we may assume that there exists a point $\mathbf{x} \notin \mathcal{S}$. If $\bar{\mathbf{x}} \in \mathcal{S}^o$, then there exists a boundary point $\mathbf{y}$ of $\mathcal{S}$ such that $\mathbf{y} \in ri([\mathbf{x}, \bar{\mathbf{x}}])$. Let $\mathcal{H} = \{\mathbf{x} \,|\, \mathbf{C}'\mathbf{x} = \alpha, \mathbf{x} \in \mathbf{R}^n\}$ be a hyperplane supporting $\mathcal{S}$ at $\mathbf{y}$. Then $\mathbf{x} \notin \mathcal{H}$ (since otherwise $\mathcal{H}$ would contain the interior point $\bar{\mathbf{x}}$ - an impossibility). If, say, $\mathcal{S} \subset [\mathcal{H}^-] = \{\mathbf{x} \,|\, \mathbf{C}'\mathbf{x} \leq \mathbf{C}'\mathbf{y}, \mathbf{x} \in \mathbf{R}^n\}$, then $\mathbf{x} \notin [\mathcal{H}^-]$. Since $\mathbf{x}$ was any point not in $\mathcal{S}$, it follows that $\mathcal{S}$ is representable as the intersection of all the closed half-spaces that contain it (by theorem 3.2.4). Thus $\mathcal{S}$ is an intersection of convex sets and consequently convex. Q.E.D.

If we couple theorem 3.2.5 with theorem 3.2.3, then a necessary and sufficient condition for a closed set with nonempty interior to be convex is

3.2.6. THEOREM. Let $\mathcal{S}$ be a closed set in $\mathbf{R}^n$ with $\mathcal{S}^o \neq \phi$. Then $\mathcal{S}$ is convex if and only if $\mathcal{S}$ has a supporting hyperplane at each of its boundary points.

The preceding separation theorems involved the existence of a hyperplane which separates a convex set $\mathcal{S}$ (possibly closed) in $\mathbf{R}^n$ and a point exterior to $\mathcal{S}$. Let us now turn to the development of a set of theorems which address the separation of two convex sets $\mathcal{S}_1$, $\mathcal{S}_2$ in $\mathbf{R}^n$.

We state first

3.2.7. STRONG SEPARATION THEOREM [Minkowski, 1911]. Let $\mathcal{S}_1$, $\mathcal{S}_2$ be two non-empty disjoint convex sets in $\mathbf{R}^n$ with $\mathcal{S}_1$ compact and $\mathcal{S}_2$ closed. Then there exists a vector $\mathbf{C}(\neq\mathbf{0})\in\mathbf{R}^n$ and a scalar $\alpha\in\mathbf{R}$ such that the following equivalent conditions hold:

(a) $\mathbf{C}'\mathbf{x}_1 < \mathbf{C}'\mathbf{x}_2$ for $\mathbf{x}_1\in\mathcal{S}_1$, $\mathbf{x}_2\in\mathcal{S}_2$ or

(b) $\inf\{\mathbf{C}'\mathbf{x}\,|\,\mathbf{x}\in\mathcal{S}_2\} > \sup\{\mathbf{C}'\mathbf{x}\,|\,\mathbf{x}\in\mathcal{S}_1\}$ or

(c) $\begin{cases}\mathbf{C}'\mathbf{x} < \alpha, \ \mathbf{x}\in\mathcal{S}_1;\\ \mathbf{C}'\mathbf{x} > \alpha, \ \mathbf{x}\in\mathcal{S}_2.\end{cases}$

PROOF. Consider the set $\mathcal{S} = \mathcal{S}_1 - \mathcal{S}_2 = \{\mathbf{x}\,|\,\mathbf{x} = \mathbf{x}_1 - \mathbf{x}_2, \mathbf{x}_1\in\mathcal{S}_1, \mathbf{x}_2\in\mathcal{S}_2\}$. As noted in chapter 2, $\mathcal{S}$ is convex and, since $\mathcal{S}_1\cap\mathcal{S}_2 = \phi$, it follows that $\mathbf{0}\notin\mathcal{S}$. Moreover, $\mathcal{S}$ is closed (as demonstrated in chapter 1). Hence, from theorem 3.2.2, there exists a $\mathbf{C}(\neq\mathbf{0})$ and a scalar α such that $\mathbf{C}'\mathbf{x} \leq \alpha < \mathbf{C}'\mathbf{0} = 0$ for all $\mathbf{x}\in\mathcal{S}$ or $\mathbf{C}'(\mathbf{x}_1-\mathbf{x}_2) < 0$, $\mathbf{x}_1\in\mathcal{S}_1$, $\mathbf{x}_2\in\mathcal{S}_2$. But this inequality implies that $\mathbf{C}'\mathbf{x}_1 < \mathbf{C}'\mathbf{x}_2$ for $\mathbf{x}_1\in\mathcal{S}_1$, $\mathbf{x}_2\in\mathcal{S}_2$ (Figure 3.7) or $\sup\{\mathbf{C}'\mathbf{x}\,|\,\mathbf{x}\in\mathcal{S}_1\} < \inf\{\mathbf{C}'\mathbf{x}\,|\,\mathbf{x}\in\mathcal{S}_2\}$. If we set $\alpha_1 = \sup\{\mathbf{C}'\mathbf{x}\,|\,\mathbf{x}\in\mathcal{S}_1\}$, $\alpha_2 = \inf\{\mathbf{C}'\mathbf{x}\,|\,\mathbf{x}\in\mathcal{S}_2\}$, then, for $\alpha = (\alpha_1+\alpha_2)/2$, $\mathbf{C}'\mathbf{x} < \alpha$, $\mathbf{x}\in\mathcal{S}_1$, and $\mathbf{C}'\mathbf{x} > \alpha$, $\mathbf{x}\in\mathcal{S}_2$. Q.E.D.

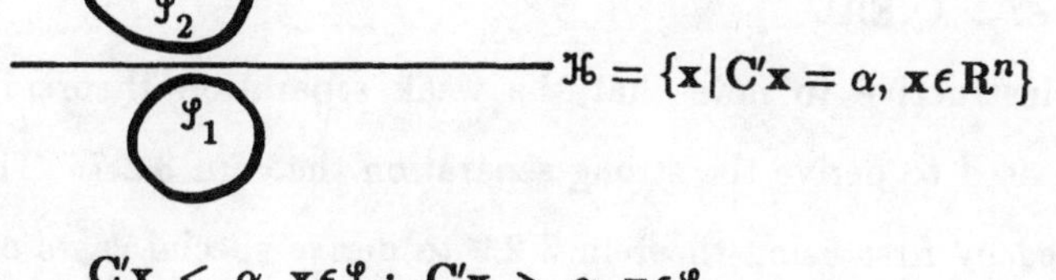

$\mathbf{C}'\mathbf{x} < \alpha$, $\mathbf{x}\in\mathcal{S}_1$; $\mathbf{C}'\mathbf{x} > \alpha$, $\mathbf{x}\in\mathcal{S}_2$

Figure 3.7

To set the stage for the next theorem which follows, theorems 3.2.2, 3.2.3.a taken together allow us to derive

3.2.8. COROLLARY. Let $\mathcal{S}$ be a nonempty convex set in $\mathbf{R}^n$ with $\bar{\mathbf{x}} \notin \mathcal{S}$. Then there exists a vector $\mathbf{C}(\neq \mathbf{0}) \in \mathbf{R}^n$ such that $\mathbf{C}'(\mathbf{x}-\bar{\mathbf{x}}) \leq 0$ or $\mathbf{C}'\mathbf{x} \leq \mathbf{C}'\bar{\mathbf{x}}$ for all $\mathbf{x} \in \bar{\mathcal{S}}$.

PROOF. If $\bar{\mathbf{x}} \notin \bar{\mathcal{S}} = \mathcal{S} \cup \partial(\mathcal{S})$, then the corollary follows from theorem 3.2.2, *i.e.*, $\mathbf{C}'\mathbf{x} \leq \mathbf{C}'\bar{\mathbf{x}} = sup\{\mathbf{C}'\mathbf{x} | \mathbf{x} \in \mathcal{S}\}$. If $\bar{\mathbf{x}} \in \partial(\mathcal{S})$, then this corollary follows from theorem 3.2.3.a so that $\mathbf{C}'\mathbf{x} \leq \mathbf{C}'\bar{\mathbf{x}} = sup\{\mathbf{C}'\mathbf{x} | \mathbf{x} \in \bar{\mathcal{S}}\}$. Q.E.D.

On the basis of this corollary we have the

3.2.9. WEAK SEPARATION THEOREM [Minkowski, 1911]. Let $\mathcal{S}_1$, $\mathcal{S}_2$ be two non-empty disjoint convex sets in $\mathbf{R}^n$. Then there exists a vector $\mathbf{C}(\neq \mathbf{0}) \in \mathbf{R}^n$ and a scalar $\alpha \in \mathbf{R}$ such that the following equivalent conditions hold:

(a) $\mathbf{C}'\mathbf{x}_1 \leq \mathbf{C}'\mathbf{x}_2$ for $\mathbf{x}_1 \in \mathcal{S}_1$, $\mathbf{x}_2 \in \mathcal{S}_2$ or

(b) $inf\{\mathbf{C}'\mathbf{x} | \mathbf{x} \in \mathcal{S}_2\} \geq sup\{\mathbf{C}'\mathbf{x} | \mathbf{x} \in \mathcal{S}_1\}$ or

(c) $\begin{cases} \mathbf{C}'\mathbf{x} \leq \alpha, \ \mathbf{x} \in \mathcal{S}_1; \\ \mathbf{C}'\mathbf{x} \geq \alpha, \ \mathbf{x} \in \mathcal{S}_2. \end{cases}$

PROOF. As stated in the proof of theorem 3.2.7 above, $\mathcal{S} = \mathcal{S}_1 - \mathcal{S}_2 = \{\mathbf{x} | \mathbf{x}=\mathbf{x}_1-\mathbf{x}_2, \ \mathbf{x}_1 \in \mathcal{S}_1, \ \mathbf{x}_2 \in \mathcal{S}_2\}$ is convex and, with $\mathcal{S}_1 \cap \mathcal{S}_2 = \phi$, $\mathbf{0} \notin \mathcal{S}$. By corollary 3.2.8, there exists a $\mathbf{C}(\neq \mathbf{0})$ such that $\mathbf{C}'(\mathbf{x}-\bar{\mathbf{x}}) = \mathbf{C}'(\mathbf{x}-\mathbf{0}) = \mathbf{C}'(\mathbf{x}_1-\mathbf{x}_2) \leq 0$ for all $\mathbf{x} \in \mathcal{S}$. Hence $\mathbf{C}'\mathbf{x}_1 \leq \mathbf{C}'\mathbf{x}_2$ for $\mathbf{x}_1 \in \mathcal{S}_1, \mathbf{x}_2 \in \mathcal{S}_2$ or $inf\{\mathbf{C}'\mathbf{x} | \mathbf{x} \in \mathcal{S}_2\} \geq sup\{\mathbf{C}'\mathbf{x} | \mathbf{x} \in \mathcal{S}_1\}$. If $\alpha_1 = sup\{\mathbf{C}'\mathbf{x} | \mathbf{x} \in \mathcal{S}_1\}$, $\alpha_2 = inf\{\mathbf{C}'\mathbf{x} | \mathbf{x} \in \mathcal{S}_2\}$, then taking $\alpha = (\alpha_1+\alpha_2)/2$ renders $\mathbf{C}'\mathbf{x} \leq \alpha$, $\mathbf{x} \in \mathcal{S}_1$, and $\mathbf{C}'\mathbf{x} \geq \alpha$, $\mathbf{x} \in \mathcal{S}_2$. Q.E.D.

It is instructive to note that the weak separation theorem 3.2.9 can actually be used to derive the strong separation theorem 3.2.7. This may be accomplished by first using theorem 3.2.9 to derive special cases of theorems 3.2.2a, 3.2.2 (wherein $\mathbf{y}=\mathbf{0}$), which in turn lead to the development of theorem 3.2.7. To this end we begin with

3.2.2.b. COROLLARY. Let $\mathcal{S}$ be a nonempty convex set in $\mathbf{R}^n$ with $\mathbf{0}\notin\overline{\mathcal{S}}$. Then there exists a vector $\mathbf{C}(\neq\mathbf{0})$ and a scalar $\alpha\in\mathbf{R}$ such that $\mathbf{C}'\mathbf{x}<\alpha<0$ for all $\mathbf{x}\in\mathcal{S}$.

PROOF. Since $\mathbf{0}\notin\overline{\mathcal{S}}$, there exists an open sphere of radius $\delta>0$ about $\mathbf{0}$, $B(\mathbf{0},\delta)$, such that $B(\mathbf{0},\delta)\cap\mathcal{S}=\phi$. Since $B(\mathbf{0},\delta)$ is also nonempty and convex, theorem 3.2.9 posits the existence of a hyperplane $\mathcal{H}=\{\mathbf{x}\,|\,\mathbf{C}'\mathbf{x}=\beta,\mathbf{x}\in\mathbf{R}^n\}$ such that

$$\mathbf{C}'\mathbf{x}\leq\beta,\ \mathbf{x}\in\mathcal{S}$$
$$\mathbf{C}'\mathbf{x}\geq\beta,\ \mathbf{x}\in B(\mathbf{0},\delta).$$

For $0<\tau\in\mathbf{R}$ sufficiently small, $B(\mathbf{0},\delta)$ must contain the vector $\tau\mathbf{C}(\tau\mathbf{C}\neq\mathbf{0})$. Let $\beta\leq-\tau\|\mathbf{C}\|^2<0$ and choose $\alpha=\beta/2$. Then $\mathbf{C}'\mathbf{x}\leq\beta<0$, $\mathbf{x}\in\mathcal{S}$. Q.E.D.

From this result we easily obtain its more restrictive version

3.2.2.c. LEMMA. Let $\mathcal{S}$ be a nonempty closed convex set in $\mathbf{R}^n$ with $\mathbf{0}\notin\mathcal{S}$. Then there exists a vector $\mathbf{C}(\neq\mathbf{0})\in\mathbf{R}^n$ and a scalar $\alpha\in\mathbf{R}$ such that $\mathbf{C}'\mathbf{x}<\alpha<0$ for all $\mathbf{x}\in\mathcal{S}$.

PROOF. Since $\mathcal{S}$ is closed and $\mathbf{0}\notin\mathcal{S}$, it follows that $\mathbf{0}\notin\overline{\mathcal{S}}(=\mathcal{S}$ if $\mathcal{S}$ is closed). Hence by the preceding corollary there exists a vector $\mathbf{C}$ and a scalar α such that $\mathbf{C}'\mathbf{x}<\alpha<0$, $\mathbf{x}\in\mathcal{S}$. Q.E.D.

If we now deem $\mathcal{S}_1$, $\mathcal{S}_2$ nonempty disjoint convex sets in $\mathbf{R}^n$ with $\mathcal{S}_1$ compact and $\mathcal{S}_2$ closed, then as noted above in the proof of theorem 3.2.5, $\mathcal{S}=\mathcal{S}_1-\mathcal{S}_2$ is closed, $\mathbf{0}\notin\mathcal{S}$, and thus the strong separation results (a)-(c) follow.

The weak separation theorem 3.2.9, while enabling us to ultimately derive the strong separation theorem 3.2.7, also serves as the foundation for a variety of additional separation results and thus is sometimes referred to as the "fundamental separation theorem." In particular, we now list, without proof, a collection of theorems which can be derived with the aid of theorem 3.2.9. Specifically,

3.2.9.a. THEOREM. Let $\mathcal{S}$ be a nonempty convex set in $\mathbf{R}^n$ which has no point in common with the nonnegative orthant. Then there exists a vector $\mathbf{0}<\mathbf{C}\in\mathbf{R}^n$ such that $\mathbf{C}'\mathbf{x}\leq0$ for all $\mathbf{x}\in\mathcal{S}$. (Here one of the sets $\mathcal{S}_1, \mathcal{S}_2$ serves as the non-negative orthant.)

The next theorem informs us that two convex sets in $\mathbf{R}^n$ need not be disjoint to be separated by a hyperplane.

3.2.9.b. THEOREM. Let $\mathcal{S}_1, \mathcal{S}_2$ be convex sets in $\mathbf{R}^n$ and such that $\mathcal{S}_1 \neq \phi, \mathcal{S}_2^o \neq \phi$. Then if $\mathcal{S}_1 \cap \mathcal{S}_2^o = \phi$ there exists a vector $\mathbf{C}(\neq\mathbf{0})\in\mathbf{R}^n$ and a scalar $\alpha\in\mathbf{R}$ such that conditions (a)-(c) of theorem 3.2.9 hold. Moreover, if $\mathcal{S}_1$, $\mathcal{S}_2$ are open, then the separation is strict.

3.2.9.c. THEOREM. Let $\mathcal{S}_1, \mathcal{S}_2$ be convex sets in $\mathbf{R}^n$ with $\mathcal{S}_1^o, \mathcal{S}_2^o \neq \phi$. If $\mathcal{S}_1^o \cap \mathcal{S}_2^o = \phi$, then there exists a vector $\mathbf{C}(\neq\mathbf{0})\in\mathbf{R}^n$ and a scalar $\alpha\in\mathbf{R}$ such that conditions (a)-(c) of theorem 3.2.9 hold.

At this point we seek to close the gap in the hierarchy of separation characteristics by briefly exploring the notion of proper separation. The first theorem involving proper separation holds for general sets in $\mathbf{R}^n$. That is,

3.2.10. PROPER SEPARATION THEOREM [Rockafellar, 1970]. Let $\mathcal{S}_1, \mathcal{S}_2$ be non-empty sets in $\mathbf{R}^n$. There exists a hyperplane separating $\mathcal{S}_1$, $\mathcal{S}_2$ properly if and only if there exists a vector $\mathbf{C}(\neq\mathbf{0})\in\mathbf{R}^n$ such that:

(a) $\mathit{inf}\{\mathbf{C}'\mathbf{x} \mid \mathbf{x}\in\mathcal{S}_2\} \geq \mathit{sup}\{\mathbf{C}'\mathbf{x} \mid \mathbf{x}\in\mathcal{S}_1\}$,

(b) $\mathit{sup}\{\mathbf{C}'\mathbf{x} \mid \mathbf{x}\in\mathcal{S}_2\} > \mathit{inf}\{\mathbf{C}'\mathbf{x} \mid \mathbf{x}\in\mathcal{S}_1\}$

PROOF: (sufficiency) Suppose $\mathbf{C}$ satisfies (a), (b). For $\alpha_1 = \mathit{inf}\{\mathbf{C}'\mathbf{x} \mid \mathbf{x}\in\mathcal{S}_2\}$ and $\alpha_2 = \mathit{sup}\{\mathbf{C}'\mathbf{x} \mid \mathbf{x}\in\mathcal{S}_1\}$, form $\alpha = (\alpha_1+\alpha_2)/2\in\mathbf{R}$ so that $\mathcal{H} = \{\mathbf{x} \mid \mathbf{C}'\mathbf{x} = \alpha\}$. Clearly $\mathcal{S}_1 \subseteq [\mathcal{H}^-] = \{\mathbf{x} \mid \mathbf{C}'\mathbf{x} \leq \alpha\}$, $\mathcal{S}_2 \subseteq [\mathcal{H}^+] = \{\mathbf{x} \mid \mathbf{C}'\mathbf{x} \geq \alpha\}$. Since (b) implies that $\mathcal{S}_1$, $\mathcal{S}_2$ are not both contained in $\mathcal{H}$, it follows that $\mathcal{H}$ separates $\mathcal{S}_1$, $\mathcal{S}_2$ properly.

(necessity) If $\mathcal{S}_1$, $\mathcal{S}_2$ are separated properly, it follows that $\mathbf{C}'\mathbf{x} \leq \alpha$ for every $\mathbf{x}\in\mathcal{S}_1$ and $\mathbf{C}'\mathbf{x} \geq \alpha$ for every $\mathbf{x}\in\mathcal{S}_2$ with strict inequality holding for at least one $\mathbf{x}\in\mathcal{S}_1$ or $\mathbf{x}\in\mathcal{S}_2$. Thus $\mathbf{C}$ meets conditions (a), (b) above. Q.E.D.

A somewhat stronger result is provided by

3.2.11. PROPER SEPARATION THEOREM [Rockafellar, 1970]. Let $\mathcal{S}_1, \mathcal{S}_2$ be non-empty convex sets in $\mathbf{R}^n$. There exists a hyperplane $\mathcal{H}$ separating $\mathcal{S}_1$, $\mathcal{S}_2$ properly if and only if $ri(\mathcal{S}_1) \cap ri(\mathcal{S}_2) = \phi$.

PROOF. (sufficiency) Let $\mathcal{S} = \mathcal{S}_1 - \mathcal{S}_2$ with $ri(\mathcal{S}) = ri(\mathcal{S}_1) - ri(\mathcal{S}_2)$. If $ri(\mathcal{S}_1) \cap ri(\mathcal{S}_2) = \phi$, then $\mathbf{0} \notin ri(\mathcal{S})$ so that there exists a hyperplane $\mathcal{H}$ containing $\mathbf{0}$ such that $ri(\mathcal{S})$ is contained in one of its associated open half-spaces; the closure of that half-space then contains $\mathcal{S}$, since $\mathcal{S} \subset \overline{ri(\mathcal{S})}$. So if $\mathbf{0} \notin ri(\mathcal{S})$, there exists a vector $\mathbf{C}(\neq \mathbf{0})$ such that

$$0 \leq inf\{\mathbf{C}'\mathbf{x} \,|\, \mathbf{x} \in \mathcal{S}\} = inf\{\mathbf{C}'\mathbf{x}_2 \,|\, \mathbf{x}_2 \in \mathcal{S}_2\} - sup\{\mathbf{C}'\mathbf{x}_1 \,|\, \mathbf{x}_1 \in \mathcal{S}_1\},$$

$$0 < sup\{\mathbf{C}'\mathbf{x} \,|\, \mathbf{x} \in \mathcal{S}\} = sup\{\mathbf{C}'\mathbf{x}_2 \,|\, \mathbf{x}_2 \in \mathcal{S}_2\} - inf\{\mathbf{C}'\mathbf{x}_1 \,|\, \mathbf{x}_1 \in \mathcal{S}_1\}.$$

Hence $\mathcal{S}_1$, $\mathcal{S}_2$ can be separated properly by virtue of theorem 3.2.10.

(necessity) The preceding two inequalities imply that $\mathbf{0} \notin ri(\mathcal{S})$ (which in turn implies that $ri(\mathcal{S}_1) \cap ri(\mathcal{S}_2) = \phi$) since they guarantee the existence of a closed half-space $[\mathcal{H}^+] = \{\mathbf{x} \,|\, \mathbf{C}'\mathbf{x} \geq 0, \mathbf{x} \in \mathbf{R}^n\} \supset \mathcal{S}$ and such that $ri[\mathcal{H}^+] = (\mathcal{H}^+)$ meets $\mathcal{S}$. Hence $ri(\mathcal{S}) \subset (\mathcal{H}^+)$, *i.e.*, $\mathcal{S}$ is contained in the closure of $[\mathcal{H}^+]$ but not entirely contained in the relative boundary of $[\mathcal{H}^+]$. Q.E.D.

Additional separation results which are at times useful are the following:

3.2.12. THEOREM. Let $\mathcal{S}$ be a nonempty convex set in $\mathbf{R}^n$ with $\mathbf{0} \notin \overline{\mathcal{S}}$. Then there exists a vector $\mathbf{C}(\neq \mathbf{0}) \in \mathbf{R}^n$ such that $\mathbf{C}'\mathbf{x} \geq \|\mathbf{C}\|^2 > 0$ for any $\mathbf{x} \in \overline{\mathcal{S}}$. (This result may be derived using theorem 3.2.1.a.)

3.2.13. THEOREM. Let $\mathcal{S}$ be a nonempty convex set in $\mathbf{R}^n$ with $\mathbf{0}$ on the boundary of $\mathcal{S}$. Then there exists a vector $\mathbf{C}(\neq \mathbf{0}) \in \mathbf{R}^n$ such that $\mathbf{C}'\mathbf{x} \leq 0$ for any $\mathbf{x} \in \overline{\mathcal{S}}$. (This result follows from the preceding theorem.)

3.2.14. THEOREM. Let $\mathcal{S}_1$, $\mathcal{S}_2$ be nonempty convex sets in $\mathbf{R}^n$ with $\mathbf{0}$ on the boundary of $\mathcal{S}_1 - \mathcal{S}_2$. Then there exists a vector $\mathbf{C}(\neq \mathbf{0}) \in \mathbf{R}^n$ and a scalar $\alpha \in \mathbf{R}$ such that $\mathbf{C}'\mathbf{x} \leq \alpha$, $\mathbf{x} \in \overline{\mathcal{S}}_1$;

$C'x \geq \alpha$, $x \in \bar{\mathscr{S}}_2$. (This theorem may be derived by using theorem 3.2.11.)

3.2.15. THEOREM. Let $\mathscr{S}_1$, $\mathscr{S}_2$ be nonempty convex sets in R^n. There exists a hyperplane $\mathscr{H}$ separating $\mathscr{S}_1$, $\mathscr{S}_2$ strongly if and only if

$$inf\{\|x_1 - x_2\| \mid x_1 \in \mathscr{S}_1, x_2 \in \mathscr{S}_2\} > 0,$$

i.e., $0 \notin \overline{\mathscr{S}_1 - \mathscr{S}_2}$.

3.2.16. THEOREM. Let $\mathscr{S}_1$, $\mathscr{S}_2$ be nonempty convex sets in R^n with $\bar{\mathscr{S}}_1 \cap \bar{\mathscr{S}}_2 = \phi$. If either set is bounded, there exists a hyperplane $\mathscr{H}$ separating $\mathscr{S}_1$, $\mathscr{S}_2$ strongly.

Finally, a set of observations concerning the existence of supporting hyperplanes for convex sets (also offered without proof) are the following:

3.2.17. THEOREM. Let $\mathscr{S}$ be a nonempty convex set in R^n and let $\mathscr{T}$ be a nonempty convex subset of $\mathscr{S}$ with $ri(\mathscr{S}) \cap \mathscr{T} = \phi$. Then there exists a proper supporting hyperplane to $\mathscr{S}$ containing $\mathscr{T}$.

3.2.18. THEOREM. A nonempty convex set $\mathscr{S}$ in R^n has a normal $C(\neq 0) \in R^n$ at each of its boundary points, *i.e.*, there exists a $C(\neq 0)$ such that $C'x = 0$ for each $x \in \partial(\mathscr{S})$.

Let us now conclude this section with some general summary comments pertaining to the existence of separating and supporting hyperplanes for convex sets. We noted above that under some fairly unrestrictive conditions: (1) given a point exterior to a convex set, a (separating) hyperplane can be constructed in a fashion such that it passes through the point and does not meet the convex set; and (2) given two disjoint convex sets, there exists a (separating) hyperplane that has one of the convex sets in one of its closed half-spaces and the other convex set in its other closed half-space. Moreover, these hyperplanes can be shown to exist in a weak as well as in a strong sense, where the latter case typically emerges if: (a) when separating a convex set and a point, we introduce the additional assumption that the set is closed (or, in weaker terms, that the point being separated from the convex set is a point of closure); and (b) when separating two disjoint convex sets, one must be compact if the other is only closed. And as a limiting case of these separation results, given a boundary point of a convex set, there is a (supporting) hyperplane that passes through the boundary point and has the convex set in one of its closed half-spaces. In fact, if the interior of a convex set is nonempty, then all boundary points are points of support.

It is important to note that separability of nonempty sets does not imply disjointness or convexity (see Figures 3.8.a, b, respectively). But if nonempty sets are convex and disjoint (or at least have no interior points in common), then they are separable.

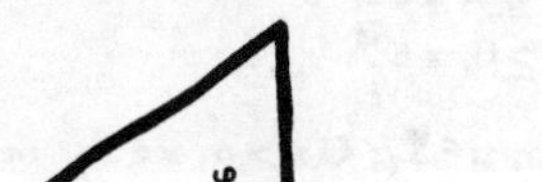

$\mathcal{S}_1, \mathcal{S}_2$ convex; $\mathcal{S}_1 \cap \mathcal{S}_2 \neq \phi$

a.

$\mathcal{S}_1 \cap \mathcal{S}_2 = \phi$; $\mathcal{S}_2$ not convex

b.

Figure 3.8

The principal separation/support theorems developed above are listed for convenience in Tables 3.1, 2 respectively.

Table 3.1

SEPARATION THEOREMS FOR CONVEX SETS

Hypotheses		Conclusions
1. $\mathcal{S}$ is nonempty, closed, convex; $\mathbf{y} \notin \mathcal{S}$	$\rightarrow$	$\mathbf{C}'\mathbf{x} \le \alpha < \mathbf{C}'\mathbf{y},\ \mathbf{x} \in \mathcal{S}$ or $\mathbf{C}'\mathbf{y} < \alpha \le \mathbf{C}'\mathbf{x},\ \mathbf{x} \in \mathcal{S}$.
2. $\mathcal{S}$ is nonempty, convex; $\mathbf{y} \notin \overline{\mathcal{S}}$	$\rightarrow$	$\mathbf{C}'\mathbf{y} > \alpha = sup\{\mathbf{C}'\mathbf{x} \mid \mathbf{x} \in \mathcal{S}\}$ or $inf\{\mathbf{C}'\mathbf{x} \mid \mathbf{x} \in \mathcal{S}\} = \alpha > \mathbf{C}'\mathbf{y}$.
3. $\mathcal{S}$ is nonempty, convex; $\overline{\mathbf{x}} \notin \mathcal{S}$	$\rightarrow$	$\mathbf{C}'\mathbf{x} \le \mathbf{C}'\overline{\mathbf{x}},\ \mathbf{x} \in \overline{\mathcal{S}}$ or $\mathbf{C}'\mathbf{x} \ge \mathbf{C}'\overline{\mathbf{x}},\ \mathbf{x} \in \overline{\mathcal{S}}$.
4. $\mathcal{S}$ is nonempty, convex; $\mathbf{0} \notin \overline{\mathcal{S}}$	$\rightarrow$	$\mathbf{C}'\mathbf{x} < \alpha < 0,\ \mathbf{x} \in \mathcal{S}$ or $\mathbf{C}'\mathbf{x} > \alpha > 0,\ \mathbf{x} \in \mathcal{S}$.
5. $\mathcal{S}$ is nonempty, closed, convex; $\mathbf{0} \notin \mathcal{S}$	$\rightarrow$	$\mathbf{C}'\mathbf{x} < \alpha < 0,\ \mathbf{x} \in \mathcal{S}$ or $\mathbf{C}'\mathbf{x} > \alpha > 0,\ \mathbf{x} \in \mathcal{S}$.
6. $\mathcal{S}$ is nonempty, convex; $\mathcal{S} \cap \{\mathbf{x} \mid \mathbf{x} \geqq \mathbf{0}\} = \phi$	$\rightarrow$	$\mathbf{C} > \mathbf{0},\ \mathbf{C}'\mathbf{x} \le 0,\ \mathbf{x} \in \mathcal{S}$.
7. $\mathcal{S}$ is nonempty, convex; $\mathbf{0} \notin \overline{\mathcal{S}}$	$\rightarrow$	$\mathbf{C}'\mathbf{x} \ge \Vert \mathbf{C} \Vert^2 > 0,\ \mathbf{x} \in \overline{\mathcal{S}}$.
8. $\mathcal{S}$ is nonempty, convex; $\mathbf{0} \in \partial(\mathcal{S})$	$\rightarrow$	$\mathbf{C}'\mathbf{x} \le 0,\ \mathbf{x} \in \overline{\mathcal{S}}$ or $\mathbf{C}'\mathbf{x} \ge 0,\ \mathbf{x} \in \overline{\mathcal{S}}$.
9. $\mathcal{S}_1, \mathcal{S}_2$ nonempty, disjoint, convex; $\mathcal{S}_1$ compact, $\mathcal{S}_2$ closed	$\rightarrow$	$\mathbf{C}'\mathbf{x} < \alpha,\ \mathbf{x} \in \mathcal{S}_1;\ \mathbf{C}'\mathbf{x} > \alpha,\ \mathbf{x} \in \mathcal{S}_2$ or $\mathbf{C}'\mathbf{x} < \alpha,\ \mathbf{x} \in \mathcal{S}_2;\ \mathbf{C}'\mathbf{x} > \alpha,\ \mathbf{x} \in \mathcal{S}_1$.
10. $\mathcal{S}_1, \mathcal{S}_2$ nonempty, disjoint, convex	$\rightarrow$	$\mathbf{C}'\mathbf{x} \le \alpha,\ \mathbf{x} \in \mathcal{S}_1;\ \mathbf{C}'\mathbf{x} \ge \alpha,\ \mathbf{x} \in \mathcal{S}_2$ or $\mathbf{C}'\mathbf{x} \le \alpha,\ \mathbf{x} \in \mathcal{S}_2;\ \mathbf{C}'\mathbf{x} \ge \alpha,\ \mathbf{x} \in \mathcal{S}_1$.
11. $\mathcal{S}_1, \mathcal{S}_2$ convex; $\mathcal{S}_1 \ne \phi,\ \mathcal{S}_2^o \ne \phi,\ \mathcal{S}_1 \cap \mathcal{S}_2^o = \phi$	$\rightarrow$	$\mathbf{C}'\mathbf{x} \le \alpha,\ \mathbf{x} \in \mathcal{S}_1;\ \mathbf{C}'\mathbf{x} \ge \alpha,\ \mathbf{x} \in \mathcal{S}_2$ or $\mathbf{C}'\mathbf{x} \le \alpha,\ \mathbf{x} \in \mathcal{S}_2;\ \mathbf{C}'\mathbf{x} \ge \alpha,\ \mathbf{x} \in \mathcal{S}_1$.
12. $\mathcal{S}_1, \mathcal{S}_2$ convex; $\mathcal{S}_1^o, \mathcal{S}_2^o \ne \phi,\ \mathcal{S}_1^o \cap \mathcal{S}_2^o = \phi$	$\rightarrow$	$\mathbf{C}'\mathbf{x} \le \alpha,\ \mathbf{x} \in \mathcal{S}_1;\ \mathbf{C}'\mathbf{x} \ge \alpha,\ \mathbf{x} \in \mathcal{S}_2$ or $\mathbf{C}'\mathbf{x} \le \alpha,\ \mathbf{x} \in \mathcal{S}_2;\ \mathbf{C}'\mathbf{x} \ge \alpha,\ \mathbf{x} \in \mathcal{S}_1$.
13. $\mathcal{S}_1, \mathcal{S}_2$ nonempty, convex; $\mathbf{0} \in \partial(\mathcal{S}_1 - \mathcal{S}_2)$	$\rightarrow$	$\mathbf{C}'\mathbf{x} \le \alpha,\ \mathbf{x} \in \overline{\mathcal{S}}_1;\ \mathbf{C}'\mathbf{x} \ge \alpha,\ \mathbf{x} \in \overline{\mathcal{S}}_2$ or $\mathbf{C}'\mathbf{x} \le \alpha,\ \mathbf{x} \in \overline{\mathcal{S}}_2;\ \mathbf{C}'\mathbf{x} \ge \alpha,\ \mathbf{x} \in \overline{\mathcal{S}}_1$.
14. $\mathcal{S}_1, \mathcal{S}_2$ nonempty, convex; $ri(\mathcal{S}_1) \cap ri(\mathcal{S}_2) = \phi$	$\rightarrow$	$inf\{\mathbf{C}'\mathbf{x} \mid \mathbf{x} \in \mathcal{S}_2\} \ge sup\{\mathbf{C}'\mathbf{x} \mid \mathbf{x} \in \mathcal{S}_1\}$, $sup\{\mathbf{C}'\mathbf{x} \mid \mathbf{x} \in \mathcal{S}_2\} \ge inf\{\mathbf{C}'\mathbf{x} \mid \mathbf{x} \in \mathcal{S}_1\}$.
15. $\mathcal{S}_1, \mathcal{S}_2$ nonempty, convex; $\mathbf{0} \notin \overline{\mathcal{S}_1 - \mathcal{S}_2}$	$\rightarrow$	$inf\{\mathbf{C}'\mathbf{x} \mid \mathbf{x} \in \mathcal{S}_2\} > sup\{\mathbf{C}'\mathbf{x} \mid \mathbf{x} \in \mathcal{S}_1\}$.
16. $\mathcal{S}_1, \mathcal{S}_2$ nonempty, convex; $\overline{\mathcal{S}}_1 \cap \overline{\mathcal{S}}_2 = \phi$, $\mathcal{S}_1$ or $\mathcal{S}_2$ bounded	$\rightarrow$	$inf\{\mathbf{C}'\mathbf{x} \mid \mathbf{x} \in \mathcal{S}_2\} > sup\{\mathbf{C}'\mathbf{x} \mid \mathbf{x} \in \mathcal{S}_1\}$.

Table 3.2
SUPPORT THEOREMS FOR CONVEX SETS

Hypotheses		Conclusion
1. $\mathcal{S}$ is nonempty, closed, convex; $\bar{\mathbf{x}} \in \partial(\mathcal{S})$	$\rightarrow$	$\begin{cases} \mathbf{C}'\mathbf{x} \leq \alpha = \mathbf{C}'\bar{\mathbf{x}} = sup\{\mathbf{C}'\mathbf{x} \mid \mathbf{x} \in \mathcal{S}\} \text{ or} \\ \mathbf{C}'\mathbf{x} \geq \alpha = \mathbf{C}'\bar{\mathbf{x}} = inf\{\mathbf{C}'\mathbf{x} \mid \mathbf{x} \in \mathcal{S}\}. \end{cases}$
2. $\mathcal{S}$ is nonempty, convex, $\mathcal{S}^o \neq \phi$; $\bar{\mathbf{x}} \in \partial(\mathcal{S})$	$\rightarrow$	$\begin{cases} \mathbf{C}'\mathbf{x} \leq \alpha = \mathbf{C}'\bar{\mathbf{x}} = sup\{\mathbf{C}'\mathbf{x} \mid \mathbf{x} \in \bar{\mathcal{S}}\} \text{ or} \\ \mathbf{C}'\mathbf{x} \geq \alpha = \mathbf{C}'\bar{\mathbf{x}} = inf\{\mathbf{C}'\mathbf{x} \mid \mathbf{x} \in \bar{\mathcal{S}}\}. \end{cases}$
3. $\mathcal{S}$ is nonempty, convex; $\mathcal{S} \supset \mathcal{T}$ is nonempty, convex; $ri(\mathcal{S}) \cap \mathcal{T} = \phi$	$\rightarrow$	there exists $\mathcal{H} = \{\mathbf{x} \mid \mathbf{C}'\mathbf{x} = \alpha\}$ such that $\mathcal{S} \not\subset \mathcal{H} \supset \mathcal{T}$.
4. $\mathcal{S}$ is nonempty, convex	$\rightarrow$	$\mathbf{C}'\mathbf{x} = 0$ for each $\mathbf{x} \in \partial(\mathcal{S})$.

If we relax the assumption that $\mathcal{S}_1, \mathcal{S}_2$ are convex sets in $\mathbf{R}^n$, then a reformulation of theorem 3.2.7 appears as

3.2.7.1. STRONG SEPARATION THEOREM. Let $\mathcal{S}_1, \mathcal{S}_2$ be non-empty compact sets in $\mathbf{R}^n$. Then there exists a vector $\mathbf{C}(\neq \mathbf{0}) \in \mathbf{R}^n$ and a scalar $\alpha \in \mathbf{R}$ such that

$$\mathbf{C}'\mathbf{x} < \alpha,\ \mathbf{x} \in \mathcal{S}_1;$$
$$\mathbf{C}'\mathbf{x} > \alpha,\ \mathbf{x} \in \mathcal{S}_2$$

if and only if $co(\mathcal{S}_1) \cap co(\mathcal{S}_2) = \phi$.

PROOF. (sufficiency) Let $co(\mathcal{S}_1) \cap co(\mathcal{S}_2) = \phi$. Since the convex hull of a compact set is compact, theorem 3.2.7 ensures the existence of a hyperplane $\mathcal{H} = \{\mathbf{x} \mid \mathbf{C}'\mathbf{x} = \alpha,\ \mathbf{x} \in \mathbf{R}^n\}$ that strongly sepa-rates $co(\mathcal{S}_1)$, $co(\mathcal{S}_2)$. But since $\mathcal{S}_1 \subseteq co(\mathcal{S}_1)$ and $\mathcal{S}_2 \subseteq co(\mathcal{S}_2)$, it follows that $\mathcal{H}$ strongly separates $\mathcal{S}_1, \mathcal{S}_2$ as well.

(necessity) Suppose $\mathcal{H}$ strongly separates $\mathcal{S}_1, \mathcal{S}_2$. Let $\mathbf{C}'\mathbf{x} < \alpha$, $\mathbf{x} \in \mathcal{S}_1$; $\mathbf{C}'\mathbf{x} > \alpha$, $\mathbf{x} \in \mathcal{S}_2$. For arbitrary $\{\mathbf{x}_1, \ldots, \mathbf{x}_r\} \subset \mathcal{S}_1$, let

$$\bar{\mathbf{x}} = \sum_{i=1}^{r} \lambda_i \mathbf{x}_i,\ \sum_{i=1}^{r} \lambda_i = 1,\ 0 \leq \lambda_i \in \mathbf{R},\ i = 1, \cdots, r.$$

Then

$$\mathbf{C}'\bar{\mathbf{x}} = \sum_{i=1}^{r} \lambda_i \mathbf{C}'\mathbf{x}_i < \alpha \sum_{i=1}^{r} \lambda_i = \alpha$$

and thus $\mathbf{C}'\mathbf{x} < \alpha$, $\mathbf{x} \in co(\mathcal{S}_1)$. Similarly, let $\hat{\mathbf{x}}$ be expressible as a convex combination of the arbitrary set of vectors $\{\mathbf{x}_1, \ldots, \mathbf{x}_s\} \subset \mathcal{S}_2$. Then in like fashion it is easily shown that $\mathbf{C}'\hat{\mathbf{x}} > \alpha$ and thus $\mathbf{C}'\mathbf{x} > \alpha$, $\mathbf{x} \in co(\mathcal{S}_2)$. Hence $\mathcal{H}$ strongly separates $co(\mathcal{S}_1)$, $co(\mathcal{S}_2)$. Then by theorem 3.2.7 it must be true that $co(\mathcal{S}_1) \cap co(\mathcal{S}_2) = \phi$. Q.E.D.

A special case of this theorem occurs if one of the sets contains but a single point. In this regard we have

3.2.7.2 THEOREM. Let $\mathcal{S}$ be a compact set in $\mathbf{R}^n$ with $\bar{\mathbf{x}} \notin \mathcal{S}$. Then there exists a vector $\mathbf{C}(\neq \mathbf{0}) \in \mathbf{R}^n$ and a scalar $\alpha \in \mathbf{R}$ such that $\mathbf{C}'\mathbf{x} < \alpha < \mathbf{C}'\bar{\mathbf{x}}$ for all $\mathbf{x} \in \mathcal{S}$ if and only if $\bar{\mathbf{x}} \notin co(\mathcal{S})$.

3.3. Separation Renders Disjoint Alternatives

An important application of "(strong) separation" is that it serves as an input in deriving what will be termed a ***theorem of the alternative.*** While chapter 6 treats such theorems in considerable detail (at this point in our discussion the reader is encouraged to examine the brief introductory section 6.1), some advanced exposure to the approach employed in developing such theorems will be of considerable value.

One of the most versatile (in fact, fundamental) theorems of the alternative is ***Farkas' theorem.*** As we shall now see, this theorem can be construed as being based upon the existence of a hyperplane that strongly separates a closed convex set and a vector exterior to the set. While Farkas' theorem appears in a variety of different forms, it is essentially concerned with the conditions underlying the existence of a nonnegative solution to a linear system of nonhomogeneous equations. Specifically,

3.3.1. FARKAS' THEOREM [Farkas, 1902]. For $\mathbf{A}$ an $(m \times n)$ matrix with $\mathbf{b} \in \mathbf{R}^n, \mathbf{x} \in \mathbf{R}^n$, and $\mathbf{y} \in \mathbf{R}^m$, a necessary and sufficient condition for the system $\mathbf{A}'\mathbf{y} = \mathbf{b}$, $\mathbf{y} \geqq \mathbf{0}$ to have a solution is that $\mathbf{b}'\mathbf{x} > 0$ for all $\mathbf{x}$ satisfying $\mathbf{A}\mathbf{x} \leqq \mathbf{0}$.

This theorem can be restated in the usual "theorem of the alternative format" as

3.3.1.a. FARKAS' THEOREM OF THE ALTERNATIVE. For any $(m \times n)$ matrix $\mathbf{A}$ and a vector $\mathbf{b} \in \mathbf{R}^n$, either

(I) $\mathbf{Ax} \leqq \mathbf{0}$, $\mathbf{b'x} > 0$ has a solution $\mathbf{x} \in \mathbf{R}^n$ or

(II) $\mathbf{A'y} = \mathbf{b}$, $\mathbf{y} \geqq \mathbf{0}$ has a solution $\mathbf{y} \in \mathbf{R}^m$,

but never both.[3]

PROOF. (II implies $\bar{\text{I}}$). If II has a solution $\mathbf{y} \geqq \mathbf{0}$ such that $\mathbf{b} = \mathbf{A'y}$, then for all $\mathbf{x}$ such that $\mathbf{Ax} \leqq \mathbf{0}$, it follows that $\mathbf{b'x} = \mathbf{y'Ax} \leq 0$ — an obvious contradiction to I. Hence I possesses no solution.

($\bar{\text{II}}$ implies I). ***Version*** **1.** Let $\mathcal{S} = \{\mathbf{u} \mid \mathbf{u} = \mathbf{A'y}, \mathbf{y} \geqq \mathbf{0}, \mathbf{u} \in \mathbf{R}^n\}$. Clearly $\mathcal{S}$ is closed and convex. Since II has no solution, $\mathbf{b} \notin \mathcal{S}$. By the strong separation theorem 3.2.2, there exists a $\mathbf{C}(\neq \mathbf{0}) \in \mathbf{R}^n$ and a scalar $\alpha \in \mathbf{R}$ such that $\mathbf{C'u} \leq \alpha < \mathbf{C'b}$ for all $\mathbf{u} \in \mathcal{S}$. Since $\mathbf{0} = \mathbf{A'0} \in \mathcal{S}$, it follows that $\alpha \geq 0$ so that $\mathbf{C'b} > 0$. In addition, $\mathbf{C'u} = \mathbf{C'A'y} = \mathbf{y'AC} \leq \alpha$ for all $\mathbf{y} \geqq \mathbf{0}$. Since $\alpha \geq 0$ and each component of $\mathbf{y}(\geqq \mathbf{0})$ can be made arbitrari-ly large, $\mathbf{y'AC} \leq \alpha$ if $\mathbf{AC} \leqq \mathbf{0}$. Combining these results we see that there exists a $\mathbf{C}(\neq \mathbf{0})$ such that $\mathbf{b'C} > 0$ and $\mathbf{AC} \leqq \mathbf{0}$. Thus I has a solution.

---

[3]The sense of the inequalities in I may be reversed if we set $\mathbf{A} = -\mathbf{C}$, $\mathbf{b} = -\mathbf{d}$. Hence an equivalent version of 3.3.1.a appears as

3.3.1.b. FARKAS' THEOREM OF THE ALTERNATIVE. For any $(m \times n)$ matrix $\mathbf{C}$ and a vector $\mathbf{d} \in \mathbf{R}^n$, either

(I') $\mathbf{Cx} \geqq \mathbf{0}$, $\mathbf{d'x} < 0$ has a solution $\mathbf{x} \in \mathbf{R}^n$ or

(II') $\mathbf{C'y} = \mathbf{d}$, $\mathbf{y} \geqq \mathbf{0}$ has a solution $\mathbf{y} \in \mathbf{R}^m$,

but never both.

Two additional modifications of theorem 3.3.1.a are at times useful. If $\mathbf{A'y} = \mathbf{b}$ in II above is replaced by $\mathbf{A'y} \leqq \mathbf{b}$, then either

$$\text{(FARKAS 3.3.1.c)}\begin{cases} (\text{I}'')\ \mathbf{Ax} \leqq \mathbf{0},\ \mathbf{b'x} > 0,\ \mathbf{x} \geqq \mathbf{0} \text{ has a solution } \mathbf{x} \in \mathbf{R}^n \text{ or} \\ (\text{II}'')\ \mathbf{A'y} \leqq \mathbf{b},\ \mathbf{y} \geqq \mathbf{0} \text{ has a solution } \mathbf{y} \in \mathbf{R}^m, \end{cases}$$

but never both. Furthermore, if the nonnegativity restriction is omitted in II'', then either

$$\text{(FARKAS 3.3.1.d)}\begin{cases} (\text{I}''')\ \mathbf{Ax} = \mathbf{0},\ \mathbf{b'x} > 0,\ \mathbf{x} \geqq \mathbf{0} \text{ has a solution } \mathbf{x} \in \mathbf{R}^n \text{ or} \\ (\text{II}''')\ \mathbf{A'y} \leqq \mathbf{b}, \text{ has a solution } \mathbf{y} \in \mathbf{R}^m, \end{cases}$$

but never both.

Version 2. Let $\mathcal{S}_1 = \{\mathbf{u} \mid \mathbf{u} = \mathbf{A}'\mathbf{y}, \mathbf{y} \geqq \mathbf{0}, \mathbf{u} \in \mathbf{R}^n\}$, $\mathcal{S}_2 = \{\mathbf{b}\}$. Clearly $\mathcal{S}_1, \mathcal{S}_2$ are nonempty, disjoint, and convex. Furthermore, $\mathcal{S}_1$ is closed and $\mathcal{S}_2$ is compact. Given that II has no solution (since $\mathcal{S}_1 \cap \mathcal{S}_2 = \phi$), the strong separation theorem 3.2.7 posits the existence of a $\mathbf{C}(\neq \mathbf{0}) \in \mathbf{R}^n$ and a scalar $\alpha \in \mathbf{R}$ such that $\mathbf{C}'\mathbf{u} = \mathbf{C}'\mathbf{A}'\mathbf{y} < \alpha$ for $\mathbf{u} \in \mathcal{S}_1$ and $\mathbf{C}'\mathbf{b} > \alpha$. Since $0 = \mathbf{A}'\mathbf{0} \in \mathcal{S}_1$, we have $\alpha < 0$ and thus $\mathbf{b}'\mathbf{C} > 0$. And by an argument similar to that used in version 1 of this proof above, we may conclude that $\mathbf{AC} \leqq \mathbf{0}$. Hence we again find that $\mathbf{C}(\neq \mathbf{0})$ is a solution to I. Q.E.D.

Since the theorem of Farkas is intimately tied to the separation result of Minkowski, it is alternatively termed the ***Minkowski-Farkas theorem*** or the ***theorem of the separating hyperplane.*** That is, if I has no solution, then there exists a hyperplane $\mathbf{b}'\mathbf{x} = 0$ which separates the sets $\{\mathbf{x} \mid \mathbf{b}'\mathbf{x} > 0, \mathbf{x} \in \mathbf{R}^n\}$ and $\{\mathbf{x} \mid \mathbf{Ax} \leqq \mathbf{0}, \mathbf{x} \in \mathbf{R}^n\}$ (Figure 3.9.a). And if II′ (see footnote 3) has no solution, there exists a vector $\mathbf{x}$ which makes an obtuse angle with $\mathbf{d}$ and a non-obtuse angle with the rows of $\mathbf{C}$. Thus the hyperplane $\mathbf{d}'\mathbf{x} = 0$ has $\mathcal{S}_1' = \{\mathbf{v} \mid \mathbf{v} = \mathbf{C}'\mathbf{y}, \mathbf{y} \geqq \mathbf{0}, \mathbf{v} \in \mathbf{R}^n\}$ on one side and $\mathcal{S}_2' = \{\mathbf{d}\}$ on the other (Figure 3.9.b).

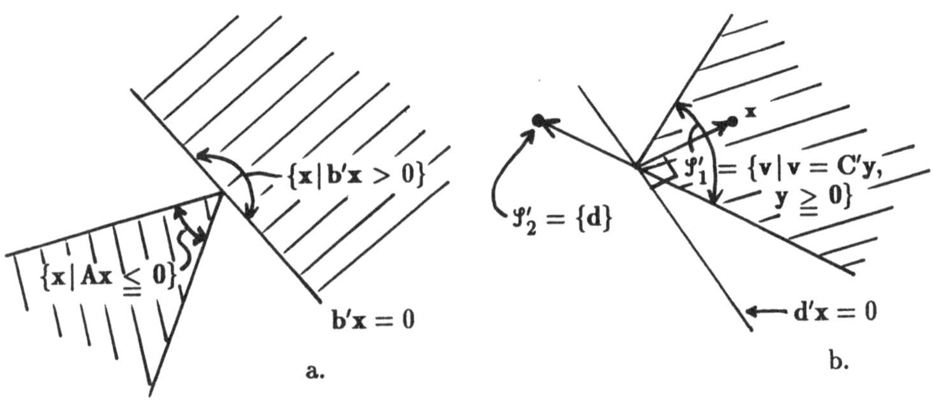

Figure 3.9

An additional theorem of the alternative, namely ***Gordan's theorem***, which considers the circumstances under which there exists a semipositive solution to a homogeneous equation system, can also be viewed as based upon the existence of a hyperplane that strictly separates a closed convex set and a point outside of the set. In this regard,

3.3.2. GORDAN'S THEOREM OF THE ALTERNATIVE [Gordan, 1873]. For any $(m \times n)$ matrix $\mathbf{A}$, either

(I) $\mathbf{Ax} > \mathbf{0}$ has a solution $\mathbf{x} \in \mathbf{R}^n$ or

(II) $\mathbf{A'y} = \mathbf{0}$, $\mathbf{y} \geq \mathbf{0}$ has a solution $\mathbf{y} \in \mathbf{R}^m$,

but never both.[4]

PROOF. (II implies $\bar{\text{I}}'$). Let $\mathbf{y} \geq \mathbf{0}$ be a solution to II. Suppose to the contrary that there exists for I′ an $\mathbf{x}$ such that $\mathbf{Cx} < \mathbf{0}$. Then, since $\mathbf{y} \neq \mathbf{0}$, $\mathbf{y'Cx} = \mathbf{x'C'y} < 0$. But since II requires $\mathbf{C'y} = \mathbf{0}$, we must have $\mathbf{x'C'y} = 0$ and thus a contradiction to I′. Hence I′ cannot have a solution.

($\overline{\text{II}}$ implies I′). Given $\mathbf{1} \in \mathbf{R}^m$, let $\mathbf{1'y} = 1$. Then the system $\mathbf{A'y} = \mathbf{0}$, $\mathbf{1'y} = 1$ may be rewritten as $\mathbf{D'y} = \mathbf{e}_{n+1}$, where $\mathbf{D} = [\mathbf{A}, \mathbf{1}]$ is of order $(m \times n{+}1)$. We may now form the nonempty, closed, and convex set $= \{\mathbf{u} \mid \mathbf{u} = \mathbf{D'y}, \mathbf{y} \geq \mathbf{0}, \mathbf{y} \in \mathbf{R}^m\} \subset \mathbf{R}^{n+1}$. Since II has no solution, $\mathbf{e}_{n+1} \notin \mathcal{S}$. By the strong separation theorem 3.2.2, there exists a $\bar{\mathbf{C}}(\neq \mathbf{0}) \in \mathbf{R}^{n+1}$ and a scalar $\alpha \in \mathbf{R}$ such that $\bar{\mathbf{C}}'\mathbf{u} \leq \alpha < \bar{\mathbf{C}}'\mathbf{e}_{n+1} = c \in \mathbf{R}$ for all $\mathbf{u} \in \mathcal{S}$. Since $\mathbf{0} = \mathbf{D'0} \in \mathcal{S}$, it follows that $0 \leq \alpha < c$. Moreover, $\bar{\mathbf{C}}'\mathbf{u} = \bar{\mathbf{C}}'\mathbf{D'y} - \mathbf{y'D}\bar{\mathbf{C}} \leq \alpha$ for all $\mathbf{y} \geqq \mathbf{0}$. Since each component of $\mathbf{y}$ can be made arbitrarily large, $\mathbf{y'D}\bar{\mathbf{C}} \leq \alpha$ if $\mathbf{D}\bar{\mathbf{C}} \leqq \mathbf{0}$. Given

$$\mathbf{D}\bar{\mathbf{C}} = [\mathbf{A}, \mathbf{1}]\,\bar{\mathbf{C}} = [\mathbf{A}, \mathbf{1}]\begin{bmatrix}\mathbf{C}\\ c\end{bmatrix} = \mathbf{AC} + c\mathbf{1} \leqq \mathbf{0},$$

[4] If we let $\mathbf{A} = -\mathbf{C}$ in I above, then the sense of this inequality may be reversed to render the equivalent specification

3.3.2.a. GORDAN'S THEOREM OF THE ALTERNATIVE. For any $(m \times n)$ matrix $\mathbf{C}$, either

(I′) $\mathbf{Cx} < \mathbf{0}$ has a solution $\mathbf{x} \in \mathbf{R}^n$ or

(II) $\mathbf{C'y} = \mathbf{0}$, $\mathbf{y} \geq \mathbf{0}$ has a solution $\mathbf{y} \in \mathbf{R}^m$,

but never both.

it follows that since $c > 0$, there exists a $\mathbf{C}(\neq \mathbf{0})$ such that $\mathbf{AC} < \mathbf{0}$. Thus I′ has a solution. Q.E.D.

It is important to note that Gordan's theorem can be construed as a corollary to Farkas' theorem. To see this we shall again let $\mathbf{1} \epsilon \mathbf{R}^m$. Then I′ in Gordan's theorem 3.3.2.a is equivalent to asserting that the system $\mathbf{Cx} \leqq -\mathbf{1}$ has an unrestricted solution $\mathbf{x} \epsilon \mathbf{R}^n$. By II‴ of Farkas' theorem 3.3.1.d, this inequality is equivalent to $\bar{\text{I}}'''$ or $\mathbf{Cx} = \mathbf{0}$, $-\mathbf{1}'\mathbf{x} > 0$, $\mathbf{x} \geqq \mathbf{0}$ has no solution. Hence no $\mathbf{y}(\neq \mathbf{0})$ can be found for which $\mathbf{C}'\mathbf{y} = \mathbf{0}$, $\mathbf{y} \geqq \mathbf{0}$ and thus $\bar{\text{II}}$ holds for Gordan's theorem 3.3.2.a.

Conversely, if there exists a $\mathbf{y}(\neq \mathbf{0})$ such that $\mathbf{A}'\mathbf{y} = \mathbf{0}$, $\mathbf{y} \geqq \mathbf{0}$ in II of Gordan's theorem 3.3.2.a, then equivalently $\bar{\text{II}}'''$ holds in Farkas' theorem 3.3.1.d or there is no $\mathbf{y}$ for which $\mathbf{A}'\mathbf{y} \leqq \mathbf{0}$. But this implies that $\bar{\text{I}}'$ prevails in Gordan's theorem 3.3.2.a so that there is no $\mathbf{x}$ such that $\mathbf{Cx} < \mathbf{0}$.

As far as the geometry of Gordan's theorem 3.3.2 is concerned, either:

(I) there exists a vector $\mathbf{x}$ which forms a strictly acute angle $(< \pi/2)$ with the rows $\boldsymbol{\alpha}_i$, $i=1,\ldots,m$, of $\mathbf{A}$; or

(II) the null vector is expressible as a nonnegative nontrivial linear combination of the columns $\boldsymbol{\alpha}_i'$, $i=1,\ldots,m$, of $\mathbf{A}'$, *i.e.*, $\mathbf{0} \epsilon \operatorname{co}(\boldsymbol{\alpha}_1',\ldots,\boldsymbol{\alpha}_m')$,

but never both. Alternatively, if $\mathcal{S}$ is any subspace spanned by the rows $\boldsymbol{\alpha}_i$, $i=1,\ldots,m$, of $\mathbf{A}$, then either

(I) there is a positive vector $\mathbf{x} \epsilon \mathcal{S}$ or

(II) there is a semipositive vector
$$\mathbf{y} \epsilon \mathcal{S}^{\perp} = \{\mathbf{y} \mid \boldsymbol{\alpha}_i\mathbf{y} = 0,\ \boldsymbol{\alpha}_i \epsilon \mathcal{S},\ i=1,\ldots,m\},$$

but never both.

The final theorem of the alternative to be presented in this section is due to Gale. It is concerned with the existence of a nonnegative solution to a system of linear inequalities and formally appears as

3.3.3. GALE'S THEOREM OF THE ALTERNATIVE [Gale, 1960]. For any $(m \times n)$ matrix $\mathbf{A}$, either

(I) $\mathbf{Ax} \leqq \mathbf{b}$, $\mathbf{x} \geqq \mathbf{0}$, has a solution $\mathbf{x} \in \mathbf{R}^n$ or

(II) $\mathbf{A'y} \geq \mathbf{0}$, $\mathbf{b'y} < 0$, $\mathbf{y} \geqq \mathbf{0}$ has a solution $\mathbf{y} \in \mathbf{R}^m$,

but never both.

PROOF. (I implies $\overline{\text{II}}$). Let $\mathbf{x} \geqq \mathbf{0}$ satisfy $\mathbf{Ax} \leqq \mathbf{b}$ or $\mathbf{x'A'} \leqq \mathbf{b'}$. Assume to the contrary that II has a solution, *i.e.,* there is a $\mathbf{y} \geqq \mathbf{0}$ for which $\mathbf{x'A'y} \leq \mathbf{b'y} < 0$. Since both $\mathbf{x} \geqq \mathbf{0}$ and $\mathbf{A'y} \geqq \mathbf{0}$, we obtain the contradiction $\mathbf{x'A'Y} < 0$ to II. Hence II has no solution.

($\overline{\text{I}}$ implies II). For $\mathbf{0} \leq \mathbf{u} \in \mathbf{R}^m$, let $\mathbf{Ax} + \mathbf{I}_m\mathbf{u} = \bar{\mathbf{A}}\bar{\mathbf{x}} = \mathbf{b}$, where $\bar{\mathbf{A}} = [\mathbf{A}, \mathbf{I}_m]$, $\bar{\mathbf{x}}' = [\mathbf{x}', \mathbf{u}']$. In addition, let $\mathcal{S} = \{\mathbf{v} \mid \mathbf{v} = \bar{\mathbf{A}}\bar{\mathbf{x}},\ \bar{\mathbf{x}} \geqq \mathbf{0},\ \mathbf{v} \in \mathbf{R}^m\}$. Since I has no solution, $\mathbf{b} \notin \mathcal{S}$. Since $\mathcal{S}$ is closed and convex, the strong separation theorem 3.2.2 informs us that there exists a $\mathbf{C}\ (\neq \mathbf{0}) \in \mathbf{R}^m$ and a scalar $\alpha \in \mathbf{R}$ such that $\mathbf{C'b} < \alpha \leq \mathbf{C'v}$ for all $\mathbf{v} \in \mathcal{S}$. With $\mathbf{0} = \bar{\mathbf{A}}\mathbf{0} \in \mathcal{S}$, we have $\alpha \leq 0$ so that $\mathbf{C'b} < 0$. Moreover, $\mathbf{C'v} = \mathbf{C'}\bar{\mathbf{A}}\bar{\mathbf{x}} = \bar{\mathbf{x}}'\bar{\mathbf{A}}'\mathbf{C} \geq \alpha$ for all $\bar{\mathbf{x}} \geqq \mathbf{0}$. Since $\alpha \leq 0$ we must have

$$\bar{\mathbf{A}}'\mathbf{C} = \begin{bmatrix} \mathbf{A}' \\ \mathbf{I}_n \end{bmatrix} \mathbf{C} \geqq \mathbf{0}$$

or $\mathbf{A'C} \geq \mathbf{0}$. Hence there exists a $\mathbf{C}(\neq \mathbf{0})$ such that $\mathbf{b'x} < 0$ and $\mathbf{A'C} \geqq \mathbf{0}$ to wit II has a solution. Q.E.D.

As was the case with Gordan's theorem of the alternative, Gale's theorem is also a corollary to the theorem of Farkas. For instance, given that I has no solution $\mathbf{x} \geqq \mathbf{0}$, it follows that $\mathbf{Ax} + \mathbf{I}_m\mathbf{u} = \mathbf{b}$ also has no solution $\mathbf{x} \geqq \mathbf{0}$, $\mathbf{0} \leq \mathbf{u} \in \mathbf{R}^m$. Then from Farkas' theorem 3.3.1.b ($\overline{\text{II}}'$ implies I′), there exists a $\mathbf{y}$ such that $\mathbf{A'y} \geqq \mathbf{0}$, $\mathbf{I}_m\mathbf{y} \geqq \mathbf{0}$, and $\mathbf{b'y} < 0$. Hence II has a solution.

A useful corollary to the preceding theorem addresses the existence of a semipositive solution to a homogeneous system of linear inequalities. Specifically,

3.3.4. COROLLARY. For any $(m \times n)$ matrix $\mathbf{A}$, either

(I) $\mathbf{Ax} \leqq \mathbf{0}$, $\mathbf{x} \geq \mathbf{0}$ has a solution $\mathbf{x} \in \mathbf{R}^n$ or

(II) $\mathbf{A'y} > \mathbf{0}$, $\mathbf{y} \geqq \mathbf{0}$ has a solution $\mathbf{y} \in \mathbf{R}^m$,

but never both.

PROOF. (I implies $\bar{\text{II}}$). Let $\mathbf{x} \geq \mathbf{0}$ satisfy $\mathbf{Ax} \leqq \mathbf{0}$. If we assume to the contrary that there exists a $\mathbf{y} \geqq \mathbf{0}$ such that $\mathbf{A'y} > \mathbf{0}$, then $\mathbf{x'A'y} \leq 0$. But since $\mathbf{x} \neq \mathbf{0}$, we must have $\mathbf{A'y} \leqq \mathbf{0}$ — a contradiction to II. Hence II cannot possess a solution.

($\bar{\text{I}}$ implies II). Since it is assumed that I admits no $\mathbf{x} \geq \mathbf{0}$ for which $\mathbf{Ax} \leqq \mathbf{0}$, it follows that the system $\mathbf{Ax} \leqq \mathbf{0}$, $\mathbf{1'x} \geqq 1$ or

$$\begin{bmatrix} \mathbf{A} \\ -\mathbf{1}' \end{bmatrix} \mathbf{x} \leqq -\mathbf{e}_{n+1}$$

does not admit an $\mathbf{x} \geq \mathbf{0}$ either. Then from II of Gale's theorem 3.3.3,

$$[\mathbf{A}', -\mathbf{1}]\bar{\mathbf{Y}} = [\mathbf{A}', -\mathbf{1}] \begin{bmatrix} \mathbf{y} \\ y \end{bmatrix} \geqq \mathbf{0}, \quad -\mathbf{e}'_{n+1}\bar{\mathbf{y}} = -\mathbf{e}'_{n+1} \begin{bmatrix} \mathbf{y} \\ y \end{bmatrix} < 0$$

has a solution $\mathbf{0} \leqq \bar{\mathbf{y}} \in \mathbf{R}^{n+1}$ or $\mathbf{0} \leqq \mathbf{y} \in \mathbf{R}^m$, $0 \leq y \in \mathbf{R}$. Hence $\mathbf{A'y} - y\mathbf{1} \geqq \mathbf{0}$, $-y < 0$ implies $\mathbf{A'y} \geqq y\mathbf{1} > \mathbf{0}$ as required so that II of this corollary obtains. Q.E.D.

3.4. Exercises

1. Given the convex set $\mathcal{S} = \{\mathbf{x} \mid \|\mathbf{x}\|^2 \leq 1\}$ in $\mathbf{R}^2$, find the equation of the supporting hyperplane at $\mathbf{x}_0' = \left(\frac{1}{\sqrt{2}}, \frac{1}{\sqrt{2}}\right)$.

2. Given $\mathcal{S} = \{\mathbf{x}' = (x_1, x_2) \mid 4x_1 - 6x_2 \leq 6\}$ in $\mathbf{R}^2$, find the equation of the supporting hyperplane at $\mathbf{x}_0' = (3, 1)$.

3. Find the equation of a separating hyperplane for the sets $\mathcal{S}, \mathcal{T}$ in $\mathbf{R}^2$, where

$$\mathcal{S} = \{\mathbf{x}' = (x_1, x_2) \mid x_1 x_2 \geq 1\},$$
$$\mathcal{T} = \{\mathbf{x}' = (x_1, x_2) \mid 3x_1 + 2x_2 \leq 5,\ x_1 + x_2 \leq 2\}.$$

4. Prove theorem 3.2.7.a.

5. Prove theorems 3.2.10, 3.2.11.

6. Prove theorem 3.2.14.

7. Verify that two hyperplanes are parallel if and only if their normals are scalar multiples of each other.

8. For $\mathbf{C}' = (3, 5)$, find an α such that $\mathbf{x}_0' = (2, 2)$ lies above the hyperplane $\mathbf{C}'\mathbf{x} = \alpha$.

CHAPTER 4
CONVEX CONES IN R^n

4.1. Convex Cones

A ***cone*** $\mathcal{C}$ in R^n is a set of points such that if $\mathbf{x} \in \mathcal{C}$, then so is every nonnegative scalar multiple of $\mathbf{x}$, *i.e.*, if $\mathbf{x} \in \mathcal{C}$, then $\lambda\mathbf{x} \in \mathcal{C}$ for $0 \leq \lambda \in R$, $\mathbf{x} \in R^n$ (see Figure 4.1.a for $\mathcal{C}$ in R^2). If we consider the set of points $\mathcal{B} = \{\mathbf{x}\}$, then the ***cone generated by*** $\mathcal{B}$ is $\mathcal{C} = \{\mathbf{y} | \mathbf{y} = \lambda\mathbf{x},\ 0 \leq \lambda \in R,\ \mathbf{x} \in \mathcal{B}\}$. And if $\mathbf{0} \notin \mathcal{B}$ and for each $\mathbf{y}(\neq \mathbf{0}) \in \mathcal{C}$ there are unique $\mathbf{x} \in \mathcal{B}$ with $\lambda > 0$ such that $\mathbf{y} = \lambda\mathbf{x}$, then $\mathcal{B}$ is termed a ***base*** of $\mathcal{C}$. Moreover, the point $\mathbf{0} \in R^n$ is termed the ***vertex*** of a cone and is an element of every cone since $\mathbf{Y} = 0\mathbf{x} = \mathbf{0} \in \mathcal{C}$. In general, $\mathcal{C}$ is "not a subspace" of R^n since the definition of $\mathcal{C}$ holds only for $\lambda \geq 0$ and not all λ.

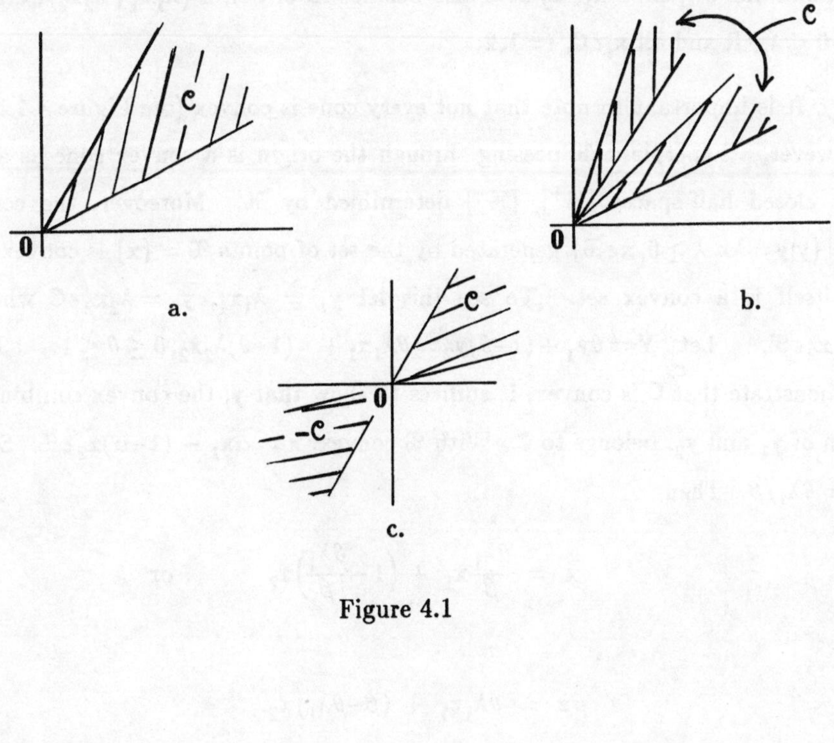

Figure 4.1

In what follows we shall consider cones which are convex. Generally speaking, a convex cone is one which is a convex set. More specifically, a cone $\mathcal{C}$ in $\mathbf{R}^n$ is termed a ***convex cone*** if and only if it is closed under the operations of addition and multiplication by a non-negative scalar, *i.e.*, $\mathcal{C}$ in $\mathbf{R}^n$ is a convex cone if and only if: (a) for $\mathbf{x}\in\mathcal{C}$, $\lambda\mathbf{x}\in\mathcal{C}$, $0\leq\lambda\in\mathbf{R}$; and (b) for $\mathbf{x}_1,\mathbf{x}_2\in\mathcal{C}$, $\mathbf{x}_1+\mathbf{x}_2\in\mathcal{C}$. In this regard, conditions (a), (b) are necessary since, if $\mathcal{C}$ is a convex cone and $\mathbf{x}\in\mathcal{C}$, then by definition, $\lambda\mathbf{x}\in\mathcal{C}$ for $\lambda\geq 0$. Moreover, if $\mathbf{v}_1,\mathbf{v}_2\in\mathcal{C}$, then both $\mathbf{x}_1 = \lambda\mathbf{v}_1$ and $\mathbf{x}_2 = (1-\lambda)\mathbf{v}_2$ are members of $\mathcal{C}$ for $0\leq\lambda\leq 1$. With $\mathcal{C}$ convex, $\lambda\mathbf{v}_1+(1-\lambda)\mathbf{v}_2 = \mathbf{x}_1+\mathbf{x}_2\in\mathcal{C}$. These conditions are sufficient in that $\lambda\mathbf{x}\in\mathcal{C}$ if $\mathbf{x}\in\mathcal{C}$ (so clearly $\mathcal{C}$ is a cone); and $\mathbf{x}_1+\mathbf{x}_2\in\mathcal{C}$ if $\mathbf{x}_1$, $\mathbf{x}_2\in\mathcal{C}$. If $\mathbf{v}_1,\mathbf{v}_2\in\mathcal{C}$, then since $\mathbf{x}_1+\mathbf{x}_2 = \lambda\mathbf{v}_1+(1-\lambda)\mathbf{v}_2\in\mathcal{C}$, $0\leq\lambda\leq 1$, it follows that $\mathcal{C}$ is convex.

Equivalently, $\mathcal{C}$ in $\mathbf{R}^n$ is a convex cone if any non-negative linear combination of points $\mathbf{x}_1$, $\mathbf{x}_2$ of $\mathcal{C}$ also belongs to $\mathcal{C}$, *i.e.*, if $(\lambda_1\mathbf{x}_1+\lambda_2\mathbf{x}_2)\in\mathcal{C}$ for all $0\leq\lambda_i\in\mathbf{R}$ and all $\mathbf{x}_i\in\mathcal{C}$, $i=1,2$.

It is important to note that not every cone is convex (see Figure 4.1.b). However, a hyperplane $\mathcal{H}$ passing through the origin is a convex cone as are the closed half-spaces $[\mathcal{H}^+]$, $[\mathcal{H}^-]$ determined by $\mathcal{H}$. Moreover, the cone $\mathcal{C}=\{\mathbf{y}|\mathbf{y}=\lambda\mathbf{x},\lambda\geq 0,\mathbf{x}\in\mathcal{S}\}$ generated by the set of points $\mathcal{S}=\{\mathbf{x}\}$ is convex if $\mathcal{S}$ itself is a convex set. To see this let $\mathbf{y}_1 = \lambda_1\mathbf{x}_1$, $\mathbf{y}_2=\lambda_2\mathbf{x}_2\in\mathcal{C}$ when $\mathbf{x}_1,\mathbf{x}_2\in\mathcal{S}$. Let $\mathbf{Y}=\theta\mathbf{y}_1+(1-\theta)\mathbf{y}_2=\theta\lambda_1\mathbf{x}_1+ (1-\theta)\lambda_2\mathbf{x}_2, 0\leq\theta\leq 1$. To demonstrate that $\mathcal{C}$ is convex, it suffices to show that $\mathbf{y}$, the convex combination of $\mathbf{y}_1$ and $\mathbf{y}_2$, belongs to $\mathcal{C}$. With $\mathcal{S}$ convex, $\mathbf{x} = \alpha\mathbf{x}_1 + (1-\alpha)\mathbf{x}_2\in\mathcal{S}$. Set $\alpha = \theta\lambda_1/\beta$. Then

$$\mathbf{x} = \frac{\theta\lambda_1}{\beta}\mathbf{x}_1 + \left(1-\frac{\theta\lambda_1}{\beta}\right)\mathbf{x}_2 \qquad \text{or}$$

$$\beta\mathbf{x} = \theta\lambda_1\mathbf{x}_1 + (\beta-\theta\lambda_1)\mathbf{x}_2.$$

For $\beta = \theta\lambda_1 + (1-\theta)\lambda_2$, the preceding expression becomes $\beta\mathbf{x} = \theta\lambda_1\mathbf{x}_1 + (1-\theta)\lambda_2\mathbf{x}_2 = \mathbf{y}$. So with $\mathbf{y}$ a multiple β of $\mathbf{x}$, it follows that $\mathbf{y}\epsilon\mathcal{C}$ and thus $\mathcal{C}$ is convex.

For a convex cone $\mathcal{C} = \{\mathbf{y} \,|\, \mathbf{y} = \lambda\mathbf{x},\ \lambda \geq 0,\ \mathbf{x}\epsilon\mathbf{R}^n\}$, the ***negative cone*** of $\mathcal{C}$ is the convex cone $-\mathcal{C} = \{-\mathbf{y} \,|\, -\mathbf{y} = \lambda(-\mathbf{x}),\ \lambda \geq 0,\ \mathbf{x}\epsilon\mathcal{C}\}$ (Figure 4.1.c) while the ***orthogonal cone*** of $\mathcal{C}$ is the convex cone $\mathcal{C}^{\perp} = \{\mathbf{v} \,|\, \mathbf{y}'\mathbf{v} = 0$ for all $\mathbf{y}\epsilon\mathcal{C}\}$. Moreover, $C^{\perp}$ is a subspace of $\mathbf{R}^n$ since if $\mathbf{v}\epsilon\mathcal{C}^{\perp}$, then $\mathbf{y}'\mathbf{v} = 0$ and $\lambda\mathbf{y}'\mathbf{v} = 0$ for "all" λ (not just $\lambda \geq 0$) so that $\lambda\mathbf{y}'\mathbf{v}\epsilon\mathcal{C}^{\perp}$ for any λ.

The ***dimension*** (or ***rank***) of a convex cone $\mathcal{C}$, *dim*($\mathcal{C}$), is the dimension of the "smallest subspace of $\mathbf{R}^n$ which contains $\mathcal{C}$" and thus amounts to the dimension of $\mathcal{C} + (-\mathcal{C})$. Equivalently, *dim(*$\mathcal{C}$*)* may be viewed as the maximum number of linearly independent vectors contained in $\mathcal{C}$.

For $\mathcal{C}$ a convex cone in $\mathbf{R}^n$:

(a) The closure $\bar{\mathcal{C}}$ of $\mathcal{C}$ is convex.

(b) $\mathcal{C}$ is termed ***solid*** if $\mathcal{C}^o \neq \phi$.

(c) $\mathcal{C}$ is never a strictly bounded set (although it may be bounded either from above or from below) except if, trivially, $\mathbf{0}$ is its only element.

A convex cone $\mathcal{C}$ in $\mathbf{R}^n$ is said to be ***pointed*** if it contains no subspace other than $\{\mathbf{0}\}$, *i.e.*, given a vector $\boldsymbol{\alpha}\epsilon\mathbf{R}^n$, $\mathcal{C}$ contains no ***line*** $\mathcal{L} = \{\mathbf{x} \,|\, \mathbf{x} = \lambda\boldsymbol{\alpha}$ for all $\lambda\epsilon\mathbf{R}$, $\mathbf{x}\epsilon\mathbf{R}^n\}$. Based upon this definition it follows that if a convex cone has a vertex, it is unique and must be located at the origin. Hence a convex cone with vertex $\mathbf{0}$ is pointed. It is also true that:

(a) A convex cone $\mathcal{C}$ in $\mathbf{R}^n$ is pointed if and only if $\mathcal{C} \cap (-\mathcal{C}) = \{\mathbf{0}\}$. (If $\mathcal{C} \cap (-\mathcal{C}) \neq \{\mathbf{0}\}$, then there exists an $\mathbf{x}\epsilon\mathcal{C}$ such that $-\mathbf{x}\epsilon\mathcal{C}$ and thus $\mathbf{0}$ can be written as the convex combination $\mathbf{0} = \frac{1}{2}\mathbf{x} + \frac{1}{2}(-\mathbf{x})$. But this implies that $\mathbf{0}$ is not a vertex of $\mathcal{C}$ since it lies between the points $\mathbf{x}$, $-\mathbf{x}$.)

(b) If $\mathcal{C}$ is a pointed convex cone in $\mathbf{R}^n$, there exists a hyperplane $\mathcal{H}$ passing through $\mathbf{0}$ which separates $\mathcal{C}$, $-\mathcal{C}$ and supports both simultaneously.

(c) A convex cone $\mathcal{C}$ in $\mathbf{R}^n$ which is also a linear subspace has no vertex and consequently is not pointed.

Fundamental operations on convex cones in $\mathbf{R}^n$ are addition, intersection, and duality, *i.e.*,

(a) if $\mathcal{C}_1$ $\mathcal{C}_2$ are convex cones, their ***sum***

$$\mathcal{C}_1 + \mathcal{C}_2 = \{\mathbf{x} \mid \mathbf{x} = \mathbf{x}_1 + \mathbf{x}_2, \mathbf{x}_1 \in \mathcal{C}_1, \mathbf{x}_2 \in \mathcal{C}_2\}$$

is a convex cone (Figure 4.2.a).

(b) if $\mathcal{C}_1$, $\mathcal{C}_2$ are convex cones, their ***intersection***

$$\mathcal{C}_1 \cap \mathcal{C}_2 = \{\mathbf{x} \mid \mathbf{x} \in \mathcal{C}_1 \text{ and } \mathbf{x} \in \mathcal{C}_2\}$$

is a convex cone (Figure 4.2.b).

(c) if $\mathcal{C}$ is a convex cone, the ***dual cone***

$$\mathcal{C}^* = \{\mathbf{y} \mid \mathbf{x}'\mathbf{y} \leq 0 \text{ for all } \mathbf{x} \in \mathcal{C}\}$$

is a convex cone (Figure 4.2.c).

Note that $\mathcal{C}^*$ consists of all vectors making a non-acute angle ($\geq \pi/2$) with all vectors of $\mathcal{C}$.

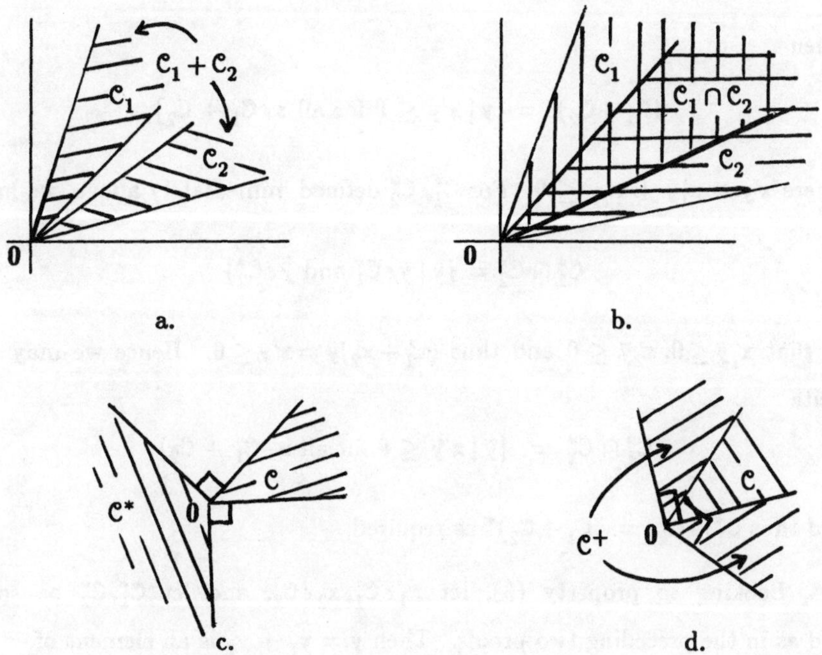

Figure 4.2

As for the properties of these operations on convex cones in $\mathbf{R}^n$ are concerned:

(1) If $\mathcal{C}_1 \subset \mathcal{C}_2$, then $\mathcal{C}_2^* \subset \mathcal{C}_1^*$.
(2) $(\mathcal{C}_1+\mathcal{C}_2)^* = \mathcal{C}_1^* \cap \mathcal{C}_2^*$.
(3) $\mathcal{C}_1^*+\mathcal{C}_2^* \subset (\mathcal{C}_1 \cap \mathcal{C}_2)^*$.
(4) $\mathcal{C} \subset (\mathcal{C}^*)^* = \mathcal{C}^{**}$.

To verify property (1), let $\mathbf{x}_1 \in \mathcal{C}_2$ whenever $\mathbf{x}_1 \in \mathcal{C}_1 (\mathcal{C}_1 \subset \mathcal{C}_2)$ and let $\mathbf{x}_2 \in \mathcal{C}_2$. Then $\mathbf{x} = \mathbf{x}_1 + \mathbf{x}_2 \in \mathcal{C}_2$. Define

$$\begin{aligned} \mathcal{C}_1^* &= \{\mathbf{y}_1 \mid \mathbf{x}_1'\mathbf{y}_1 \leq 0 \text{ for all } \mathbf{x}_1 \in \mathcal{C}_1\}, \\ \mathcal{C}_2^* &= \{\mathbf{y}_2 \mid \mathbf{x}_2'\mathbf{y}_2 \leq 0 \text{ for all } \mathbf{x}_2 \in \mathcal{C}_2\}. \end{aligned}$$

Since $\mathbf{x} \in \mathcal{C}_2$, $\mathbf{x}'\mathbf{y}_2 = \mathbf{x}_1'\mathbf{y}_2 + \mathbf{x}_2'\mathbf{y}_2 \leq 0$ for all $\mathbf{x}_1, \mathbf{x}_2 \in \mathcal{C}_2$. But $\mathbf{x}_1'\mathbf{y}_2 = \mathbf{x}_1'\mathbf{y}_1$ since $\mathbf{x}_1 \in \mathcal{C}_1$. Hence $\mathbf{x}_1'\mathbf{y}_2 + \mathbf{x}_2'\mathbf{y}_2 = \mathbf{x}_1'\mathbf{y}_1 + \mathbf{x}_2'\mathbf{y}_2 \leq \mathbf{x}_1'\mathbf{y}_1$ so that $\mathcal{C}_2^* \subset \mathcal{C}_1^*$.

As for property (2), let $\mathbf{x}_1 \in \mathcal{C}_1$, $\mathbf{x}_2 \in \mathcal{C}_2$, and $\mathbf{x} = \mathbf{x}_1 + \mathbf{x}_2 \in \mathcal{C}_1 + \mathcal{C}_2$.

Then

$$(\mathcal{C}_1+\mathcal{C}_2)^* = \{\mathbf{y}\,|\,\mathbf{x}'\mathbf{y} \leq 0 \text{ for all } \mathbf{x}\in\mathcal{C}_1 + \mathcal{C}_2\},$$

where $\mathbf{x}'\mathbf{y} = \mathbf{x}_1'\mathbf{y} + \mathbf{x}_2'\mathbf{y} \leq 0$. For $\mathcal{C}_1^*$, $\mathcal{C}_2^*$ defined immediately above, we have

$$\mathcal{C}_1^* \cap \mathcal{C}_2^* = \{\bar{\mathbf{y}}\,|\,\bar{\mathbf{y}}\in\mathcal{C}_1^* \text{ and } \bar{\mathbf{y}}\in\mathcal{C}_2^*\}$$

so that $\mathbf{x}_1'\bar{\mathbf{y}} \leq 0, \mathbf{x}_2'\bar{\mathbf{y}} \leq 0$ and thus $(\mathbf{x}_1'+\mathbf{x}_2')\bar{\mathbf{y}} = \mathbf{x}'\bar{\mathbf{y}} \leq 0$. Hence we may also write

$$\mathcal{C}_1^* \cap \mathcal{C}_2^* = \{\bar{\mathbf{y}}\,|\,\mathbf{x}'\bar{\mathbf{y}} \leq 0 \text{ for all } \mathbf{x}\in\mathcal{C}_1 + \mathcal{C}_2\}$$

and thus $\mathcal{C}_1^* \cap \mathcal{C}_2^* = (\mathcal{C}_1+\mathcal{C}_2)^*$ as required.

Looking to property (3), let $\mathbf{x}_1\in\mathcal{C}_1, \mathbf{x}_2\in\mathcal{C}_2$, and let $\mathcal{C}_1^*, \mathcal{C}_2^*$ be specified as in the preceding two proofs. Then $\mathbf{y} = \mathbf{y}_1 + \mathbf{y}_2$ is an element of

$$\mathcal{C}_1^* + \mathcal{C}_2^* = \{\mathbf{y}\,|\,\mathbf{x}'\mathbf{y} \leq 0 \text{ for all } \mathbf{x}\in\mathcal{C}\},$$

where $\mathcal{C}$ is the cone whose dual is $\mathcal{C}_1^* + \mathcal{C}_2^*$. For $\bar{\mathbf{x}}\in\mathcal{C}_1 \cap \mathcal{C}_2$,

$$(\mathcal{C}_1 \cap \mathcal{C}_2)^* = \{\bar{\mathbf{y}}\,|\,\bar{\mathbf{x}}'\bar{\mathbf{y}} \leq 0 \text{ for all } \bar{\mathbf{x}}\in\mathcal{C}_1 \cap \mathcal{C}_2\}.$$

From the specification of $\bar{\mathbf{x}}$, it follows that $\bar{\mathbf{x}}'\mathbf{y}_1 \leq 0, \bar{\mathbf{x}}'\mathbf{y}_2 \leq 0$ and thus $\bar{\mathbf{x}}'(\mathbf{y}_1+\mathbf{y}_2) = \bar{\mathbf{x}}'\mathbf{y} \leq 0$. Hence $\bar{\mathbf{x}}'\mathbf{y}+\bar{\mathbf{x}}'\bar{\mathbf{y}} = \bar{\mathbf{x}}'(\bar{\mathbf{y}}+\mathbf{y}) \leq 0$ for $\bar{\mathbf{x}}\in\mathcal{C}_1 \cap \mathcal{C}_2$. Since $\mathbf{y}\in\mathcal{C}_1^*+\mathcal{C}_2^*$ and $\bar{\mathbf{y}}+\mathbf{y}\in(\mathcal{C}_1 \cap \mathcal{C}_2)^*$, it follows that $\mathcal{C}_1^*+\mathcal{C}_2^* \subset (\mathcal{C}_1 \cap \mathcal{C}_2)^*$.

To verify property (4), let $\mathbf{x}\in\mathcal{C}$. Then

$$\mathcal{C}^* = \{\mathbf{y}\,|\,\mathbf{x}'\mathbf{y} \leq 0 \text{ for all } \mathbf{x}\in\mathcal{C}\}.$$

Let

$$\mathcal{C}^{**} = \{\bar{\mathbf{x}}\,|\,\mathbf{y}'\bar{\mathbf{x}} \leq 0 \text{ for all } \mathbf{y}\in\mathcal{C}^*\}.$$

Hence $\mathbf{y}'\mathbf{x}+\mathbf{y}'\bar{\mathbf{x}} = \mathbf{y}'(\mathbf{x}+\bar{\mathbf{x}}) \leq 0$ for $\mathbf{y}\in\mathcal{C}^*$. Since $\mathbf{x}\in\mathcal{C}$ and $\mathbf{x}+\bar{\mathbf{x}}\in\mathcal{C}^{**}$, $\mathcal{C}\subset\mathcal{C}^{**}$.

Relative to fundamental operation (c) above we may take the negative of the dual cone $\mathcal{C}^*$ of $\mathcal{C}$ in $\mathbf{R}^n$ so as to form the ***polar cone*** $\mathcal{C}^+ = \{\mathbf{y} \mid \mathbf{x}'\mathbf{y} \geq 0$ for all $\mathbf{x} \in \mathcal{C}\}$ (Figure 4.2.d) consisting of all vectors making a non-obtuse angle ($\leq \pi/2$) with all vectors of $\mathcal{C}$. Clearly $\mathcal{C}^+$ is a convex cone if $\mathcal{C}$ is. For convex cones in $\mathbf{R}^n$, the polarity operator possess the following properties:

(1) If $\mathcal{C}_1 \subset \mathcal{C}_2$, then $\mathcal{C}_2^+ \subset \mathcal{C}_1^+$

(2) $(\mathcal{C}_1 + \mathcal{C}_2)^+ = \mathcal{C}_1^+ \cap \mathcal{C}_2^+$.

(3) $\mathcal{C}_1^+ + \mathcal{C}_2^+ \subset (\mathcal{C}_1 \cap \mathcal{C}_2)^+$.

(4) $\mathcal{C} \subset (\mathcal{C}^+)^+ = \mathcal{C}^{++}$.

The proofs of these properties are similar to those given in support of the preceding set of assertions involving the duality operator.

4.2. Finite Cones

A convex cone $\mathcal{C}$ in $\mathbf{R}^n$ is termed a ***finite cone*** if it consists of the set of all non-negative linear combinations of a finite set of vectors, *i.e.*, for points $\mathbf{x}_j \in \mathbf{R}^n$, $j=1,\ldots,m$,

$$\mathcal{C} = \left\{\mathbf{x} \mid \mathbf{x} = \sum_{j=1}^{m} \lambda_j \mathbf{x}_j,\ 0 \leq \lambda_j \in \mathbf{R}\right\}. \tag{4.1}$$

Here $\mathcal{C}$ is said to be ***generated*** or ***spanned*** by the points $\mathbf{x}_j$. Hence any vector which can be expressed as a non-negative linear combination of a finite set of vectors $\mathbf{x}_j$, $j=1,\ldots,m$, lies in the finite cone spanned by those vectors. An alternative way of structuring $\mathcal{C}$ in (4.1) is to state that a convex cone $\mathcal{C}$ in $\mathbf{R}^n$ is a finite cone if for some $(n \times m)$ matrix $\mathbf{A}' = [\mathbf{a}_1,\ldots,\mathbf{a}_m]$, every $\mathbf{x} \in \mathcal{C}$ is a non-negative linear combination of the $(n \times 1)$ column vectors $\mathbf{a}_j$, $j=1,\ldots,m$, of $\mathbf{A}'$, *i.e.*,

$$\mathcal{C} = \left\{ \mathbf{x} \,|\, \mathbf{x} = \mathbf{A}'\boldsymbol{\lambda} = \sum_{j=1}^{m} \lambda_j \mathbf{a}_j,\ \mathbf{0} \leq \boldsymbol{\lambda} \epsilon \mathbf{R}^m,\ \mathbf{x} \epsilon \mathbf{R}^m \right\}. \tag{4.1.1}$$

A cone such as (4.1.1) is sometimes referred to as a "closed" convex cone. As we shall see later on, "finite cones" (those generated by a finite set of vectors) are synononomous with what will be termed "polyhedral convex cones," the latter begin formed as the intersection of a finite number of half-planes, each of whose associated hyperplanes pass through the origin. Polyhedral convex cones are typically written as the set $\{\mathbf{x} \,|\, \mathbf{A}\mathbf{x} \leqq \mathbf{0}\}$ in $\mathbf{R}^n$.

The essential operations on finite cones are:

(a) if $\mathcal{C}_1, \mathcal{C}_2$ are finite cones in $\mathbf{R}^n$, their ***sum*** $\mathcal{C}_1 + \mathcal{C}_2 = \{\mathbf{x} \,|\, \mathbf{x} = \mathbf{x}_1 + \mathbf{x}_2,\ \mathbf{x}_1 \epsilon \mathcal{C}_1,\ \mathbf{x}_2 \epsilon \mathcal{C}_2\}$ is a finite cone;

(b) if $\mathcal{C}_1$, $\mathcal{C}_2$ are finite cones in $\mathbf{R}^n$, their ***intersection*** $\mathcal{C}_1 \cap \mathcal{C}_2$ is a finite cone.

A special case of (4.1), (4.1.1) occurs when there is but a single generator. That is, a convex cone $\mathcal{C}$ in $\mathbf{R}^n$ generated by a single vector $\mathbf{a}(\neq \mathbf{0})$ is termed a ***ray*** or ***half-line*** (denoted $(\mathbf{a})$). So for $\mathbf{a} \epsilon \mathbf{R}^n$,

$$(\mathbf{a}) = \{\mathbf{x} \,|\, \mathbf{x} = \lambda \mathbf{a},\ 0 \leq \lambda \epsilon \mathbf{R},\ \mathbf{x} \epsilon \mathbf{R}^n\}$$

(Figure 4.3.a). Given $(\mathbf{a}) = \{\mathbf{x} \,|\, \mathbf{x} = \lambda \mathbf{a}, \mathbf{a} \neq \mathbf{0}, \lambda \geq 0, \mathbf{x} \epsilon \mathbf{R}^n\}$, the ***orthogonal cone to (a)*** is the hyperplane $(\mathbf{a})^{\perp} = \mathcal{H} = \{\mathbf{y} \,|\, \mathbf{a}'\mathbf{y} = 0\}$ passing through the origin (Figure 4.3.b) while the ***dual of (a)*** is the half-space $(\mathbf{a})^* = [\mathcal{H}^-] = \{\mathbf{y} \,|\, \mathbf{a}'\mathbf{y} \leq 0\}$; it consists of all vectors making a non-acute angle with $\mathbf{a}$ (Figure 4.3.c). Additionally, the ***polar*** of $(\mathbf{a})$ is the half-space $(\mathbf{a})^+ = [\mathcal{H}^+] = \{\mathbf{y} \,|\, \mathbf{a}'\mathbf{y} \geq 0\}$; it consists of all vectors making a non-obtuse angle with $\mathbf{a}$ (Figure 4.3.d).

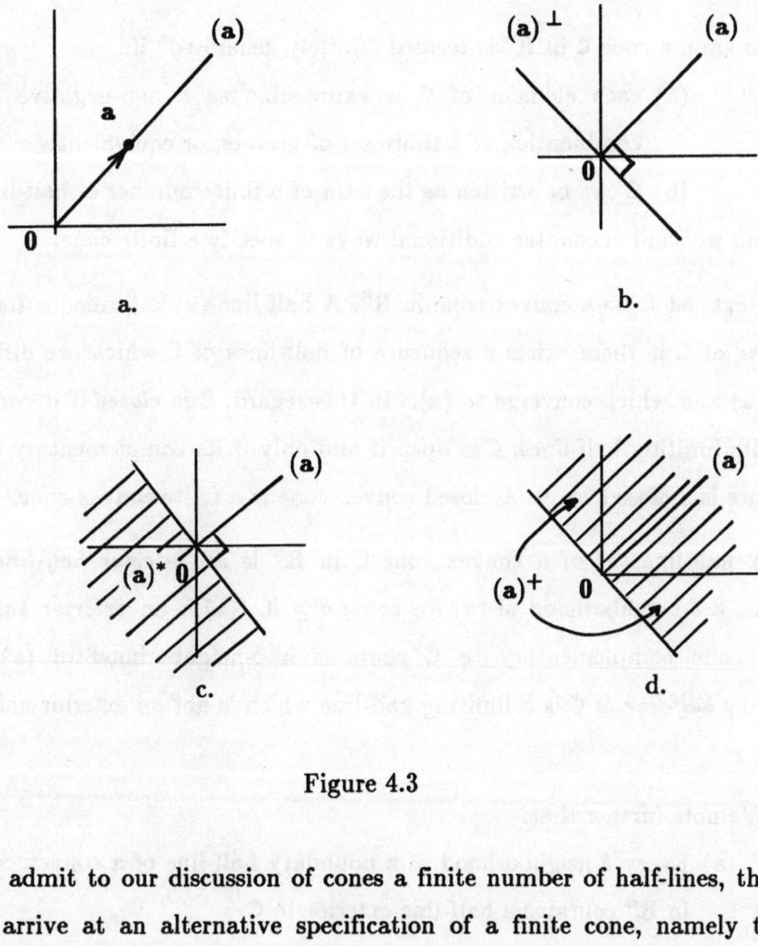

Figure 4.3

If we admit to our discussion of cones a finite number of half-lines, then we may arrive at an alternative specification of a finite cone, namely that a convex cone $\mathcal{C}$ in $\mathbf{R}^n$ is termed a finite cone if there exists a finite number of half-lines $(\mathbf{a}_j)$ in $\mathbf{R}^n$, $j=1,\ldots,m$, such that $\mathcal{C} = \Sigma_{j=1}^m (\mathbf{a}_j)$. Here $\mathcal{C}$ is to be interpreted as the sum of the m cones $(\mathbf{a}_j)$ and may be represented as

$$\mathcal{C} = \left\{ \mathbf{x} \,|\, \mathbf{x} = \sum_{j=1}^{m} \mathbf{x}_j,\ \mathbf{x}_j \in (\mathbf{a}_j),\ \mathbf{x} \in \mathbf{R}^n \right\} \tag{4.1.2}$$

by property (a) immediately above. But since $\mathbf{x} = \Sigma_{j=1}^m \mathbf{x}_j = \Sigma_{j=1}^m \lambda_j \mathbf{a}_j = \mathbf{A}'\boldsymbol{\lambda}$, $\boldsymbol{\lambda} \geqq \mathbf{0}$, equation (4.1.2) is just $\mathcal{C} = \{\mathbf{x} \,|\, \mathbf{x} = \mathbf{A}'\boldsymbol{\lambda},\ \boldsymbol{\lambda} \geqq \mathbf{0},\ \mathbf{x} \in \mathbf{R}^n\}$, thus establishing the equivalence of (4.1), (4.1.1), and (4.1.2).

In sum, a cone $\mathcal{C}$ in $\mathbf{R}^n$ is termed "finitely generated" if:

(a) each element of $\mathcal{C}$ is expressible as a non-negative linear combination of a finite set of vectors; or equivalently,

(b) $\mathcal{C}$ can be written as the sum of a finite number of half-lines.

Later on we shall encounter additional ways to specify a finite cone.

Next, let $\mathcal{C}$ be a convex cone in $\mathbf{R}^n$. A half-line $(\mathbf{a})$ is termed a ***limiting half-line*** of $\mathcal{C}$ if there exists a sequence of half-lines of $\mathcal{C}$ which are different from $(\mathbf{a})$ and which converge to $(\mathbf{a})$. In this regard, $\mathcal{C}$ is ***closed*** if it contains all of its limiting half-lines; $\mathcal{C}$ is open if and only if its complementary set of half-lines is a closed cone. A closed convex cone is a finite convex cone.

A half-line $(\mathbf{a})$ of a convex cone $\mathcal{C}$ in $\mathbf{R}^n$ is an ***interior half-line*** if $\mathcal{C}$ contains a δ-neighborhood of $(\mathbf{a})$ for some $\delta \geq 0$. $(\mathbf{a})$ is an ***exterior half-line*** if the cone complementary to $\mathcal{C}$ contains a δ-neighborhood of $(\mathbf{a})$. A ***boundary half-line*** of $\mathcal{C}$ is a limiting half-line which is not an exterior half-line of $\mathcal{C}$.

We note further that:

(a) Every δ-neighborhood of a boundary half-line of a convex cone $\mathcal{C}$ in $\mathbf{R}^n$ contains a half-line exterior to $\mathcal{C}$.

(b) A convex cone $\mathcal{C}$ in $\mathbf{R}^n$ and its exterior have the same boundary half-lines.

In defining a finite cone no mention was made of the linear independence of the spanning set $\{\mathbf{x}_j, j=1,\ldots,m\}$. Indeed, the vectors $\mathbf{x}_j$ need not be linearly independent. However, if some subset of the vectors generating a finite cone $\mathcal{C}$ is linearly independent, then we may note that the ***dimension*** of $\mathcal{C}$ is the maximum number of linearly independent vectors in $\mathcal{C}$. So if $\mathcal{C} = \{\mathbf{x} \mid \mathbf{x} = \mathbf{A}'\boldsymbol{\lambda}, \boldsymbol{\lambda} \geqq \mathbf{0}, \mathbf{x} \in \mathbf{R}^n\}$, then $dim(\mathcal{C}) = \rho(\mathbf{A}')$.

It was mentioned earlier that any hyperplane $\mathcal{H}$ through the origin is a convex cone as are $[\mathcal{H}^+]$, $[\mathcal{H}^-]$. In addition, $\mathcal{H}$, $[\mathcal{H}^+]$, and $[\mathcal{H}^-]$ are all finite cones. In this light we note briefly that:

(a) If $\mathbf{A}' = [\mathbf{a}_1,\ldots,\mathbf{a}_m]$ is $(n\times m)$, the solution set $\mathcal{C}$ of the system of homogeneous linear inequalities $\mathbf{A}'\boldsymbol{\lambda} \leqq \mathbf{0}$ is a finite cone. ($\mathbf{A}'\boldsymbol{\lambda} \leqq \mathbf{0}$ consists of the n inequalities $\boldsymbol{\alpha}_i\boldsymbol{\lambda} \leq 0$, $i=1,\ldots,n$, each specifying a closed half-plane of the form $[\mathcal{H}_i^-]=(\boldsymbol{\alpha}_i)^*=\{\boldsymbol{\lambda} \mid \boldsymbol{\alpha}_i\boldsymbol{\lambda} \leq 0, \boldsymbol{\lambda}\in\mathbf{R}^m\}$, where $\boldsymbol{\alpha}_i$ is the ith row of $\mathbf{A}'$. Since each $(\boldsymbol{\alpha}_i)^*$ is a cone, $\mathcal{C}=\cap_{i=1}^n(\boldsymbol{\alpha}_i)^*$ is a finite cone by property (b), p. 4.) $\mathcal{C}$ in $\mathbf{R}^2$ is illustrated in Figure 4.4. In like fashion it can be rationalized that the solution set $\mathcal{C}$ of $\mathbf{A}'\boldsymbol{\lambda} \geqq \mathbf{0}$ is a finite cone since $(\boldsymbol{\alpha}_i)^+=\{\boldsymbol{\lambda} \mid \boldsymbol{\alpha}_i\boldsymbol{\lambda} \geq 0, \boldsymbol{\lambda}\in\mathbf{R}^m\}$, $i=1,\ldots,n$, is a finite cone and $\mathcal{C}=\cap_{i=1}^n(\boldsymbol{\alpha}_i)^+$.

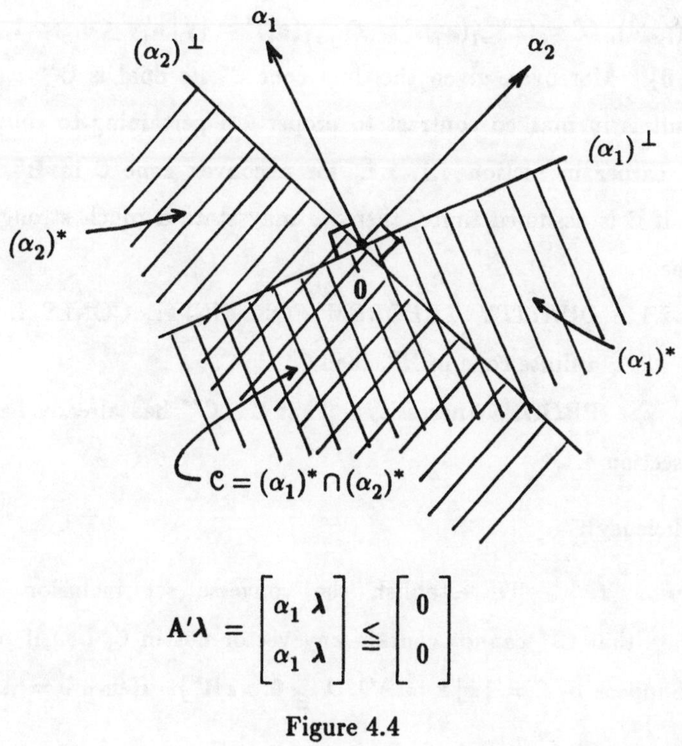

$$\mathbf{A}'\boldsymbol{\lambda} = \begin{bmatrix} \alpha_1\ \boldsymbol{\lambda} \\ \alpha_1\ \boldsymbol{\lambda} \end{bmatrix} \leqq \begin{bmatrix} 0 \\ 0 \end{bmatrix}$$

Figure 4.4

(b) If $\mathbf{A}' = [\mathbf{a}_1, \ldots, \mathbf{a}_m]$ is $(n \times m)$, the set of non-negative solutions of the homogeneous linear system $\mathbf{A}'\boldsymbol{\lambda} = \mathbf{0}$ forms a finite cone. (Since each of the n hyperplanes $\mathcal{H}_i = (\boldsymbol{\alpha}_i)^{\perp} = \{\boldsymbol{\lambda} \mid \boldsymbol{\alpha}_i \boldsymbol{\lambda} = 0, \boldsymbol{\lambda} \in \mathbf{R}^m\}$ constituting $\mathbf{A}'\boldsymbol{\lambda} = \mathbf{0}$ is a finite cone, as is the non-negative orthant $\mathcal{N} = \{\mathbf{x} \mid \mathbf{x} \geqq \mathbf{0}, \mathbf{x} \in \mathbf{R}^m\}$, it follows that $\mathcal{N} \cap (\cap_{i=1}^{n} (\boldsymbol{\alpha}_i)^{\perp})$ is a finite cone by property (b) above.)

For $\mathcal{C}$ in $\mathbf{R}^n$ a finite cone, the ***dual cone*** $\mathcal{C}^*$ is the finite cone consisting of all vectors making a non-acute angle ($\geq \pi/2$) with the vectors of $\mathcal{C}$, *i.e.*, if $\mathbf{A}' = [\mathbf{a}_1, \ldots, \mathbf{a}_m]$ is $(n \times m)$ and $\mathcal{C} = \{\mathbf{x} \mid \mathbf{x} = \mathbf{A}'\boldsymbol{\lambda}, \boldsymbol{\lambda} \geqq \mathbf{0}, \mathbf{x} \in \mathbf{R}^n\}$ then $\mathcal{C}^* = \{\mathbf{y} \mid \mathbf{A}\mathbf{y} \leqq \mathbf{0}, \mathbf{y} \in \mathbf{R}^n\}$ (Figure 4.5.a). Here $\mathcal{C}^*$ is the intersection of finitely many half-spaces $\mathbf{a}_j'\boldsymbol{\lambda} \leq 0$, $j = 1, \ldots, m$, whose boundary hyperplanes $(\mathbf{a}_j)^{\perp} = \{\mathbf{y} \mid \mathbf{a}_j'\mathbf{y} = 0\}$ pass through the origin. That is, for half-line $(\mathbf{a}_j)$, its dual $(\mathbf{a}_j)^*$ is the half-space $\mathbf{a}_j'\mathbf{y} \leq 0$. Then, since $\mathcal{C} = \Sigma_{j=1}^{m} (\mathbf{a}_j)$ and, by duality property 2 above (p. 4), $\mathcal{C}^* = (\Sigma_{j=1}^{m} (\mathbf{a}_j))^* = \cap_{j=1}^{m} (\mathbf{a}_j)^* = \{\mathbf{y} \mid \mathbf{a}_j'\mathbf{y} \leq 0, j = 1, \ldots, m\} = \{\mathbf{y} \mid \mathbf{A}\mathbf{y} \leqq \mathbf{0}\}$. Moreover, given the dual cone $\mathcal{C}^*$ its dual is $\mathcal{C}^{**} = \mathcal{C}$. This latter result is in marked contrast to property 4 pertaining to convex cones presented earlier in section 4.1, *i.e.*, for a convex cone $\mathcal{C}$ in $\mathbf{R}^n$, $\mathcal{C} \subset \mathcal{C}^{**}$. However, if $\mathcal{C}$ is assumed finite, then we may state a much stronger result, namely the

4.2.1. DUALITY THEOREM FOR FINITE CONES I. If $\mathcal{C}$ is a finite cone in $\mathbf{R}^n$, then $\mathcal{C}^{**} = \mathcal{C}$.

PROOF. (necessity) That $\mathcal{C} \subset \mathcal{C}^{**}$ has already been established in section 4.1.

(sufficiency):

Version 1. To establish the converse set inclusion we shall demonstrate that $\mathcal{C}^{**}$ cannot contain any vector not in $\mathcal{C}$, *i.e.*, if $\mathbf{b} \notin \mathcal{C}$, then $\mathbf{b} \notin \mathcal{C}^{**}$. Suppose $\mathbf{b} \notin \mathcal{C} = \{\mathbf{x} \mid \mathbf{x} = \mathbf{A}'\boldsymbol{\lambda}, \boldsymbol{\lambda} \geqq \mathbf{0}, \mathbf{x} \in \mathbf{R}^n\}$. Hence $\mathbf{b} = \mathbf{A}'\boldsymbol{\lambda}$ has no non-negative solution and thus, by Farkas' theorem 3.3.1.a, there exists a $\mathbf{y}\epsilon\mathbf{R}^n$ such that $\mathbf{Ay}\leq\mathbf{0}$, $\mathbf{b'y}>0$. The first inequality implies that $\mathbf{y}\epsilon\mathcal{C}^*$ while the second implies, for $\mathbf{y}\epsilon\mathcal{C}^*$, that $\mathbf{b}\notin\mathcal{C}^{**}$.

Version 2. It is most instructive to alternatively prove sufficiency by the direct application of the strong separation theorem 3.2.2. We again endeavor to demonstrate that if $\mathbf{b}\notin\mathcal{C}$, then $\mathbf{b}\notin\mathcal{C}^{**}$. Suppose $\mathbf{b}\notin\mathcal{C}$. Since $\mathcal{C}$ is a closed convex set (*i.e.*, $\mathcal{C}$ is a finite cone formed as the intersection of n closed half-spaces), there exists a vector $\mathbf{p}(\neq\mathbf{0})\epsilon\mathbf{R}^n$ and a scalar $\alpha\epsilon\mathbf{R}$ such that $\mathbf{p'x}\leq\alpha<\mathbf{p'b}$ for all $\mathbf{x}\epsilon\mathcal{C}$. By the definition of $\mathcal{C}$, $\mathbf{x}\epsilon\mathcal{C}$ implies $\lambda\mathbf{x}\epsilon\mathcal{C}, \lambda>0$. Then $\mathbf{p'}(\lambda\mathbf{x})\leq\alpha$, $\mathbf{p'x}\leq\alpha/\lambda$, and thus $\mathbf{p'x}\leq 0$ as $\lambda\rightarrow\infty$. Hence $\mathbf{p}\epsilon\mathcal{C}^*$. And since $\mathbf{0}\epsilon\mathcal{C}$, $\mathbf{p'0}=0\leq\alpha<\mathbf{p'b}$. But then $\mathbf{p'b}>0$ for $\mathbf{p}\epsilon\mathcal{C}^*$ implies that $\mathbf{b}\epsilon\mathcal{C}^{**}$. Q.E.D.

This theorem has established that the relationship between $\mathcal{C}$ and $\mathcal{C}^{**}$ is symmetric.

In what follows we shall find it convenient to structure the relationship between $\mathcal{C}$, $\mathcal{C}^*$, and $\mathcal{C}^{**}$ as:

$$\begin{aligned}\mathcal{C} &= \{\mathbf{x}\,|\,\mathbf{x}=\mathbf{A}'\boldsymbol{\lambda},\ \boldsymbol{\lambda}\geqq\mathbf{0},\ \mathbf{x}\epsilon\mathbf{R}^n\},\\ \mathcal{C}^* &= \{\mathbf{y}\,|\,\mathbf{x'y}\leq 0,\ \mathbf{x}\epsilon\mathcal{C},\ \mathbf{y}\epsilon\mathbf{R}^n\}\\ &= \{\mathbf{y}\,|\,\boldsymbol{\lambda}'\mathbf{Ay}\leq 0,\ \boldsymbol{\lambda}\geqq\mathbf{0}\}\\ &= \{\mathbf{y}\,|\,\mathbf{A'y}\leqq\mathbf{0}\},\text{ and}\\ \mathcal{C}^{**} &= \{\mathbf{x}\,|\,\mathbf{y'x}\leq 0,\ \mathbf{y}\epsilon\mathcal{C}^*\}.\end{aligned}$$

The essential duality properties of finite cones in $\mathbf{R}^n$ are:

(1) $(\mathcal{C}^*)^*=\mathcal{C}$.

(2) If $\mathcal{C}_1\subset\mathcal{C}_2$, then $\mathcal{C}_2^*\subset\mathcal{C}_1^*$.

(3) $(\mathcal{C}_1+\mathcal{C}_2)^*=\mathcal{C}_1^*\cap\mathcal{C}_2^*$.

(4) $(\mathcal{C}_1\cap\mathcal{C}_2)^*=\mathcal{C}_1^*+\mathcal{C}_2^*$.

It is easily seen that property (1) is simply a restatement of the duality theorem 4.1 for finite cones given earlier. Properties (2) and (3) have already been proved above for the case of arbitrary convex cones. Looking to property (4), it is true from properties (1), (3) that $\mathcal{C}_1^* + \mathcal{C}_2^* = (\mathcal{C}_1^* + \mathcal{C}_2^*)^{**} = (\mathcal{C}_1^{**} \cap \mathcal{C}_2^{**})^* = (\mathcal{C}_1 \cap \mathcal{C}_2)^*$.

We may now observe that a cone $\mathcal{C}$ in $\mathbf{R}^n$ is ***closed*** if and only if $\mathcal{C}^{**} = \mathcal{C}$, where $\mathcal{C}^{**}$ is the ***closure*** $\bar{\mathcal{C}}$ of $\mathcal{C}$. Here $\mathcal{C} \subseteq \mathcal{C}^{**} \subseteq \overline{\mathcal{C}^{**}} = \bar{\mathcal{C}}$, which implies $\mathcal{C}^{**} = \bar{\mathcal{C}}$ since $\mathcal{C}^{**}$ is an intersection of half-spaces and consequently closed.

Next, with $\mathcal{C}$ in $\mathbf{R}^n$ a finite cone, the ***polar cone*** $\mathcal{C}^+$ is the finite cone consisting of all vectors making a non-obtuse angle ($\leq \pi/2$) with the vectors of $\mathcal{C}$, *i.e.*, if $\mathbf{A}'$ is $(n \times m)$ and $\mathcal{C} = [\mathbf{x} \mid \mathbf{x} = \mathbf{A}'\boldsymbol{\lambda},\ \boldsymbol{\lambda} \geqq \mathbf{0}, \mathbf{x} \in \mathbf{R}^n\}$, then $\mathcal{C}^+ = \{\mathbf{y} \mid \mathbf{AY} \geqq \mathbf{0}, \mathbf{y} \in \mathbf{R}^n\}$ (Figure 4.5.b).

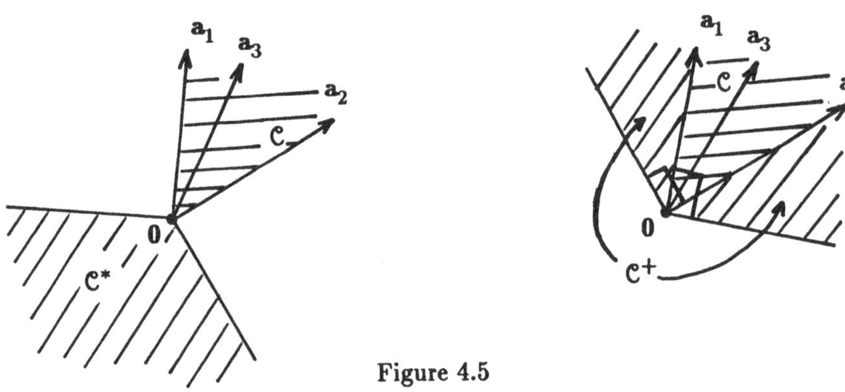

Figure 4.5

Important properties of polars of finite cones in $\mathbf{R}^n$ are:

(1) $(\mathcal{C}^+)^+ = \mathcal{C}$.

(2) If $\mathcal{C}_1 \subset \mathcal{C}_2$, then $\mathcal{C}_2^+ \subset \mathcal{C}_1^+$.

(3) $(\mathcal{C}_1 + \mathcal{C}_2)^+ = \mathcal{C}_1^+ \cap \mathcal{C}_2^+$.

(4) $(\mathcal{C}_1 \cap \mathcal{C}_2)^+ = \mathcal{C}_1^+ + \mathcal{C}_2^+$.

In fact, using property 2, one can easily rationalize the structure of $\mathcal{C}^+$ in a fashion similar to that employed in determining the form of $\mathcal{C}^*$ above.

As far as some of the salient features of convex/finite cones are concerned, we note briefly that:

(a) Any linear subspace $\mathcal{V}$ in $\mathbf{R}^n$ (including the whole space and the origin) is a convex cone. Moreover, the dual cone $\mathcal{V}^*$ is the orthogonal complement or dual subspace of $\mathcal{V}$, *i.e.*, $\mathcal{V}^* = \mathcal{V}^\perp$. (To see this let $\mathbf{y} \in \mathcal{V}^\perp$. Then $\mathbf{x}'\mathbf{y} = 0$ for all $\mathbf{x} \in \mathcal{V}$ and thus $\mathbf{y} \in \mathcal{V}^*$. Conversely, if $\mathbf{y} \in \mathcal{V}^*$, then for all $\mathbf{x}$, $-\mathbf{x} \in \mathcal{V}$, it follows that $\mathbf{x}'\mathbf{y} \leq 0$, $(-\mathbf{x})'\mathbf{y} \leq 0$ or $\mathbf{x}'\mathbf{y} = 0$ so that $\mathbf{y} \in \mathcal{V}^\perp$. Hence $\mathcal{V}^* = \mathcal{V}^\perp$.) It is also true that a linear subspace $\mathcal{V}$ in $\mathbf{R}^n$ is a finite cone. (Let $\{\mathbf{x}_1, \ldots, \mathbf{x}_r\}$ be a basis for $\mathcal{V}$. Then for any $\mathbf{x} \in \mathcal{V}$, there exist $\lambda_i \in \mathbf{R}$ such that $\mathbf{x} = \Sigma_{i=1}^r \lambda_i \mathbf{x}_i$. Take $\lambda_i = \lambda_i' - \lambda_i''$, where $\lambda_i', \lambda_i'' \geq 0$. Then

$$\begin{aligned}\mathbf{x} &= \sum_{i=1}^{r} \lambda_i' \cdot \mathbf{x}_i + \sum_{i=1}^{r} \lambda_i'' (-\mathbf{x}_i) \\ &= \sum_{i=1}^{r} (\mathbf{x}_i) + \sum_{i=1}^{r} (-\mathbf{x}_i),\end{aligned}$$

i.e., any $\mathbf{x}$ in $\mathcal{V}$ is expressible as a positively weighted sum of a finite number of half-lines.)

(b) The intersection of any linear subspace $\mathcal{V}$ in $\mathbf{R}^n$ with the non-negative orthant $\mathcal{N} = \{\mathbf{x} \mid \mathbf{x} \geqq \mathbf{0}, \mathbf{x} \in \mathbf{R}^n\}$ is a finite cone. (Let $\{\mathbf{a}_1, \ldots, \mathbf{a}_r\}$ be a basis for $\mathcal{V}^*$ and let $\mathcal{C}$ represent the set of all non-negative solutions of the equations $\mathbf{a}_i'\mathbf{x} = 0$, $i=1,\ldots,r$. Clearly $\mathcal{C}$ is a finite cone and, if $\mathbf{x} \in \mathcal{V} \cap \mathcal{N}$, then $\mathbf{x} \in \mathcal{C}$. However, if $\mathbf{x} \in \mathcal{C}$, then $\mathbf{a}_i'\mathbf{x} = 0$, $i=1,\ldots,r$. Hence $\mathbf{x} \in \mathcal{V}^{**} = \mathcal{V}$ and thus $\mathcal{C} = \mathcal{V} \cap \mathcal{N}$.)

(c) All convex cones in $\mathbf{R}^n$ intersect at the origin $\mathbf{0}$. Trivially, $\mathbf{0}$ can be thought of as the convex cone $(\mathbf{0}) = \{\mathbf{x} \mid \mathbf{x} = \lambda \mathbf{0},\ 0 \leq \lambda \in \mathbf{R}, \mathbf{x} \in \mathbf{R}^n\}$.

(d) $(\mathbf{0})^* = \mathbf{R}^n$. $(\mathbf{R}^n)^* = (\mathbf{0})$.

(e) For $\mathcal{C}$ a finite cone in $\mathbf{R}^n$, $\mathcal{C} \oplus \mathcal{C}^* = \mathbf{R}^n$. (Clearly $\mathcal{C} \cap \mathcal{C}^* = (\mathbf{0})$ since, if $\mathbf{x} \in \mathcal{C} \cap \mathcal{C}^*$, then $\mathbf{x}'\mathbf{x} \leq 0$ only if $\mathbf{x} = \mathbf{0}$. Taking $(\mathcal{C} + \mathcal{C}^*)^* = (\mathbf{0})^*$ we obtain $\mathcal{C}^* + \mathcal{C} = \mathbf{R}^n = \mathcal{C} \oplus \mathcal{C}^*$ as required.)

(f) The dual cone $\mathcal{C}$ in $\mathbf{R}^n$ is sometimes referred to as the ***negative polar cone*** since $\mathcal{C}^* = -\mathcal{C}^+$.

(g) The "smallest linear space" containing a cone $\mathcal{C}$ in $\mathbf{R}^n$ is $\mathcal{C} + (-\mathcal{C})$; it consists of all vectors expressed as the difference of vectors in $\mathcal{C}$. The "largest linear space" containing $\mathcal{C}$ is $\mathcal{C} \cap (-\mathcal{C})$; it is made up of all vectors $\mathbf{x}$ such that both x, $-\mathbf{x} \in \mathcal{C}$. The dimension of $\mathcal{C} \cap (-\mathcal{C})$ is called the ***linearity of*** $\mathcal{C}$ (denoted *lin*($\mathcal{C}$)). In this regard, if $\mathcal{C} = \{\mathbf{x} \mid \mathbf{x} = \mathbf{A}'\boldsymbol{\lambda},\ \boldsymbol{\lambda} \geqq \mathbf{0},\ \mathbf{x} \in \mathbf{R}^n\}$, where $\mathbf{A}$ is of order $(n \times m)$, then the linearity of $\mathcal{C}$ is $n - \rho(A)$. If $\mathcal{C}$ is a finite cone in $\mathbf{R}^n$ and $lin(\mathcal{C}) = 0$, then $\mathcal{C}$ is pointed. Moreover, $dim(\mathcal{C}) + lin(\mathcal{C}^+) = lin(\mathcal{C}) + dim(\mathcal{C}^+) = n$.

(h) Let $\mathcal{C}$ be a finite cone in $\mathbf{R}^n$. Then $\mathcal{C}$ is pointed if and only if $\mathcal{C}^*$ is solid.

(i) A cone $\mathcal{C}$ in $\mathbf{R}^n$ is termed ***full*** if it is finite (*i.e.*, closed and convex), pointed, and solid.

(j) Let $\mathcal{S}$ be a convex set in $\mathbf{R}^n$. The ***barrier cone*** of $\mathcal{S}$ is the convex cone $\mathcal{C} = \{\mathbf{y} \mid \mathbf{x}'\mathbf{y} \leq \beta \text{ for all } \mathbf{x} \in \mathcal{S},\ \beta \in \mathbf{R}\}$.

(k) Let $\mathcal{C}$ be a finite cone in $\mathbf{R}^n$. If $\mathcal{C}$ contains no semi-positive $\mathbf{x} \geq \mathbf{0}$, then $\mathcal{C}^*$ contains a positive $\mathbf{y} > \mathbf{0}$, *i.e.*, if $\mathcal{C}$ has no semi-positive points, then $\mathcal{C}$ is contained in the half-space determined by $\mathbf{x}'\mathbf{y} \leq 0$ for some $\mathbf{y} > \mathbf{0}$.

(l) Let $\mathcal{C}$ be a convex cone in $\mathbf{R}^n$. Then:

(i) if $\mathcal{C}^+$ has an interior point, $\mathcal{C}$ is pointed;

(ii) if $\mathcal{C}$ is closed and pointed, $\mathcal{C}^+$ has an interior point;

(iii) if $\mathcal{C}$ is closed and pointed, there is some vector $\boldsymbol{\alpha}$ such that $\boldsymbol{\alpha}'\mathbf{x} > 0$ for all $\mathbf{x}(\neq \mathbf{0}) \in \mathcal{C}$; and

(iv) for any $\boldsymbol{\alpha}$ in (iii) and any $\beta \in \mathbf{R}$, the set $\{\mathbf{x} \mid \boldsymbol{\alpha}'\mathbf{x} \leq \beta,\ \mathbf{x} \in \mathcal{C}\}$ is bounded.

(m) A vector $\mathbf{y} \in \mathbf{R}^n$ is said to be ***normal*** to a convex set $\mathcal{S}$ in $\mathbf{R}^n$ at a point $\mathbf{x}_0 \notin \mathcal{S}$ if $\mathbf{y}$ does not make an acute angle with any line segment connecting any $\mathbf{x} \in \mathcal{S}$ and $\mathbf{x}_0$. The convex set of all vectors $\mathbf{y}$ normal to $\mathcal{S}$ at $\mathbf{x}_0$ is termed the ***normal cone*** to $\mathcal{S}$ at $\mathbf{x}_0$ and appears as $\mathcal{C}_N = \{\mathbf{y} \mid \mathbf{y}'(\mathbf{x}-\mathbf{x}_0) \leq 0,\ \mathbf{x} \in \mathcal{S}\}$ (Figure 4.6). That is, $\mathcal{C}_N$ is the set of all vectors $\mathbf{y}$ such that the associated supporting closed half-spaces $[\mathcal{H}_s^-] = \{\mathbf{x} \mid \mathbf{y}'(\mathbf{x}-\mathbf{x}_0) \leq 0, \mathbf{x} \in \mathcal{S}\}$ contain $\mathcal{S}$. If no supporting hyper-plane of $\mathcal{S}$ passes through $\mathbf{x}_0$, then trivially $\mathcal{C}_N = \{\mathbf{0}\}$. The dual of $\mathcal{C}_N$ is called the ***support cone*** of $\mathcal{S}$

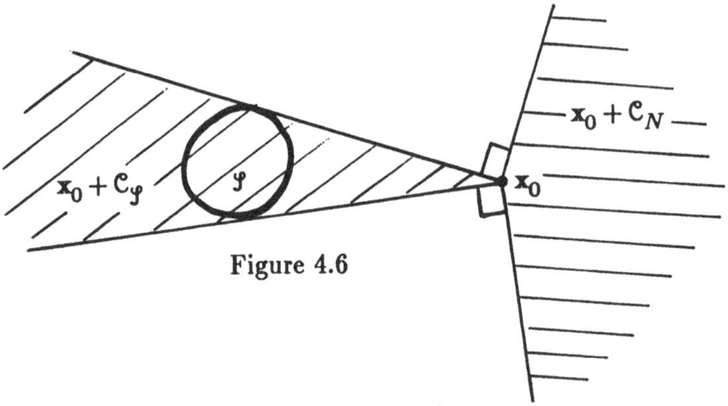

Figure 4.6

at $\mathbf{x}_0$ and will be denoted as $\mathcal{C}_N^* = \mathcal{C}_s$; it amounts to the intersection of all closed half-spaces $\{\mathbf{x} \mid \mathbf{y}'\mathbf{x} \leq 0, \mathbf{y} \in \mathcal{C}_N\}$ (Figure 4.6). If the collection of supporting closed half-spaces of $\mathcal{S}$ at $\mathbf{x}_0$ is empty, then $\mathcal{C}_s = \mathbf{R}^n$.

(n) The set of interior points $\mathcal{C}^o$ of a finite cone $\mathcal{C}$ in $\mathbf{R}^n$ is, in general, not a cone since it may be the case that $\mathbf{0} \notin \mathcal{C}^o$. However, $\mathbf{0} \in \mathcal{C}^o$ if and only if $\mathcal{C}$ equals the set of all linear combinations of points of $\mathcal{C}$.

4.3. Conical Hull

In the preceding section we indicated that a convex cone is finite if it consists of the set of all non-negative linear combinations of a finite set of vectors. This section now considers an amplification as well as some extensions of this concept.

For a set of vectors $\{\mathbf{a}_1, \ldots, \mathbf{a}_m\}$ in $\mathbf{R}^n$, their ***conical (non-negative linear) combination*** is represented as the point

$$\mathbf{x} = \sum_{j=1}^{m} \lambda_j \mathbf{a}_j, \; 0 \leq \lambda_j \in \mathbf{R} \text{ for all } j.$$

And for $\mathcal{S}$ a set in $\mathbf{R}^n$, the ***conical hull*** of $\mathcal{S}$ is the collection of all conical combinations of vectors from $\mathcal{S}$ or

$$coni(\mathcal{S}) = \left\{\mathbf{x} \mid \mathbf{x} = \sum_{j=1}^{m} \lambda_j \mathbf{x}_j,\ \mathbf{x}_j \in \mathcal{S},\ 0 \leq \lambda_j \in R, \text{ and } m \text{ is an arbitrary positive integer}\right\}.$$

Here the conical hull of any set $\mathcal{S}$ in R^n is a convex cone "spanned" by $\mathcal{S}$ with vertex at the origin (Figure 4.7.a); it is the smallest convex cone with vertex 0 containing $\mathcal{S}$. (Geometrically, a convex cone with vertex at the origin is a convex set containing all the half-lines from the origin through each of the points of the set.) In fact, $\mathcal{S}$ itself is a convex cone with vertex the origin if and only if $\mathcal{S} = coni(\mathcal{S})$.

Next, the conical hull "spanned or generated" by a finite collection of points $\{\mathbf{a}_1, \ldots, \mathbf{a}_m\}$ in R^n is a finite convex cone (Figure 4.7.b) and is expressed as the sum of a finite number of half-lines or

$$coni\{\mathbf{a}_1, \ldots, \mathbf{a}_m\} = \left\{\mathbf{x} \mid \mathbf{x} = \sum_{j=1}^{m} (\mathbf{a}_j) = \sum_{j=1}^{m} \lambda_j \mathbf{a}_j,\ 0 \leq \lambda_j \in R\right\}$$

$$= \{\mathbf{x} \mid \mathbf{x} = A'\lambda,\ \mathbf{0} \leqq \lambda \in R^m\}.$$

Hence a cone generated by $\{\mathbf{a}_1, \ldots, \mathbf{a}_m\}$ is the set $coni\{\mathbf{a}_1, \ldots, \mathbf{a}_m\}$.

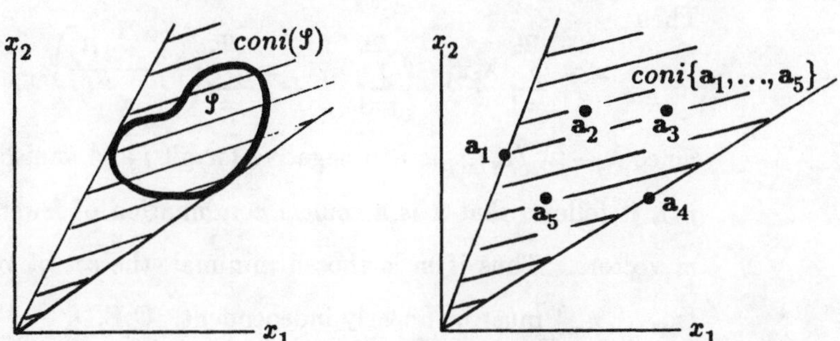

Figure 4.7

For a cone $\mathfrak{C}$ in $\mathbf{R}^n$, the conical hull of $\mathfrak{C}$, *coni*($\mathfrak{C}$), is the intersection of all convex cones containing $\mathfrak{C}$; it thus is the smallest convex cone containing $\mathfrak{C}$. Since any vector $\mathbf{x}$ of *coni*($\mathfrak{C}$) can be written as the conical combination of a finite number of vectors in $\mathfrak{C}$, it follows that the conical hull of $\mathfrak{C}$ is the set of all conical or non-negative linear combinations of points from $\mathfrak{C}$ or

$$coni(\mathfrak{C}) = \left\{ \mathbf{x} \mid \mathbf{x} = \sum_{j=1}^{m} \lambda_j \mathbf{x}_j, \mathbf{x}_j \in \mathfrak{C},\ 0 \leq \lambda_j \in \mathbf{R}, \text{ and } m \text{ is an arbitrary positive integer} \right\}.$$

Moreover, if $dim(\mathfrak{C}) = p$, then any vector $\mathbf{x}(\neq \mathbf{0})$ in *coni*($\mathfrak{C}$) is expressible as the conical combination of p vectors of $\mathfrak{C}$, *i.e.*,

4.3.1 THEOREM [Fenchel, 1953]. Any vector $\mathbf{x}(\neq \mathbf{0}) \in coni(\mathfrak{C})$ in $\mathbf{R}^n$ is a conical combination of linearly independent vectors of $\mathfrak{C}$.

PROOF. Let $\mathbf{x} \in coni(\mathfrak{C})$, *i.e.*, $\mathbf{x} = \Sigma_{j=1}^{m} \lambda_j \mathbf{x}_j \in \mathfrak{C}$, $0 \leq \lambda_j \in \mathbf{R}$. If the vectors $\mathbf{x}_1, \ldots, \mathbf{x}_m$ are linearly dependent, then there exist coefficients $\mu_j \neq 0$, $j = 1, \ldots, m$, such that $\Sigma_{j=1}^{m} \mu_j \mathbf{x}_j = \mathbf{0}$ (where it is assumed that at least one $\mu_j > 0$). Let

$$\theta = \lambda_t / \mu_t = \min_j \left\{ \lambda_j / \mu_j \mid \mu_j > 0 \right\}.$$

Then

$$\mathbf{x} = \sum_{j=1}^{m} \lambda_j \mathbf{x}_j - \theta \sum_{j=1}^{m} \mu_j \mathbf{x}_j = \sum_{j=1}^{m} \left(\lambda_j - \frac{\lambda_t \mu_j}{\mu_t} \right) \mathbf{x}_j.$$

Since $\lambda_j - (\lambda_t / \mu_t) \mu_j$ is non-negative for all j and vanishes for $j = t$, it follows that $\mathbf{x}$ is a conical combination of fewer than m vectors. Thus if m is chosen minimal, the set of vectors $\{\mathbf{x}_1, \ldots, \mathbf{x}_m\}$ must be linearly independent. Q.E.D.

We may carry this theorem a step further by noting that, for $\mathcal{S}$ a subset of $\mathbf{R}^n$, there exists an upper limit to the number of linearly independent vectors from $\mathcal{S}$ needed to construct a conical combination. Specifically, we have

4.3.2. CARATHÉODORY'S THEOREM FOR CONES [Stoer, Witzgall, 1970]. Let $\mathcal{S}$ be any subset of $\mathbf{R}^n$ with $dim(coni(\mathcal{S})) = r$. If $\mathbf{x} \epsilon\, coni$, then there exists a set of r linearly independent vectors $\{\mathbf{x}_1, \ldots, \mathbf{x}_r\}$ in $\mathcal{S}$ such that $\mathbf{x} = coni\{\mathbf{x}_1, \ldots, \mathbf{x}_r\}$.

PROOF. For $\mathbf{x} \epsilon\, coni(\mathcal{S})$, $\mathbf{x} = \Sigma_{j=1}^{k} \lambda_j \mathbf{x}_j, \mathbf{x}_j \epsilon\, \mathcal{S}, 0 \leq \lambda_j \epsilon\, \mathbf{R}$. If $k \leq r$, the assertion is true by virtue of the preceding theorem. If $k > r$ (the number of vectors in $\mathcal{S}$ exceeds the dimension of $coni(\mathcal{S})$), then the vectors in $\mathcal{S}$ must be linearly dependent and thus the remainder of this proof mirrors the one given in support of theorem 4.3.1. The process of elimination continues until $k \leq r$. Q.E.D.

A consequence of this theorem is

4.3.3 COROLLARY. For $\mathcal{S}$ a subset of $\mathbf{R}^n$, if $\mathbf{x} \epsilon\, co(\mathcal{S})$, then $\mathbf{x} \epsilon\, co\{\mathbf{x}_0, \mathbf{x}_1, \ldots, \mathbf{x}_r\}$ for $\{\mathbf{x}_0, \mathbf{x}_1, \ldots, \mathbf{x}_r\}$ a set of affinely independent vectors in $\mathcal{S}$.

PROOF. Let $\mathcal{T} = \left\{\begin{pmatrix}\mathbf{t}\\1\end{pmatrix} \middle| \mathbf{t} \epsilon\, \mathcal{S}\right\}$. Then $\mathbf{x} \epsilon\, co(\mathcal{S})$ if and only if $\begin{pmatrix}\mathbf{x}\\1\end{pmatrix} \epsilon\, coni(\mathcal{T})$. Moreover, vectors $\mathbf{x}_0, \mathbf{x}_1, \ldots, \mathbf{x}_r$ are affinely independent if and only if $\begin{pmatrix}\mathbf{x}_0\\1\end{pmatrix}, \begin{pmatrix}\mathbf{x}_1\\1\end{pmatrix}, \ldots, \begin{pmatrix}\mathbf{x}_r\\1\end{pmatrix}$ are linearly independent. The desired result thus follows immediately from the preceding theorem. Q.E.D.

We note briefly that:

(a) For any cone $\mathcal{C}$ in $\mathbf{R}^n$, $coni(\overline{\mathcal{C}}) \subset \overline{coni(\mathcal{C})}$. Moreover, $coni(\overline{\mathcal{C}}) = \overline{coni(\mathcal{C})}$ if $\mathcal{C}$ consists of a finite number of half-lines.

(b) For $\mathcal{C}$ a convex cone in $\mathbf{R}^n$, both $\overline{\mathcal{C}}$, $\overline{coni(\mathcal{C})}$ are finite cones and $\overline{\mathcal{C}} = \overline{coni(\mathcal{C})} = \mathcal{C}^{**}$.

(c) For set $\mathcal{S}$ in $\mathbf{R}^n$, $coni(\mathcal{S})$ is the intersection of all convex cones with vertex $\mathbf{0}$ that contain $\mathcal{S}$.

(d) Given an arbitrary set $\mathcal{S}$ $(\neq \phi)$ in $\mathbf{R}^n$, $\mathcal{S} \subseteq co(\mathcal{S}) \subseteq coni(\mathcal{S}) \subseteq \ell(S)$.

4.4. Extreme Vectors, Half-Lines, and Half-Spaces

A vector $\mathbf{a}(\neq \mathbf{0})$ within a convex cone $\mathcal{C}$ in $\mathbf{R}^n$ is termed an ***extreme vector*** if $\mathbf{a}$ cannot be expressed as a conical (non-negative linear) combination of any two linearly independent vectors from $\mathcal{C}$, *i.e.*, it is not possible to write $\mathbf{a} = \lambda\mathbf{a}_1 + (1-\lambda)\mathbf{a}_2,\ 0 < \lambda \epsilon R,$ for linearly independent vectors $\mathbf{a}_1, \mathbf{a}_2 \epsilon \mathcal{C}$. (Figure 4.8.a exhibits $\mathbf{a}_1, \mathbf{a}_2$ as extreme vectors in $\mathcal{C}$.) As we shall now see, extreme vectors may be used to generate a pointed finite cone. (It is important to note that not every finite cone is spanned by its extreme vectors, *e.g.*, neither a subspace nor a half-space in $\mathbf{R}^n$ for $n > 2$ possesses extreme vectors.)

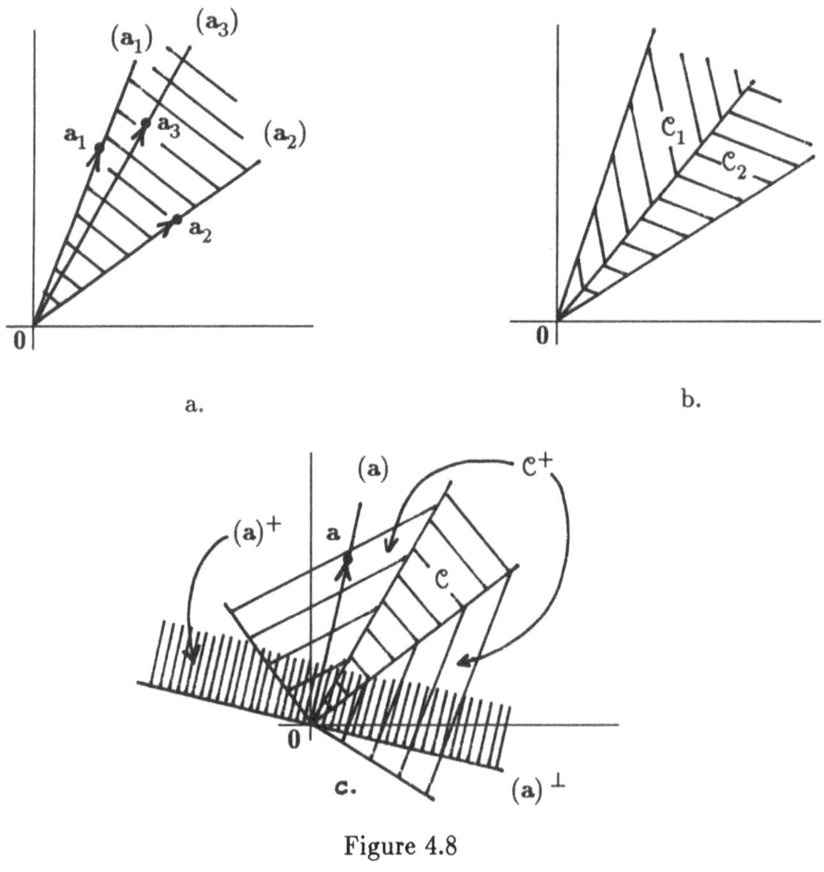

Figure 4.8

Under what conditions can a set of extreme vectors be used to obtain a finite cone? To answer this let us start with the following definition. Specifically, a set of vectors $\{\mathbf{a}_1, \ldots, \mathbf{a}_m\}$ in $\mathbf{R}^n$ arranged as the columns of an $(n \times m)$ matrix $\mathbf{A}'$ is said to be ***semi-positively independent*** if the homogeneous linear system $\mathbf{A}'\boldsymbol{\lambda} = \mathbf{0}$ has no semi-positive solution $\boldsymbol{\lambda} \geq \mathbf{0}$. Geometrically, the set $\{\mathbf{a}_1, \ldots, \mathbf{a}_m\}$ is semi-positively independent if the cone they generate is pointed, *i.e.*, contains no subspace other than $\{\mathbf{0}\}$. (Note that ***semi-positive dependence*** holds if $\mathbf{A}'\boldsymbol{\lambda} = \mathbf{0}, \boldsymbol{\lambda} \geqq \mathbf{0}$, and $\boldsymbol{\lambda}$ has at least one strictly positive component.) Next, a half-line $(\mathbf{a})$ of a finite cone $\mathcal{C}$ in $\mathbf{R}^n$ is an ***extreme half-line*** if $\mathbf{a}$ is an extreme vector. Based upon these notions we have

4.4.1. THEOREM [Gale, 1960]. If the set of vectors $\{\mathbf{a}_1, \ldots, \mathbf{a}_m\}$ in $\mathbf{R}^n$ is semi-positively independent, then the finite cone $\mathcal{C}$ spanned by these vectors is:

(1) pointed; and

(2) the sum of its extreme half-lines or $\mathcal{C} = \Sigma_{j=1}^{m}(a_j)$. (Equivalently, $\mathcal{C}$ is the conical combination of its extreme vectors.)

PROOF. Since both vectors $a, -a$ cannot belong to $\mathcal{C}$ unless $a = \mathbf{0}$, it follows that $\mathcal{C}$ must be pointed.

From $\{a_1, \ldots, a_m\}$ remove, one at a time, any vector which is expressible as a conical combination of the remain-ing vectors. By continuing this process we ultimately obtain a new (possibly renumbered) set of vectors $\{\boldsymbol{a}_1,\ldots,\boldsymbol{a}_r\}$, $r \leqq m$, such that: (1) each remaining vector in this new set is extreme; and (2) $\mathcal{C} = \Sigma_{j=1}^{r} (\boldsymbol{a}_j)$. Suppose to the contrary that, say, $\boldsymbol{a}_1$ is not extreme. Then $\boldsymbol{a}_1 = \Sigma_{j=1}^{r} \lambda_j \boldsymbol{a}_j$, $0 \leq \lambda_j \epsilon \boldsymbol{R}$. If $\lambda_1 \geq 1$ we obtain

$$(\lambda_1 - 1)\, \mathbf{a}_1 + \sum_{j=2}^{r} \lambda_j \boldsymbol{a}_j = \mathbf{0},$$

a violation of the assumption that the original set $\{\mathbf{a}_1, \ldots, \boldsymbol{a}_m\}$ is semi-positively independent. If $\lambda_1 < 1$, then

$$\mathbf{a}_1 = (1-\lambda_1)^{-1} \sum_{j=2}^{r} \lambda_j \boldsymbol{a}_j.$$

But, contrary to requirement (1) immediately above, this means that $\mathbf{a}_1$ is not an extreme vector since it is a linear combination, with non-negative coefficients $\lambda_j(1-\lambda_1)^{-1}$, of the vectors $\mathbf{a}_j$, $j=2,\ldots,r$. Q.E.D.

A glance back at Figure 4.8.a reveals that although $\mathcal{C}$ may be written as $(\mathbf{a}_1)+(\mathbf{a}_2)+(\mathbf{a}_3)$, the half-line $(\mathbf{a}_3)$ is not extreme and thus redundant. Hence we need only write $\mathcal{C} = (\mathbf{a}_1) + (\mathbf{a}_2)$ to completely generate this cone.

A collection of pointed cones $\mathcal{C}_1,\ldots,\mathcal{C}_p$ in $\mathbf{R}^n$ is said to be ***semi-positively independent*** if, for $\mathbf{x}_j \epsilon\, \mathcal{C}_j$, $j=1,\ldots,p$, $\Sigma_{j=1}^{p} \mathbf{x}_j = \mathbf{0}$ implies $\mathbf{x}_j = \mathbf{0}$ for all j. So for semi-positive independent cones $\mathcal{C}_1,\ldots,\mathcal{C}_p$ with $\mathbf{x}_j \epsilon\, \mathcal{C}_j$, $j=1,\ldots,p$, $\Sigma_{j=1}^{p} \mathbf{x}_j \neq \mathbf{0}$ unless $\mathbf{x}_j = \mathbf{0}$ for all j. (Figure 4.8.b displays two semi-positively independent cones $\mathcal{C}_1$, $\mathcal{C}_2$.)

Next, if $\mathcal{C}$ is a finite cone in $\mathbf{R}^n$ with the vector $\mathbf{a} \epsilon\, \mathcal{C}^+$, then $(\mathbf{a})^+$ is a ***supporting half-space*** for $\mathcal{C}$ if $\mathcal{C}$ is contained in $(\mathbf{a})^+$ while $(\mathbf{a})^\perp$, the "generator" of $(\mathbf{a})^+$, is called a ***supporting hyperplane*** for $\mathcal{C}$ (Figure 4.8.c). Let $\{\mathbf{a}_1,\ldots,\mathbf{a}_m\}$ be a set of vectors in $\mathbf{R}^n$. For $\mathcal{C} = \Sigma_{j=1}^{m} (\mathbf{a}_j)$ with $dim(\mathcal{C}) = n$, a supporting half-space $(\mathbf{a})^+$ is called an ***extreme supporting half-space*** if $(\mathbf{a})^\perp$ contains $n-1$ linearly independent vectors from among the set of generators $\{\mathbf{a}_1, \ldots, \mathbf{a}_m\}$, *i.e.*, $\mathbf{a}'\mathbf{a}_j = 0$ for any $n-1$ of the indices $1, \ldots, m$. Here $(\mathbf{a})^{\perp}$, the generator of extreme $(\mathbf{a})^{+}$, is termed an ***extreme supporting hyperplane*** to $\mathcal{C}$. Geomet-rically, a supporting half-space $(\mathbf{a})^{+}$ for a finite cone $\mathcal{C}$ in $\mathbf{R}^n$ is an extreme supporting half-space if $(\mathbf{a})$ is an extreme half-line of $\mathcal{C}^{+}$.

We observe further that:

(a) If $\mathcal{C}$ is a convex cone in $\mathbf{R}^n$ and $(\mathbf{a})$ is a half-line exterior to $\mathcal{C}$, then there exists a supporting half-space of $\mathcal{C}$ which does not contain $(\mathbf{a})$.

(b) A convex cone which is not the whole of $\mathbf{R}^n$ has a supporting half-space.

(c) If $(\mathbf{a})$ is a boundary half-line of the convex cone $\mathcal{C}$ in $\mathbf{R}^n$ and there is a supporting half-space $(\mathbf{b})^{+} = \{\mathbf{x} \mid \mathbf{b}'\mathbf{x} \geq 0, \mathbf{x} \in \mathcal{C}\}$ to $\mathcal{C}$ such that $\mathbf{a}'\mathbf{b} = 0$, then $\mathbf{a}$ is on the boundary of $(\mathbf{b})^{+}$.

(d) The closure $\overline{coni(\mathcal{C})}$ of the conical hull of a cone $\mathcal{C}$ in $\mathbf{R}^n$ is the intersection of all supporting hyperplanes of $\mathcal{C}$.

(e) If $\mathcal{H}$ is a supporting hyperplane to a cone $\mathcal{C}$ in $\mathbf{R}^n$, then $coni(\mathcal{C} \cap \mathcal{H}) = coni(\mathcal{C}) \cap \mathcal{H}$.

(f) A half-line which is the only half-line in the intersection of a supporting hyperplane and a finite cone must be an extreme half-line.

4.5. Extreme Solutions of Homogeneous Linear Inequalities

We noted in section 4.2 above that for an $(n \times m)$ matrix $\mathbf{A}' = [\mathbf{a}_1, \ldots, \mathbf{a}_m]$, the solution set $\mathcal{C}$ of the system of homogeneous linear inequalities $\mathbf{A}'\boldsymbol{\lambda} \leqq \mathbf{0}$ is a finite cone, it is also termed a ***polyhedral covex cone*** since it is the intersection of a finite collection of closed half-spaces whose boundary hyperplanes pass through the origin. While our rationalization of this statement was purely geometric in nature (Figure 4.4), we shall now explore this result in a more formal fashion, *i.e.*, in order to determine the structure of $\mathcal{C}$, let us examine a theorem which essentially states that "every polyhedral convex cone has a finite set of generators." That is to say, if $\mathbf{A}'$ is any $(n \times m)$ matrix, then the set $\{\boldsymbol{\lambda} \mid \mathbf{A}'\boldsymbol{\lambda} \leqq \mathbf{0}, \boldsymbol{\lambda} \epsilon R^m\}$ is a polyhedral convex cone. More specifically, we look to

4.5.1. MINKOWSKI'S THEOREM [Minkowski, 1910; Gale, 1960]. Given an $(n \times m)$ matrix $\mathbf{A}' = [\mathbf{a}_1, \ldots, a_m]$ and a vector $\boldsymbol{\lambda} \epsilon R^m$, the set of solutions of the system of homogeneous linear inequalities $\mathbf{A}'\boldsymbol{\lambda} \leqq \mathbf{0}$ is a finite or polyhedral convex cone $\mathcal{C} = \Sigma_{i=1}^{k}(\hat{\boldsymbol{\lambda}}_i)$.

PROOF. Consider the subspace $\mathcal{L} = \{\mathbf{x} \mid \mathbf{x} = \mathbf{A}'\boldsymbol{\lambda}, \boldsymbol{\lambda} \epsilon \mathbf{R}^m\}$ in $\mathbf{R}^n$. Since we know that $\mathcal{L} \cap \mathcal{N}, \mathcal{N} = \{\mathbf{x} \mid \mathbf{x} \geqq \mathbf{0}, \mathbf{x} \epsilon \mathbf{R}^n\}$, is a finite cone, we may write $\mathcal{L} \cap \mathcal{N} = \Sigma_{i=1}^{r}(\mathbf{x}_i)$. Let us choose $\boldsymbol{\lambda}_i$ such that $-\mathbf{x}_i = \mathbf{A}'\boldsymbol{\lambda}_i$, $i = 1, \ldots, r$, and let $\bar{\mathcal{L}}$ be the linear subspace of solutions of the homogeneous system $\mathbf{A}'\boldsymbol{\lambda} = \mathbf{0}$. Since $\bar{\mathcal{L}}$ is a finite cone, it is expressible as $\bar{\mathcal{L}} = \Sigma_{i=1}^{s}(\bar{\boldsymbol{\lambda}}_i)$ for some finite set of vectors $\{\bar{\boldsymbol{\lambda}}_1, \ldots, \bar{\boldsymbol{\lambda}}_s\}$.

If $\mathbf{A}'\boldsymbol{\lambda} \leqq \mathbf{0}$, then $-\mathbf{A}'\boldsymbol{\lambda} \epsilon \mathcal{L} \cap \mathcal{N}$. Hence there exists $0 \leq \eta_i \epsilon R$ such that $-\mathbf{A}'\boldsymbol{\lambda} = \Sigma_{i=1}^{r} \eta_i \mathbf{x}_i$ or $\mathbf{A}'\boldsymbol{\lambda} = -\Sigma_{i=1}^{r} \eta_i \mathbf{x}_i = \Sigma_{i=1}^{r} {}_i\mathbf{A}'\boldsymbol{\lambda}_i$ so that $\mathbf{A}'(\boldsymbol{\lambda} - \Sigma_{i=1}^{r} \eta_i \boldsymbol{\lambda}_i) = \mathbf{0}$. Hence $(\boldsymbol{\lambda} - \Sigma_{i=1}^{r} \eta_i \boldsymbol{\lambda}_i) \epsilon \bar{\mathcal{L}}$ and thus $\boldsymbol{\lambda} - \Sigma_{i=1}^{r} \eta_i \boldsymbol{\lambda}_i = \Sigma_{i=1}^{s} \mu_i \bar{\mathbf{x}}_i$ for $0 \leq \mu_i \epsilon \mathbf{R}$. Clearly

$$\boldsymbol{\lambda} = \sum_{i=1}^{r} \eta_i \boldsymbol{\lambda}_i + \sum_{i=1}^{s} \mu_i \bar{\boldsymbol{\lambda}}_i = \sum_{i=1}^{k} \zeta_i \hat{\boldsymbol{\lambda}}_i$$

$$= \sum_{i=1}^{k} (\hat{\boldsymbol{\lambda}}_i), \; 0 \leq \zeta_i \epsilon \mathbf{R}.$$

Thus the set $\mathcal{C}$ of solutions of $\mathbf{A}'\boldsymbol{\lambda} \leqq \mathbf{0}$ is expressible as the sum of a finite number of half-lines and thus constitutes a finite cone. Q.E.D.

Sometimes Minkowski's theorem appears as: for every homogeneous system $\mathbf{A}'\boldsymbol{\lambda} \leqq \mathbf{0}$ there exists a finite set of vectors $\{\mathbf{u}_1, \ldots, \mathbf{u}_t\}$ such that $\boldsymbol{\lambda}^*$ satisfies $\mathbf{A}'\boldsymbol{\lambda}^* \leqq \mathbf{0}$ if and only if $\boldsymbol{\lambda}^* = \Sigma_{i=1}^{t} \theta_i \mathbf{u}_i$, $0 \leq \theta_i \epsilon \mathbf{R}$.

In sum, Minkowski's theorem posits that the intersection of finitely many half-spaces is a conical combination of finitely many generators or, equivalently, the sum of finitely many half-lines. Looked at in another fashion, given finite sets of vectors $\mathcal{A} = \{\mathbf{a}_1,\ldots,\mathbf{a}_p\}$, $\mathcal{B} = \{\mathbf{b}_1,\ldots,\mathbf{b}_q\}$ in $\mathbf{R}^n$, with $\mathbf{A}' = [\mathbf{a}_1,\ldots,\mathbf{a}_p]$ and $\mathbf{B}' = [\mathbf{b}_1\ldots,\mathbf{b}_q]$, , Minkowski's theorem states that

$$\mathcal{A}^* = coni(\mathcal{B})$$

or

$$\{\mathbf{y} \mid \mathbf{A}\mathbf{y} \leqq \mathbf{0}\} = coni\{\mathbf{b}_1,\ldots,\mathbf{b}_q\} = \{Y \mid Y = B'V,\, \mathbf{v} \geqq \mathbf{0}\}.$$

Let $\mathcal{C}$ in $\mathbf{R}^m$ represent the finite cone of solutions to $\mathbf{A}'\boldsymbol{\lambda} \leqq \mathbf{0}$. Then $\bar{\boldsymbol{\lambda}} \epsilon \mathbf{R}^m$ is an ***extreme solution*** of $\mathbf{A}'\boldsymbol{\lambda} \leqq \mathbf{0}$ if $\bar{\boldsymbol{\lambda}}$ is an extreme vector of $\mathcal{C}$. Why are extreme solutions important? If we are interested in finding all solutions to $\mathbf{A}'\boldsymbol{\lambda} \leqq \mathbf{0}$, then, as will be demonstrated below, we need only concentrate on determining a finite set of extreme solutions to this inequality system. All remaining solutions are constructed by simply taking conical combinations of the extreme solutions.

In what follows an inequality in system $\mathbf{A}'\boldsymbol{\lambda} \leqq \mathbf{0}$ which holds as an equality ("$=$") for some $\bar{\boldsymbol{\lambda}} \epsilon \mathcal{C} = \{\boldsymbol{\lambda} \mid \mathbf{A}' \leqq \mathbf{0}\}$ will be termed ***binding***, and any inequality which holds as a strict inequality ("$<$") at a $\bar{\boldsymbol{\lambda}} \epsilon \mathcal{C}$ is deemed ***non-binding***. Moreover, let $\mathcal{I} = \{i \mid \boldsymbol{\alpha}_i\boldsymbol{\lambda} = 0\}$ depict the index set of binding inequalities and let $\mathcal{I}^c = \{i \mid \boldsymbol{\alpha}_i\boldsymbol{\lambda} < 0\}$ represent the (complementary) index set of non-binding inequalities.

How may we characterize an extreme solution to $\mathbf{A}'\boldsymbol{\lambda} \leqq \mathbf{0}$? Specifically, a vector $\bar{\boldsymbol{\lambda}} \epsilon \mathbf{R}^m$ is an extreme solution of $\mathbf{A}'\boldsymbol{\lambda} \leqq \mathbf{0}$ if it satisfies exactly $m-1$ linearly independent equations $\boldsymbol{\alpha}_i\bar{\boldsymbol{\lambda}} = 0$, where $\mathbf{A}' = [\mathbf{a}_1,\ldots,\mathbf{a}_m]$ is of order $(n \times m)$ and $\boldsymbol{\alpha}_i$ is the ith row of $\mathbf{A}'$, $i = 1,\ldots,n$. Geometrically, $\bar{\boldsymbol{\lambda}}$ lies on $m-1$ independent hyperplanes determined by the $\boldsymbol{\alpha}_i$ and passing through the origin. Alternatively, we may describe $\bar{\boldsymbol{\lambda}}$ and $m-1$ of the $\boldsymbol{\alpha}_i$'s as being orthogonal. This characterization may be stated algebraically as

4.5.2 THEOREM [Gale, 1960]. A solution $\bar{\lambda} \epsilon \mathbf{R}^m$ of the system of homogeneous linear inequalities $\mathbf{A}'\lambda \leqq \mathbf{0}$ is extreme if and only if the column matrix $\mathbf{A}_1'$ containing the α_i's for which $\alpha_i\lambda = 0$ has rank $m-1$.

PROOF. Let $\mathbf{A}_1'$ denote the coefficient matrix for the system of binding equalities $\alpha_i\bar{\lambda} = 0$, $i \epsilon \mathfrak{I}$, and let $\rho(\mathbf{A}_1') = m-1$. If $\bar{\lambda} = \theta\lambda_1 + (1-\theta)\lambda_2$, $0 < \theta \epsilon R$, is not an extreme vector, then $\mathbf{A}_1'\bar{\lambda} = \mathbf{A}_1'(\theta\lambda_1) + \mathbf{A}_1'((1-\theta)\lambda_2)$, which implies that $\mathbf{A}_1'\lambda_1 = \mathbf{A}_1'\lambda_2 = \mathbf{0}$. From theorem 1.2.2, the solution sets for these latter two sets of homogeneous equations are a linear subspaces of dimension one so that the set $\{\lambda_1, \lambda_2\}$ is linearly dependent -- a contradiction of the assumption that $\bar{\lambda}$ is not extreme.

Conversely, suppose $\bar{\lambda} \epsilon \mathfrak{C} = \{\lambda \mid \mathbf{A}'\lambda \leqq \mathbf{0}, \lambda \epsilon \mathbf{R}^m\}$. Let us partition $\mathbf{A}'$ as $\mathbf{A}' = [\mathbf{A}_1', \mathbf{A}_2']$, where $\mathbf{A}_1'\bar{\lambda} = \mathbf{0}$, $\mathbf{A}_2'\bar{\lambda} < \mathbf{0}$. If $\bar{\lambda} \neq \mathbf{0}$, then $\rho(\mathbf{A}_1') < m$. If $\rho(\mathbf{A}_1') < m-1$, then there exists a vector $\hat{\lambda} \epsilon \mathbf{R}^m$ such that $\mathbf{A}_1'\hat{\lambda} = \mathbf{0}$ with $\{\bar{\lambda}, \hat{\lambda}\}$ linearly independent. Hence we can find a $0 \neq \delta \epsilon \mathrm{R}$ sufficiently small such that the vectors $\lambda_1 = (\bar{\lambda} + \delta\hat{\lambda})/2$, $\lambda_2 = (\bar{\lambda} - \delta\hat{\lambda})/2$ are linearly independent, satisfy $\mathbf{A}_2'\lambda < \mathbf{0}$, and ultimately satisfy $\mathbf{A}'\lambda \leqq \mathbf{0}$. But since $\bar{\lambda} = \lambda_1 + \lambda_2$, $\bar{\lambda}$ cannot be an extreme vector. Hence it must be true that $\rho(\mathbf{A}_1') = m-1$. Q.E.D.

We note briefly that a cone of solutions $\mathfrak{C} = \{\lambda \mid \mathbf{A}'\lambda \leqq \mathbf{0}\}$ in $\mathbf{R}^m$ is termed ***non-degenerate*** if the columns of $\mathbf{A}'$ are linearly independent or, equivalently, $\rho(\mathbf{A}') = m$.

As we shall now verify, a non-degenerate polyhedral convex cone in $\mathbf{R}^m$ corresponding to the solution set of a system of homogeneous linear inequalities is generated by its set of extreme vectors. To this end we have

4.5.3. THEOREM [Gale, 1960]. Let $\mathfrak{C} = \{\lambda \mid \mathbf{A}'\lambda \leqq \mathbf{0}\}$ be a polyhedral convex cone in $\mathbf{R}^m$ with $\rho(\mathbf{A}') = m$. Then every solution of the homogeneous linear system $\mathbf{A}'\lambda \leqq \mathbf{0}$ is a conical combination of extreme solutions, *i.e.*, of extreme vectors of $\mathcal{C}$.

PROOF. For a finite set of generators $\{\mathbf{b}_1,\ldots,\mathbf{b}_r\}$ in $\mathbf{R}^m$, let $\mathcal{C} = \Sigma_{i=1}^r (\mathbf{b}_i)$ be the cone of all solutions of $\mathbf{A}'\boldsymbol{\lambda} \leqq \mathbf{0}$. If the $\mathbf{b}_i$'s are semi-positively dependent, then $\Sigma_{i=1}^r \theta_i \mathbf{b}_i = \mathbf{0}$, $\theta_i \geq 0$, $i = 1,\ldots,r$, and $\theta_i > 0$ for at least one i, say, $i = 1$. So with $\theta_1 > 0$, we can solve for $-\mathbf{b}_1 = \theta_1^{-1} \Sigma_{i=2}^r \theta_i \mathbf{b}_i$ and thus $-\mathbf{A}'\mathbf{b}_1 = \theta_1^{-1} \Sigma_{i=2}^r \theta_i \mathbf{A}'\mathbf{b}_i \leqq \mathbf{0}$ or $\mathbf{A}'\mathbf{b}_1 \geqq \mathbf{0}$. But since $\mathbf{b}_1 \in \mathcal{C}$ (*i.e.*, $\mathbf{A}'\mathbf{b}_1 \leqq \mathbf{0}$), this implies that $\mathbf{A}'\mathbf{b}_1 = \mathbf{0}$ so that the columns of $\mathbf{A}'$ are linearly dependent, contrary to the assumption that $\rho(\mathbf{A}') = m$. Hence the $\mathbf{b}_i$, $i = 1,\ldots,r$, must be semi-positively independent and thus, via theorem 4.4.1, $\mathcal{C}$ is the sum of its extreme half-lines and thus a conical combination of its extreme vectors. Q.E.D.

4.6 Sum Cone and Intersection Cone Equivalence

We noted above that a finite cone in $\mathbf{R}^n$ can be formed as a sum of half-lines. As we shall now see, a given finite cone can also be generated from a set of half-spaces. To develop this point in greater detail let $\{\mathbf{b}_1,\ldots,\mathbf{b}_q\}$ be a set of vectors in $\mathbf{R}^n$. Then a finite cone $\mathcal{C}$ can be written as the intersection of a finite number q of half-spaces $(\mathbf{b}_j)^* = \{\mathbf{x} \mid \mathbf{b}_j'\mathbf{x} \leq 0\}$, $j = 1,\ldots,q$, as

$$\mathcal{C} = \bigcap_{j=1}^{q} (\mathbf{b}_j)^*.$$

Here $\mathcal{C}$ consists of all vectors $\mathbf{x}$ making non-acute angles ($\geq \pi/2$) with each of the vectors $\mathbf{b}_1,\ldots,\mathbf{b}_q$. If $\mathbf{B}' = [\mathbf{b}_1,\ldots,\mathbf{b}_q]$ is of order $(n \times q)$, then the preceding cone may be rewritten as $\mathcal{C} = \{\mathbf{x} \mid \mathbf{B}\mathbf{x} \leqq \mathbf{0},\ \mathbf{x} \in \mathbf{R}^n\}$. (Note that the jth row of $\mathbf{B}$ is the normal to the supporting hyperplane defining the jth half-space.) The justification for this specification is a result due to H. Weyl (1935, 1950). In what follows we shall assume that the finite cones (or their finite sets of generators) under consideration are non-degenerate and have at least one extreme supporting hyperplane. Specifically,

4.6.1. WEYL'S THEOREM [Weyl, 1935, 1950.; Fenchel, 1953; Gale, 1957; Gerstenhaber, 1957; Simonnard, 1966]. If $\mathcal{C}$ is a finite collection of points $\mathbf{x}_i \epsilon \mathbf{R}^n$, $i=1,\ldots,p \geq n$, and $\rho[\mathbf{x}_1,\ldots,\mathbf{x}_p]=n$, then the set $\mathcal{C}_1$ of conical combinations of points of $\mathcal{C}$ equals the set $\mathcal{C}_2$ formed as the intersection of extreme half-spaces of $\mathcal{C}$, *i.e.*, $\mathcal{C}_2$ is a polyhedral convex cone.

PROOF. (Simonnard, 1966). We demonstrate first that if $\mathbf{x} \epsilon \mathcal{C}_2$, then $\mathbf{x} \epsilon \mathcal{C}_1$, *i.e.*, $\mathcal{C}_2 \subset \mathcal{C}_1$. To this end let $(\mathbf{a})^*$ be an extreme supporting half-space of $\mathcal{C}$ and suppose $\mathbf{a}'\mathbf{x}_1 < 0$. The set of extreme supporting half-spaces may be partitioned into two classes;

$$\mathcal{J} = \{(\mathbf{a}_j)^* \mid \mathbf{a}_j'\mathbf{x}_1 = 0\},$$

$$\mathcal{J}^c = \{(\mathbf{a}_j)^* \mid \mathbf{a}_j'\mathbf{x}_1 < 0\}.$$

Let $\mathbf{x}$ be a point belonging to the intersection of all the extreme supporting half-spaces and consider the point $\bar{\mathbf{x}} = \mathbf{x} - \lambda\mathbf{x}_1$. To determine which extreme supporting hyperplane contains $\bar{\mathbf{x}}$, let us form $\mathbf{a}_j'\bar{\mathbf{x}} = \mathbf{a}_j'\mathbf{x} - \lambda\mathbf{a}_j'\mathbf{x}_1 \leq 0$, $j \epsilon \mathcal{J} \cup \mathcal{J}^c$. Clearly we must choose λ so that, for some $j \epsilon \mathcal{J}, \mathbf{a}_j'\bar{\mathbf{x}} = 0$. So if we let

$$\lambda = \min_{j \epsilon \mathcal{J}^c} \left\{\mathbf{a}_j'\mathbf{x}/\mathbf{a}_j'\mathbf{x}_1 \mid \mathbf{a}_j'\mathbf{x}_1 < 0\right\} = \mathbf{a}_k'\mathbf{x}/\mathbf{a}_k'\mathbf{x}_1,$$

then $\mathbf{a}_k'\bar{\mathbf{x}} = 0$ and thus $\bar{\mathbf{x}}$ lies on the extreme supporting hyperplane $\mathbf{a}_k'\mathbf{x} = 0$.

Under a suitable changes of basis, $\mathbf{a}_k$ may be replaced by the unit column vector $\mathbf{e}_n$. The extreme supporting half-space $(\mathbf{a}_k)^*$ consequently can be rewritten as $\mathbf{a}_k'\mathbf{x} = \mathrm{x}_n \leq 0$, *i.e.*, $\{\mathrm{x}_n \leq 0\}$ is an equivalent extreme supporting half-space.

We now proceed to prove the theorem by induction on n. The theorem is obvious in $\mathbf{R}^1$ since $\mathcal{C}_1$, $\mathcal{C}_2$ trivially coincide. Next, assume that the theorem is true in $\mathbf{R}^{n-1}$. Since all points $\mathbf{x}_i \epsilon \mathcal{C}$ are such that $\mathrm{x}_{in} \leq 0$, we may partition $\mathcal{C}$ into the subsets

$$\mathcal{C}' = \{\mathbf{x}_i \,|\, x_{in} = 0\},$$
$$\mathcal{C}'' = \{\mathbf{x}_i \,|\, x_{in} < 0\},$$

where $\mathcal{C}'$ contains $n-1$ linearly independent points $\mathbf{x}_i$ given that $\{x_n \leq 0\}$ is an extreme supporting half-space. The equality $x_n = 0$ defines the subspace $\mathbf{R}^{n-1}$ which contains $\bar{\mathbf{x}}$ as well as points $\mathbf{x}_i \in \mathcal{C}'$.

Let $(\bar{\mathbf{a}})^*$ be an extreme supporting half-space of $\mathcal{C}'$ in $\mathbf{R}^{n-1}$ so that, for all $\mathbf{x} \in \mathcal{C}'$,

$$\bar{\mathbf{a}}'\mathbf{x} = \bar{a}_1 x_1 + \cdots + \bar{a}_{n-1} x_{n-1} \leq 0. \tag{4.2}$$

The half-space

$$\bar{a}_1 x_1 + \cdots + \bar{a}_{n-1} x_{n-1} - \alpha x_n \leq 0 \tag{4.3}$$

in $\mathbf{R}^n$ contains all elements of $\mathcal{C}'$ and contains all points of $\mathcal{C}''$ if α is chosen so that

$$\alpha = \min_{\mathbf{x}_i \in \mathcal{C}''} \left[x_{in}^{-1} \sum_{j=1}^{n-1} \bar{a}_j \, x_{ij} \right].$$

Since $(\bar{\mathbf{a}})^*$ is an extreme supporting half-space of $\mathcal{C}'$, (4.3) is satisfied as a strict equality by $n-2$ linearly independent points of $\mathcal{C}'$, and by a single point of $\mathcal{C}''$; the half-space (4.3) is thus an extreme supporting half-space of $\mathcal{C}$. Hence $\bar{\mathbf{x}}$ satisfies (4.3) and, since $\bar{x}_n = 0$, also satisfies (4.2). By hypothesis, $\bar{\mathbf{x}}$ is thus representable as a conical combination of the elements of $\mathcal{C}'$; $\mathbf{x}$ is then representable as a conical combination of the points of $\mathcal{C}$ from $\bar{\mathbf{x}} = \mathbf{x} - \lambda \mathbf{x}_1$ above.

Conversely, to demonstrate that $\mathcal{C}_1 \subset \mathcal{C}_2$, let $\mathbf{x}_0 \in \mathcal{C}_1$ and, if $(\mathbf{a}_j)^*$ is an arbitrary extreme supporting half-space of $\mathcal{C}_2$, then $\mathbf{a}_j'\mathbf{x}_0 = \sum_{i=1}^{p} \lambda_{0i}(\mathbf{a}_j'\mathbf{x}_i) \leq 0$ so that $\mathbf{x}_0 \in (\mathbf{a}_j)^*$ and consequently $\mathbf{x}_0 \in \mathcal{C}_2$. Q.E.D.

This theorem has demonstrated that the set $\mathcal{C}$ of conical combinations of a finite set of vectors (or sum of finitely many half-lines) in $\mathbf{R}^n$

correpsonds to the intersection of a finite number of extreme supporting half-spaces of $\mathcal{C}$, *i.e.*, "every convex cone with finitely many generators is polyhedral." If we again let $\mathcal{A} = \{\mathbf{a}_1,\ldots,\mathbf{a}_p\}$, $\mathcal{B} = \{\mathbf{b}_1,\ldots, \mathbf{b}_q\}$ be sets of vectors in $\mathbf{R}^n$ with $\mathbf{A}' = [\mathbf{a}_1,\ldots,\mathbf{a}_p]$, $\mathbf{B}' = [\mathbf{b}_1,\ldots,\mathbf{b}_q]$, then the theorem of Weyl informs us that

$$coni(\mathcal{A}) = \mathcal{B}^*$$

or

$$coni\{\mathbf{A}_1,\ldots,\mathbf{A}_p\} = \{\mathbf{x} \mid \mathbf{x} = \mathbf{A}'\boldsymbol{\lambda},\ \boldsymbol{\lambda} \geqq \mathbf{0}\} = \{\mathbf{x} \mid \mathbf{B}\mathbf{x} \leqq \mathbf{0}\}.$$

The converse proposition, namely that every intersection of a finite number of extreme supporting half-spaces may be considered as the set of conical combinations of a finite number of vectors, is incorporated in an indirect proof of Minkowski's theorem 4.5.1 and appears as

4.6.2. THEOREM [Simonnard, 1966]. Given a finite set of closed half-spaces $(\mathbf{b}_j)^* = \{\mathbf{x} \mid \mathbf{b}_j'\mathbf{x} \leq 0\}$, $j = 1,\ldots,q$, there exists a finite set of points $\mathbf{a}_i$, $i = 1,\ldots,p$, such that the intersection of the half-spaces is equal to the set of conical combinations of the points $\mathbf{a}_i$, $i = 1,\ldots,p$.

PROOF. Let $\mathcal{C}' = coni\{\mathbf{b}_1,\ldots,\mathbf{b}_q\}$, let $\mathbf{a}_i'\mathbf{x} \leq 0$, $i = 1,\ldots,p$, be the extreme supporting half-spaces of set $\{\mathbf{b}_1,\ldots,\mathbf{b}_q\}$, and let $\mathcal{C} = coni\{\mathbf{a}_1,\ldots,\mathbf{a}_p\}$. Since $\mathcal{C} = coni\{\mathbf{a}_1,\ldots,\mathbf{a}_p) = \{\mathbf{x} \mid \mathbf{x} = \mathbf{A}'\boldsymbol{\lambda},\ \boldsymbol{\lambda} \geqq \mathbf{0}\}$, we have, by the preceding theorem,

$$\mathcal{C}' = \{\mathbf{y} \mid \mathbf{a}_i'\mathbf{y} \leq 0,\ i = 1,\ldots,p\} = \mathcal{C}^*.$$

Taking the "dual" of both sides of this expression renders

$$\begin{aligned}(\mathcal{C}')^* &= \left(coni\{\mathbf{b}_1,\ldots,\mathbf{b}_q\}\right)^* \\ &= \{\mathbf{x} \mid \mathbf{b}_j'\mathbf{x} \leq 0,\ j = 1,\ldots,q\} \\ &= (\mathcal{C}^*)^* = \mathcal{C}. \quad \text{Q.E.D.}\end{aligned}$$

The importance of theorems 4.6.1, 4.6.2 is that they establish an equivalence between two fundamental ways of defining an n-dimensional cone, *i.e.*, a finite or polyhedral convex cone written as a finite sum of half-lines can also be represented as the intersection of a finite number of supporting half-spaces, each of which has its defining hyperplane passing through the origin. This consequence of Weyl's theorem can be formalized as

4.6.3. COROLLARY. Let $\{\mathbf{a}_1,\ldots,\mathbf{a}_p\}$, $\{\mathbf{b}_1,\ldots,\mathbf{b}_q\}$ represent sets of vectors in $\mathbf{R}^n$. If $\mathcal{C}$ is a finite cone in $\mathbf{R}^n$ with $dim(\mathcal{C}) = n$ and $\mathcal{C}$ is expressed as a finite sum of half-lines $\mathcal{C} = coni\{\mathbf{a}_1,\ldots,\mathbf{a}_p\} = \Sigma_{i=1}^p(\mathbf{a}_i)$, then it can also be written as the intersection of a finite number of half-spaces $\mathcal{C} = \cap_{j=1}^q(\mathbf{b}_j)^*$.

Weyl's theorem may be expressed in matrix form as

4.6.4. THEOREM [Gale, 1957]. Given an $(n \times p)$ matrix $\mathbf{A}' = [\mathbf{a}_1,\ldots,\mathbf{a}_p]$ with $\rho(\mathbf{A}) = n$, let finite cone $\mathcal{C} = \{\mathbf{x} \mid \mathbf{x} = \mathbf{A}'\boldsymbol{\lambda}, \boldsymbol{\lambda} \geqq \mathbf{0}\}$. Then there exists a $(n \times q)$ matrix $\mathbf{B} = [\mathbf{b}_1,\ldots,\mathbf{b}_q]$ such that: (1) $\mathcal{C} = \{\mathbf{x} \mid \mathbf{B}\mathbf{x} \leqq \mathbf{0}\}$; and (2) each column of $\mathbf{B}$ is orthogonal to $n-1$ linearly independent columns of $\mathbf{A}$.

Following Gale (1957) we may actually demonstrate directly the equivalence between the "sum cone" and the "intersection cone" expressions given in the preceding corollary by starting with the intersection form, *i.e.*,

$$\begin{aligned}
\mathcal{C} &= \cap_{j=1}^q (\mathbf{b}_j)^* \\
&= \left(\Sigma_{j=1}^q (\mathbf{b}_j)\right)^* && [\text{by } \mathcal{C}_1^* \cap \mathcal{C}_2^* = (\mathcal{C}_1 + \mathcal{C}_2)^*] \\
&= \left(\cap_{i=1}^p (\mathbf{a}_i)^*\right)^* && [\text{by Weyl's theorem}] \\
&= \left[\left(\Sigma_{i=1}^p (\mathbf{a}_i)\right)^*\right]^* && [\text{by } \mathcal{C}_1^* \cap \mathcal{C}_2^* = (\mathcal{C}_1 + \mathcal{C}_2)^*] \\
&= \Sigma_{i=1}^p (\mathbf{a}_i). && [\text{by } \mathcal{C}^{**} = \mathcal{C} \text{ for finite cones}]
\end{aligned}$$

All of the results/proofs provided in this section have been framed in terms of dual cones. It is important to note that the same set of theorems holds true if one chooses to rework the material using polar cones.

4.7. Additional Duality Results for Finite Cones

Theorem 4.2.1 of section 4.2 above addressed the fundamental duality property for finite cones using the dual operator $(*)$. Let us now examine this duality property using the polar operation $(+)$. While the basic approach employed in the following proof is the same as that used in the verification of theorem 4.2.1 (*i.e.*, we directly engaged Farkas' theorem or, what amounts to essentially the same thing, a strong separation theorem), we shall additionally invoke Weyl's theorem. To this end we now demonstrate that the relation between a cone and its polar is symmetric.

4.7.1. DUALITY THEOREM FOR FINITE CONES II. If $\mathcal{C}$ is a finite cone in $\mathbf{R}^n$, then $\mathcal{C}^{++} = \mathcal{C}$.

PROOF. If $\mathbf{x} \in \mathcal{C}$, then for any $\mathbf{y} \in \mathcal{C}^+$ we have $\mathbf{x}'\mathbf{y} \geq 0$ so that $\mathbf{x} \in \mathcal{C}^{++}$. Hence $\mathcal{C} \subset \mathcal{C}^{++}$.

To establish the converse set inclusion we note, by Weyl's theorem 4.6.1, that $\mathcal{C}$ is expressible as the intersection of a finite number of half-spaces. If a vector $\mathbf{u} \notin \mathcal{C}$, then for some $\boldsymbol{\alpha} \in \mathcal{C}$, there exists a supporting half-space $(\boldsymbol{\alpha})^+$ such that $\mathcal{C} \subset (\boldsymbol{\alpha})^+$, $\mathbf{u} \notin (\mathbf{a})^+$. With $\mathcal{C} \subset (\boldsymbol{\alpha})^+$, $\mathbf{x}'\boldsymbol{\alpha} \geq 0$ for all $\mathbf{x} \in \mathcal{C}$ and thus $\boldsymbol{\alpha} \in \mathcal{C}^+$. Since $\mathbf{u}\,'\mathbf{y} < 0$, it follows that $\mathbf{u}$ is not in $\mathcal{C}^{++}$. Q.E.D.

In the proof of duality theorem 4.2.1 we used Farkas' theorem of the alternative 3.3.1.a to verify the set inclusion statement $\mathcal{C}^{**} \subset \mathcal{C}$. As we shall now see, the reverse process is also valid, *i.e.*, we may easily establish that Farkas' theorem is a consequence of (or is equivalent to) the duality relation $\mathcal{C}^{**} = \mathcal{C}$ for finite cones. In fact, this duality property is often referred to as "Farkas' theorem." To demonstrate this let $\mathbf{A}'$ in a matrix of order $(n \times m)$ with $\mathcal{C} = \{\mathbf{x} \mid \mathbf{x} = \mathbf{A}'\boldsymbol{\lambda},\ \boldsymbol{\lambda} \geqq \mathbf{0}\}$. By definition,

$$\begin{aligned}\mathcal{C}^* &= \{\mathbf{y} \mid \mathbf{x}'\mathbf{y} \leq 0 \text{ for all } \mathbf{x} \in \mathcal{C}\} \\ &= \{\mathbf{y} \mid \boldsymbol{\lambda}'\mathbf{A}\mathbf{y} \leq 0,\ \boldsymbol{\lambda} \geqq \mathbf{0}\} \\ &= \{\mathbf{y} \mid \mathbf{A}\mathbf{y} \leqq \mathbf{0}\}.\end{aligned}$$

Then the dual of this dual is

$$\begin{aligned}\mathcal{C}^{**} &= \{\mathbf{x} \mid \mathbf{y} \in \mathcal{C}^* \text{ implies } \mathbf{x}'\mathbf{y} \leq 0\} \\ &= \{\mathbf{x} \mid \mathbf{A}\mathbf{y} \leqq \mathbf{0} \text{ implies } \mathbf{x}'\mathbf{y} \leq 0\}.\end{aligned}$$

The duality result $\mathcal{C}^{**} = \mathcal{C}$ may be interpreted as "$\mathbf{b} \in \mathcal{C}^{**}$ if and only if $\mathbf{b} \in \mathcal{C}$," *i.e.*,

(I) $\mathbf{A}\mathbf{y} \leqq \mathbf{0}$ implies $\mathbf{b}'\mathbf{y} \leq 0$ ($\mathbf{b} \in \mathcal{C}^{**}$);

(II) $\mathbf{A}'\boldsymbol{\lambda} = \mathbf{b},\ \boldsymbol{\lambda} \geqq \mathbf{0}$ ($\mathbf{b} \in \mathcal{C}$).

Hence $\mathcal{C}^{**} = \mathcal{C}$ means that (I) has a solution if and only if (II) has a non-negative solution.

The preceding system may be rewritten in "theorem of the alternative form" as

(I′) $\mathbf{A}\mathbf{y} \leqq \mathbf{0},\ \mathbf{b}'\mathbf{y} > 0$ ($\mathbf{b} \notin \mathcal{C}^{**}$);

(II′) $\mathbf{A}'\boldsymbol{\lambda} = \mathbf{b},\ \boldsymbol{\lambda} \geqq \mathbf{0}$ ($\mathbf{b} \in \mathcal{C}$)

((I′) is legitimate since Farkas' theorem 3.3.1.a can be viewed as a separation theorem for a finite cone $\mathcal{C}$ in $\mathbf{R}^n$and a point $\mathbf{b}$). So if $\mathbf{b} \in \mathcal{C}^{**}$ ((I′) has no solution), then (II′) has a non-negative solution ($\mathbf{b} \in \mathcal{C}$); and if $\mathbf{b} \in \mathcal{C}$ ((II′) has a non-negative solution), then (I′) has no solution ($\mathbf{b} \in \mathcal{C}^{**}$). (Note that in proving $\mathcal{C}^{**} \subset \mathcal{C}$ in theorem 4.2.1 we assumed that $\mathbf{b} \notin \mathcal{C}$ or (II′) has no solution. Hence (I′) has a solution and thus $\mathbf{b} \notin \mathcal{C}^{**}$ either.)

Similarly, we can easily demonstrate that Farkas' theorem 3.3.1.b. is a consequence of (or equivalent to) duality theorem 4.7.1 or $\mathcal{C}^{++} = \mathcal{C}$. Briefly, let

$$\mathcal{C}^+ = \{\mathbf{y} \mid \mathbf{x}'\mathbf{y} \geq 0 \text{ for all } \mathbf{x} \in \mathcal{C}\} = \{\mathbf{y} \mid \mathbf{A}\mathbf{y} \geqq \mathbf{0}\}.$$

Then the polar of the polar is

$$\mathcal{C}^{++} = \{\mathbf{x} \mid \mathbf{y} \in \mathcal{C}^{+}(\text{ or } \mathbf{A}\mathbf{y} \geqq \mathbf{0}) \text{ implies } \mathbf{x}'\mathbf{y} \geq 0\}.$$

Since $\mathcal{C}^{++} = \mathcal{C}$ means that "$\mathbf{b} \in \mathcal{C}^{++}$ if and only if $\mathbf{b} \in \mathcal{C}$," it follows that

$$(\bar{\mathrm{I}}) \quad \mathbf{A}\mathbf{y} \geqq \mathbf{0} \text{ implies } \mathbf{b}'\mathbf{y} \geq 0 \qquad (\mathbf{b} \in \mathcal{C}^{++});$$

$$(\bar{\mathrm{II}}) \quad \mathbf{A}'\boldsymbol{\lambda} = \mathbf{b},\ \boldsymbol{\lambda} \geqq \mathbf{0} \qquad (\mathbf{b} \in \mathcal{C})$$

or

$$(\bar{\mathrm{I}}') \quad \mathbf{A}\mathbf{y} \geqq \mathbf{0},\ \mathbf{b}'\mathbf{y} < 0 \qquad (\mathbf{b} \notin \mathcal{C}^{++});$$

$$(\bar{\mathrm{II}}') \quad \mathbf{A}'\boldsymbol{\lambda} = \mathbf{b},\ \boldsymbol{\lambda} \geqq \mathbf{0} \qquad (\mathbf{b} \in \mathcal{C}).$$

Given that $\mathbf{b} \in \mathcal{C}^{++}((\bar{\mathrm{I}}')$ has no solution), then $(\bar{\mathrm{II}}')$ has a non-negative solution ($\mathbf{b} \in \mathcal{C}$); and for $\mathbf{b} \in \mathcal{C}$ (($\bar{\mathrm{II}}'$) has a non-negative solution), it follows that ($\bar{\mathrm{I}}'$) has no solution ($\mathbf{b} \in \mathcal{C}^{++}$).

As observed by Stoer and Witzgall (1970), Weyl's theorem implies the duality theorem for finite cones $\mathcal{C}^{**} = \mathcal{C}$ (or Farkas' theorem) given the equivalence between the statements "$\mathcal{C} = \mathcal{C}^{**}$" and "$\mathcal{C}$ is the intersection of half-planes." That is, for finite $\mathcal{A} = \{\mathbf{a}_1, \ldots, \mathbf{a}_p\}$ in $\mathbf{R}^n$ and $\mathbf{A}' = [\mathbf{a}_1, \ldots, \mathbf{a}_p]$, Farkas' theorem may be expressed as $\mathcal{A}^{**} = coni(\mathcal{A})$. But this equality is rationalized as: for $\mathcal{A}^* = \{\mathbf{y} \mid \mathbf{A}\mathbf{y} \leqq \mathbf{0}\}$, we have $\mathcal{A}^{**} = \{\mathbf{x} \mid \mathbf{x}'\mathbf{y} \leq 0$ whenever $\mathbf{A}\mathbf{y} \leqq \mathbf{0}\} = \{\mathbf{x} \mid \mathbf{x} = \mathbf{A}'\boldsymbol{\lambda},\ \boldsymbol{\lambda} \geqq \mathbf{0}\} = coni(\mathcal{A})$.

Additionally, Weyl's theorem also implies Minkowski's theorem 4.5.1. To see this let $\mathcal{C} = \{\mathbf{y} \mid \mathbf{A}\mathbf{y} \leqq \mathbf{0}\}$ be a polyhedral convex cone in $\mathbf{R}^n$ with $\mathbf{A}' = [\mathbf{a}_1, \ldots, \mathbf{a}_p]$. Then $\{\mathbf{y} \mid \mathbf{A}\mathbf{y} \leqq \mathbf{0}\} = (coni\{\mathbf{a}_1, \ldots, \mathbf{a}_p\})^*$ yields $\mathcal{C} = coni\{\mathbf{a}_1, \ldots, \mathbf{a}_p\}^*$. From Weyl's theorem, $coni\{\mathbf{a}_1, \ldots, \mathbf{a}_p\} = \{\mathbf{x} \mid \mathbf{B}\mathbf{x} \leqq \mathbf{0}\}$ and thus $\{\mathbf{y} \mid \mathbf{A}\mathbf{y} \leqq \mathbf{0}\}^* = coni\{\mathbf{a}_1, \ldots, \mathbf{a}_p\}$ gives, under a second application of Weyl's theorem, $\mathcal{C} = \{\mathbf{x} \mid \mathbf{B}\mathbf{x} \leqq \mathbf{0}\}^* = coni\{\mathbf{b}_1, \ldots, \mathbf{b}_q\}$, where $\mathbf{B} = [\mathbf{b}_1, \ldots, \mathbf{b}_q]$. By virtue of this argument, Weyl's theorem ($coni(\mathcal{A}) = \mathcal{B}^*$) is often called the "dual" of Minkowski's theorem ($\mathcal{A}^* = coni(\mathcal{B})$).

While theorem 4.6.1 is supported by a "direct" proof of Weyl's theorem, the theorems of Minkowski and Farkas may be combined to yield, in an "indirect" fashion, the theorem of Weyl (Goldman and Tucker, 1956). That is, to demonstrate that every convex cone with finitely many generators is polyhedral, let us note that for the finitely generated cone $\mathcal{C} = coni\{\mathbf{a}_1,\ldots,\mathbf{a}_p\}$ in $\mathbf{R}^n$, its dual is the polyhedral convex cone $\mathcal{C}^* = \{\mathbf{y} \mid \mathbf{A}\mathbf{y} \leq \mathbf{0}\} = (coni\{\mathbf{a}_1,\ldots,\mathbf{a}_p\})^*$. But $\mathcal{C}^*$ has a finite set of generators $\{\mathbf{b}_1,\ldots,\mathbf{b}_q\}$ (by Minkowski's theorem 4.5.1) and its dual $\mathcal{C}^{**}$ is also a polyhedral convex cone since, by Farkas' (duality) theorem, $\mathcal{C}^{**} = \mathcal{C}$.

Based upon this discussion we may formulate a summary theorem which serves as an amalgamation of the theorems of Farkas (1902), Minkowski (1910), and Weyl (1935, 1950). That is,

4.7.2. THEOREM. A convex cone $\mathcal{C}$ in $\mathbf{R}^n$ is polyhedral if and only if it is finitely generated.

PROOF. (sufficiency) That each finitely generated cone is polyhedral follows from Weyl's theorem.

(necessity) That a polyhedral cone is finitely generated follows from the theorems of Minkowski and Farkas. Let $\mathcal{C} = \{\mathbf{x} \mid \mathbf{A}\mathbf{x} \leqq \mathbf{0},\ \mathbf{x} \in \mathbf{R}^n\}$ be a polyhedral convex cone, where $\mathbf{A}' = [\mathbf{a}_1,\ldots,\mathbf{a}_m]$ is $(n \times m)$. Since each finitely generated cone is polyhedral, there exist vectors $\mathbf{b}_1,\ldots,\mathbf{b}_r$ such that $coni\{\mathbf{a}_1,\ldots,\mathbf{a}_m\} = \{\mathbf{x} \mid \mathbf{B}\mathbf{x} \leqq \mathbf{0}, \mathbf{x} \in \mathbf{R}^n\}$, where $\mathbf{B}' = [\mathbf{b}_1,\ldots,\mathbf{b}_r]$ is $(n \times r)$. Hence we must demonstrate that $\mathcal{C} = coni\{\mathbf{b}_1,\ldots,\mathbf{b}_r\}$, *i.e.*, $\mathcal{C}$ is finitely generated. Since $\mathbf{b}_j \in \mathcal{C}$, $j = 1,\ldots,r$ (*i.e.*, $\mathbf{b}_j'\mathbf{a}_i \leq 0$, $i = 1,\ldots,m$; $j = 1,\ldots,r$), it follows that $coni\{\mathbf{b}_1,\ldots,\mathbf{b}_r\} \subseteq \mathcal{C}$.

To establish the reverse set inclusion, suppose $\mathbf{y} \in \mathcal{C}$ is such that $\mathbf{y} \notin coni\{\mathbf{b}_1,\ldots,\mathbf{b}_r\}$. With $coni\{\mathbf{b}_1,\ldots,\mathbf{b}_r\}$ polyhedral, there exists, by virtue of Farkas' theorem, a vector $\mathbf{u} \in \mathbf{R}^n$ such that $\mathbf{u}'\mathbf{b}_j \leq 0$, $j = 1,\ldots,r$, and $\mathbf{u}'\mathbf{y} > 0$. Hence $\mathbf{u} \in coni\{\mathbf{a}_1,\ldots,\mathbf{a}_m\}$ and thus $\mathbf{u}'\mathbf{x} \leq 0$ for all $\mathbf{x} \in \mathcal{C}$. But this contradicts the supposition that $\mathbf{y} \in \mathcal{C}$ and $\mathbf{u}'\mathbf{y} > 0$. Q.E.D.

4.8. Separation of Cones

It should be intuitively clear that, for a vector $\mathbf{C}(\neq \mathbf{0}) \in \mathbf{R}^n$, a hyperplane $(\mathbf{C})^{\perp} = \mathcal{H} = \{\mathbf{x} \mid \mathbf{C}'\mathbf{x} = 0, \mathbf{x} \in \mathbf{R}^n\}$ ***separates cones*** $\mathcal{C}_1, \mathcal{C}_2$ in $\mathbf{R}^n$ if $\mathcal{C}_1 \subseteq [\mathcal{H}^-] = \{\mathbf{x} \mid \mathbf{C}'\mathbf{x} \leq 0\}$ and $\mathcal{C}_2 \subseteq [\mathcal{H}^+] = \{\mathbf{x} \mid \mathbf{C}'\mathbf{x} \geq 0\}$. This general characterization of separation of cones will serve as the basic foundation for interpreting the theorems (and their consequences) which follow.

We first have a

4.8.1. PROPER SEPARATION THEOREM [Rockafellar 1970]. Let $\mathcal{S}_1$, $\mathcal{S}_2$ be non-empty subsets of $\mathbf{R}^n$, at least one of which is a cone. If there exists a hyperplane which properly separates $\mathcal{S}_1$ and $\mathcal{S}_2$, then the hyperplane passes through the origin.

PROOF. Assume $\mathcal{S}_1$ is a cone. If $\mathcal{S}_1, \mathcal{S}_2$ can be separated properly, then there exists a vector $\mathcal{C}(\neq \mathbf{0}) \in \mathbf{R}^n$ such that:

(a) $inf\{\mathbf{C}'\mathbf{x} \mid \mathbf{x} \in \mathcal{S}_2\} \geq sup\{\mathbf{C}'\mathbf{x} \mid \mathbf{x} \in \mathcal{S}_1\}$; and

(b) $sup\{\mathbf{C}'\mathbf{x} \mid \mathbf{x} \in \mathcal{S}_2\} > inf\{\mathbf{C}'\mathbf{x} \mid \mathbf{x} \in \mathcal{S}_1\}$.

And as shown in the proof of weak separation theorem 3.2.7, the hyperplane which separates $\mathcal{S}_1, \mathcal{S}_2$ properly can be written as $\mathcal{H} = \{\mathbf{x} \mid \mathcal{C}'\mathbf{x} = \alpha, \ \mathbf{x} \in \mathbf{R}^n\}$, where $\alpha = sup\{\mathbf{C}'\mathbf{x} \mid \mathbf{x} \in \mathcal{S}_1\}$. Since $\mathcal{S}_1$ is a cone, $\lambda\mathbf{C}'\mathbf{x} = \mathbf{C}'(\lambda\mathbf{x}) \leq \alpha$ for $\lambda > 0$ and all $\mathbf{x} \in \mathcal{S}_1$. But this implies that $\alpha \geq 0$ and $\mathbf{C}'\mathbf{x} \leq 0$ for every $\mathbf{x} \in \mathcal{S}_1$. Hence, for $\alpha = 0$, $\mathbf{0} \in \mathcal{H}$. Q.E.D.

On the basis of this theorem we have:

4.8.2. COROLLARY. A non-empty closed convex cone $\mathcal{C}$ in $\mathbf{R}^n$ is the intersection of the set of homogeneous closed half-spaces which contain it.

(Use theorem 4.8.1 to appropriately adjust the procedure used in the proof given in support of theorem 3.2.4.)

4.8.3. COROLLARY. Let $\mathcal{C}$ be a convex cone in $\mathbf{R}^n$ but not $\mathbf{R}^n$ itself. Then $\mathcal{C}$ is contained in some homogeneous closed half-space of $\mathbf{R}^n$, *i.e.*, there exists a vector $\mathbf{C}(\neq \mathbf{0}) \epsilon \mathbf{R}^n$ such that $\mathbf{C}'\mathbf{x} \leq 0$ for every $\mathbf{x} \epsilon \mathcal{C}$.

A refinement of theorem 3.2.9 to general cones appears as

4.8.4 PROPER SEPARATION THEOREM. Cones $\mathcal{C}_1, \mathcal{C}_2$ in $\mathbf{R}^n$ can be separated properly by a hyperplane if and only if they have no relative interior points in common, *i.e.*, $ri(\mathcal{C}_1) \cap ri(\mathcal{C}_2) = \phi$.

We next establish the notion that Farkas' theorem may be cast directly in terms of polyhedral convex cones and then interpreted as a separation theorem for either a cone and an individual vector or a cone and an open half-space. To see this let us consider the following version of Farkas' theorem.

4.8.5 FARKAS' THEOREM FOR CONES. For $\{\mathbf{a}_1, \ldots, \mathbf{a}_m\}$ a set of vectors in $\mathbf{R}^n$ and $\mathbf{A}' = [\mathbf{a}_1, \ldots, \mathbf{a}_m]$ of order $(n \times m)$, the half-space $(\mathbf{b})^* = \mathbf{y} \mid \mathbf{b}'\mathbf{y} \leq 0, \mathbf{y} \epsilon \mathbf{R}^n\}$ contains the polyhedral convex cone $\mathcal{C}^* = \{\mathbf{y} \mid \mathbf{A}\mathbf{y} \leq \mathbf{0},\ \ \mathbf{y} \epsilon \mathbf{R}^n\}$ if and only if $\mathbf{b} \epsilon\, coni\{\mathbf{a}_1, \ldots, \mathbf{a}_m\}$.

PROOF. (sufficiency) If $\mathbf{b} \epsilon\, coni\{\mathbf{a}_1, \ldots, \mathbf{a}_m\}$, then $\mathbf{A}'\boldsymbol{\lambda} = \mathbf{b}$, $\boldsymbol{\lambda} \geqq \mathbf{0}$. Hence $\boldsymbol{\lambda}'\mathbf{A} = \mathbf{b}'$ and $\boldsymbol{\lambda}'\mathbf{A}\mathbf{y} = \mathbf{b}'\mathbf{y} \leq 0$ forall $\mathbf{y} \epsilon \mathcal{C}^*$.

(necessity) Given $(\mathbf{b})$, $(\mathbf{b})^{\perp}$, and $(\mathbf{b})^*$ (Figure 4.9), let $\mathbf{y} \epsilon (\mathbf{b})^*$ whenever $\mathbf{y} \epsilon \mathcal{C}^*$. So with $\mathcal{C}^* \subseteq (\mathbf{b})^*$, $(\mathbf{b}) \subseteq coni\{\mathbf{a}_1, \ldots, \mathbf{a}_m\}$ or $\mathbf{b} \epsilon\, coni\{\mathbf{a}_1, \ldots, \mathbf{a}_m\}$. Q.E.D.

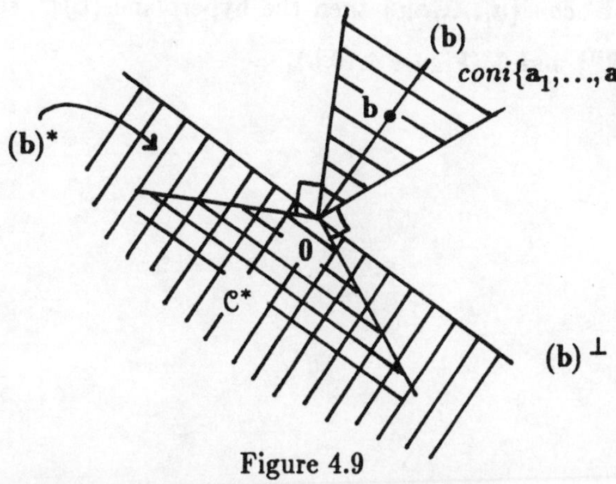

Figure 4.9

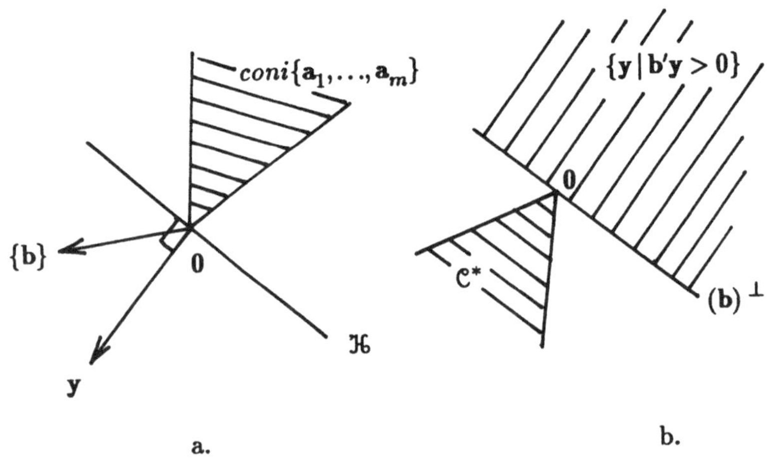

Figure 4.10

For $\mathcal{H}$ a hyperplane passing through the origin and orthogonal to $\mathbf{y} \epsilon \mathbf{R}^n$, if $\mathbf{b} \not\epsilon \, coni\{\mathbf{a}_1, \ldots, \mathbf{a}_m\}$, then $\mathbf{y}$ makes non-acute angles $(\geq \pi/2)$ with the columns of $\mathbf{A}'(\mathbf{Ay} \leqq \mathbf{0})$ and a strictly acute angle $(< \pi/2)$ with $\mathbf{b}(\mathbf{b}'\mathbf{y} > 0)$. Hence $\mathcal{H}$ has $coni\{\mathbf{a}_1, \ldots, \mathbf{a}_m\}$, on one side and $\{\mathbf{b}\}$ on the other (Figure 4.10.a. And if $\mathbf{b} \epsilon \, coni\{\mathbf{a}_1, \ldots, \mathbf{a}_m\}$ then the hyperplane $(\mathbf{b})^{\perp}$ separates sets $\{\mathbf{y} \mid \mathbf{b}'\mathbf{y} > 0, \mathbf{y} \epsilon \mathbf{R}^n\}$ and $\mathcal{C}^*$(Figure 4.10.b).

Additional remarks concerning the separation of cones are:

(a) Let $\mathcal{C}$ be a finite cone in $\mathbf{R}^n$ having no point other than $\mathbf{0}$ in common with the nonnegative orthant (*i.e.*, $\mathcal{C}$ has no semi-positive point $\mathbf{x} \geq \mathbf{0}$). Then there exists a vector $\mathbf{0} < \mathbf{p} \in \mathbf{R}^n$ such that $\mathbf{p}'\mathbf{x} \leq 0$ for all $\mathbf{x} \in \mathcal{C}$, *i.e.*, $\mathbf{0} < \mathbf{p} \in \mathcal{C}^*$. Geometrically, this proposition asserts that there exists a separating hyperplane $(\mathbf{p})^{\perp} = \{\mathbf{x} \mid \mathbf{p}'\mathbf{x} = 0, \mathbf{x} \in \mathbf{R}^n\}$ with a strictly positive normal such that the half-space $(\mathbf{p})^* = \{\mathbf{x} \mid \mathbf{p}'\mathbf{x} \leq 0, \mathbf{x} \in \mathbf{R}^n\}$ contains $\mathcal{C}$.

Next, if a finite cone $\mathcal{C}$ in $\mathbf{R}^n$ contains no vector $\mathbf{x} < \mathbf{0}$, then $\mathcal{C}^+$ contains a vector $\mathbf{y} \geq \mathbf{0}$. Geometrically, if $\mathcal{C}$ does not intersect the interior of the negative orthant $-\mathcal{N}$, then there exists a hyperplane $(\mathbf{y})^{\perp}$ separating $\mathcal{C}$ from $-\mathcal{N}$, with $\mathcal{C}$ contained in the half-space $(\mathbf{y})^+$ determined by $(\mathbf{y})^{\perp}$.

Furthermore, if the finite cone $\mathcal{C}$ in $\mathbf{R}^n$ contains no vector $\mathbf{x} \leq \mathbf{0}$, then $\mathcal{C}^+$ contains a vector $\mathbf{y} > \mathbf{0}$. In this instance we assert that if $\mathcal{C}$ does not intersect the negative orthant $-\mathcal{N}$, then there exists a hyperplane $(\mathbf{y})^{\perp}$ with normal $\mathbf{y}$ interior to $\mathcal{N}$ which separates $\mathcal{C}$ and the negative orthant. Again $\mathcal{C}$ is contained in $(\mathbf{y})^+$.

(b) Let $\mathcal{C}$ be a finite cone in $\mathbf{R}^n$ with $\mathbf{b} \in \mathbf{R}^n$ a vector exterior to $\mathcal{C}$. Then there exists a vector $\mathbf{p}(\neq \mathbf{0}) \in \mathbf{R}^n$ such that $\mathbf{p}'\mathbf{b} > 0$ and

$\mathbf{p}'\mathbf{x} \leq 0$ for $\mathbf{x} \epsilon \mathcal{C}$, *i.e.*, there exists a hyperplane $(\mathbf{p})^{\perp} = \{\mathbf{y} \mid \mathbf{p}'\mathbf{y} = 0, \mathbf{y} \epsilon \mathbf{R}^n\}$ passing through the origin $\mathbf{0}$ which strictly separates $\mathbf{b}$ and $\mathcal{C} \subset (\mathbf{p})^* = \{\mathbf{y} \mid \mathbf{p}'\mathbf{y} \leq 0, \ \mathbf{y} \epsilon \mathbf{R}^n\}$. (By Farkas' theorem 3.3.1.a, 4.8.5, and 4.2.1, $\mathcal{C} = \{\mathbf{x} \mid \mathbf{x} = \mathbf{A}'\boldsymbol{\lambda}, \ \boldsymbol{\lambda} \geqq \mathbf{0}\} = \{\mathbf{x} \mid \mathbf{y}'\mathbf{x} \leq 0, \ \mathbf{y} \epsilon \mathcal{C}^*\} = \mathcal{C}^{**}$. Then $\mathbf{b} \not\epsilon \mathcal{C}$ if and only if there is a vector $\mathbf{y} \epsilon \mathcal{C}^* = \{\mathbf{y} \mid \mathbf{A}\mathbf{y} \leqq \mathbf{0}\}$ such that $\mathbf{y}'\mathbf{b} > 0$; clearly $\mathbf{y}'\mathbf{x} \leq 0$ for all $\mathbf{x} \epsilon \mathcal{C} = \mathcal{C}^{**}$.)

(c) Let $\mathcal{C}^o$ represent the interior of a convex cone $\mathcal{C}$ in $\mathbf{R}^n$ and let $\mathcal{S}$ be a convex set in $\mathbf{R}^n$ with $\mathcal{C}^o \cap \mathcal{S} = \phi$. then there exists a hyperplane passing through the origin $\mathbf{0}$ which separates $\mathcal{C}^o, \mathcal{S}$.

4.9. Exercises

1. Is the vector $\mathbf{x}' = (1,1,1)$ an element of the cone spanned by the vectors

$$\mathbf{x}_1 = \begin{bmatrix} 2 \\ 0 \\ \frac{1}{2} \end{bmatrix}, \quad \mathbf{x}_2 = \begin{bmatrix} 1 \\ 1 \\ 4 \end{bmatrix}, \quad \mathbf{x}_3 = \begin{bmatrix} \frac{1}{2} \\ 1 \\ \frac{1}{2} \end{bmatrix}?$$

2. Sketch the rays or half-lines determined by the points $\mathbf{x}_1' = (4,2)$, $\mathbf{x}_2' = (2,6)$, and $\mathbf{x}_3' = (-2,4)$. Form the cone $\mathcal{C}$ as the sum of these half-lines. Also, sketch $-\mathcal{C}, \mathcal{C}^{\perp}, \mathcal{C}^*$, and $\mathcal{C}^+$.

3. Demonstrate that $\mathcal{C} + (-\mathcal{C})$ is the smallest subspace containing $\mathcal{I}$, *i.e.*, it is the intersection of all subspaces containing $\mathcal{C}$.

4. What is the dimension of the cone determined by the vectors $\mathbf{x}_1' = (1,3)$, and $\mathbf{x}_2' = (3,5)$? Show that $\mathcal{C}^* = -\mathcal{C}^+$.

5. Does the set $\mathcal{I} = \{\mathbf{x}' = (x_1, x_2, x_3) \mid \|\mathbf{x}\|^2 \leq 1, x_3 = 1\}$ in $\mathbf{R}^3$ constitute a cone?

6. For $\mathcal{I} = \{\mathbf{x}_1' = (1,0,0),\ \mathbf{x}_2' = (0,1,0)\}$ in $\mathbf{R}^3$, verify that *coni*($\mathcal{I}$) is the nonnegative orthant of the x_1, x_2-plane.

7. For $\mathbf{a}' = (1,2)$, is the point $\mathbf{x}_1' = (-2,1)$ a member of $(\mathbf{a})^{\perp}$?
Does $\mathbf{x}_2' = (-1,-1)$ lie in $(\mathbf{a})^*$?
Does $\mathbf{x}_3' = (1, -\frac{1}{2})$ lie in $(\mathbf{a})^+$?

8. Let $\mathbf{A} = \begin{bmatrix} 1 & 3 & 2 \\ 0 & -1 & 0 \\ 4 & 1 & 6 \end{bmatrix}$.

What is the dimension of the cone generated by **A**?

9. Prove that if $\mathcal{C}_1 \subset \mathcal{C}_2$, then $\mathcal{C}_2^+ \subset \mathcal{C}_1^+$.

10. Given $\mathbf{x}_1, \mathbf{x}_2$, and $\mathbf{x}_3$ in problem 2, demonstrate that $\mathbf{x}_2$ is not an extreme vector.

11. Are the vectors

$$\mathbf{x}_1 = \begin{bmatrix} 2 \\ 1 \end{bmatrix}, \ \mathbf{x}_2 = \begin{bmatrix} 1 \\ 1 \end{bmatrix}, \ \mathbf{x}_3 = \begin{bmatrix} 3 \\ -1 \end{bmatrix}$$

semi-positively independent?

12. Does the set of vectors

$$\mathbf{x}_1 = \begin{bmatrix} 2 \\ 1 \end{bmatrix}, \ \mathbf{x}_2 = \begin{bmatrix} -2 \\ -2 \end{bmatrix}, \ \mathbf{x}_3 = \begin{bmatrix} 1 \\ 1 \end{bmatrix}$$

generate a finite cone $\mathcal{C}$? Under what conditions can $\mathcal{C}$ be expressed as a conical combination of its extreme vectors? Determine such an expression if it exists. Find a supporting half-space for $\mathcal{C}$. Also, find a supporting hyperplane for $\mathcal{C}$.

13. Prove corollaries 4.8.2, 4.8.3.

CHAPTER 5
EXISTENCE THEOREMS FOR LINEAR SYSTEMS

5.1. Dual Homogeneous Linear Relations

This chapter is concerned with the existence of solutions to dual homogeneous linear systems. In this regard, we shall deal with pairs of finite systems of homogeneous linear equations and/or inequalities in which the variables are either nonnegative or unrestricted in sign. Specifically, these systems are structured in a fashion such that there is a one-to-one correspondence between unrestricted variables in one system and equations in the other; and between nonnegative variables in one system and inequalities in the other. Moreover, the coefficient matrix in one system is the negative transpose of the coefficient matrix of the other.

The importance of the aforementioned correspondence is that is enables us to pass from a given system of homogeneous linear equalities and/or inequalities involving nonnegative and/or unrestricted variables to a second such system and conversely. If, say, system (I) had system (II) as its dual, then obviously system (II) has system (I) as dual.

For example, systems (I), (II) which immediately follow represent "typical" general dual homogeneous systems.

$$
\left\{\begin{array}{lcl}
\left.\begin{array}{l} a_{11}x_1 + a_{12}x_2 + b_{11}y_1 + b_{12}y_2 = 0 \\ a_{21}x_1 + a_{22}x_2 + b_{21}y_1 + b_{22}y_2 = 0 \end{array}\right\} & \leftrightarrow & \left\{\begin{array}{l} u_1 \text{ unrestricted} \\ u_2 \text{ unrestricted} \end{array}\right. \\
\left.\begin{array}{l} c_{11}x_1 + c_{12}x_2 + d_{11}y_1 + d_{12}y_2 \geq 0 \\ c_{21}x_1 + c_{22}x_2 + d_{21}y_1 + d_{22}y_2 \geq 0 \end{array}\right\} & \leftrightarrow & \left\{\begin{array}{r} v_1 \geq 0 \\ v_2 \geq 0 \end{array}\right. \\
\left.\begin{array}{l} x_1 \geq 0 \\ x_2 \geq 0 \end{array}\right\} & \leftrightarrow & \left\{\begin{array}{l} -a_{11}u_1 - a_{21}u_2 - c_{11}v_1 - c_{21}v_2 \geq 0 \\ -a_{12}u_1 - a_{22}u_2 - c_{12}v_1 - c_{22}v_2 \geq 0 \end{array}\right. \\
\left.\begin{array}{l} y_1 \text{ unrestricted} \\ y_2 \text{ unrestricted} \end{array}\right\} & \leftrightarrow & \left\{\begin{array}{l} -b_{11}u_1 - b_{21}u_2 - d_{11}v_1 - d_{21}v_2 = 0 \\ -b_{12}u_1 - b_{22}u_2 - d_{12}v_1 - d_{22}v_2 = 0 \end{array}\right.
\end{array}\right\}
$$

(I) (II)

In matrix form, these dual homogeneous systems appear as

$$
(\mathrm{I}')\left\{\begin{array}{lcl}
\mathbf{Ax}+\mathbf{By}=\mathbf{0} & \longleftrightarrow & \mathbf{u} \text{ unrestricted} \\
\mathbf{Cx}+\mathbf{Dy}\geqq\mathbf{0} & \longleftrightarrow & \mathbf{v}\geqq\mathbf{0} \\
\mathbf{x}\geqq\mathbf{0} & \longleftrightarrow & -\mathbf{A}'\mathbf{u}-\mathbf{C}'\mathbf{v}\geqq\mathbf{0} \\
\mathbf{y} \text{ unrestricted} & \longleftrightarrow & -\mathbf{B}'\mathbf{u}-\mathbf{D}'\mathbf{v}=\mathbf{0}
\end{array}\right\}(\mathrm{II}')
$$

By deeming certain matrices in (I'),(II') nonvacuous, we obtain the various dual homogeneous linear systems that will be considered in the existence theorems and theorems of the alternative which follow.

5.2. Existence Theorems

5.1. TUCKER'S LEMMA [Tucker, 1956]. For any $(m\times n)$ matrix **A**, the dual systems

(I) $\mathbf{Ax}\geqq\mathbf{0}$ and

(II) $\mathbf{A}'\mathbf{y}=\mathbf{0},\ \mathbf{y}\geqq\mathbf{0}$

possess solutions $\mathbf{x}\in\mathbf{R}^n$, $\mathbf{y}\in\mathbf{R}^m$ satisfying $\boldsymbol{\alpha}_1\mathbf{x}+y_1>0$, where $\boldsymbol{\alpha}_1$ is the first row of **A** and y_1 is the first component in **y**.

PROOF (by induction on the rows of **A** or on m). For $m=1$, if $\boldsymbol{\alpha}_1=\mathbf{0}'$, set $y_1=1$, $\mathbf{x}=\mathbf{0}$; if $\boldsymbol{\alpha}_1\neq\mathbf{0}'$, set $y_1=0$, $\mathbf{x}=\boldsymbol{\alpha}_1'$.

Assume the lemma holds for a matrix **A** having m rows (in which case we get $\mathbf{x},\mathbf{y}$ satisfying (I),(II) above) and prove it for a matrix

$$\bar{\mathbf{A}}=\begin{bmatrix}\mathbf{A}\\ \alpha_{m+1}\end{bmatrix}$$

having $m+1$ rows.

To this end we may apply the lemma to $\bar{\mathbf{A}}$ by considering two mutually exclusive possibilities for the $m{+}1^{st}$ row. First, if $\boldsymbol{\alpha}_{m+1}\ \mathbf{x}\geq 0$, let $\bar{\mathbf{y}}'=(\mathbf{y}',0)$. Then

$$\bar{\mathbf{A}}\mathbf{x} \geqq \mathbf{0},\ \bar{\mathbf{A}}'\bar{\mathbf{y}} = \mathbf{0},\ \bar{\mathbf{y}} \geqq \mathbf{0},\ \alpha_1\mathbf{x} + y_1 > 0.$$

Next, if $\alpha_{m+1}\mathbf{x} < 0$, apply the lemma again to the new coefficient matrix

$$\mathbf{B} = \begin{bmatrix} \beta_1 \\ \vdots \\ \beta_m \end{bmatrix} = \begin{bmatrix} \alpha_1 + \lambda_1\alpha_{m+1} \\ \vdots \\ \alpha_m + \lambda_m\alpha_{m+1} \end{bmatrix},$$

where λ_i, $i=1,\ldots,m$, is the ith component of

$$\lambda = \frac{\mathbf{A}\mathbf{x}}{-\alpha_{m+1}\mathbf{x}} \geqq \mathbf{0}$$

so that $\mathbf{B}\mathbf{x} = \mathbf{A}\mathbf{x} + \lambda\alpha_{m+1}\mathbf{x} = \mathbf{0}$. Invoking the lemma a second time yields solutions $\mathbf{u} \in \mathbf{R}^n, \mathbf{v} \in \mathbf{R}^m$ satisfying

$$\mathbf{B}\mathbf{u} \geqq \mathbf{0},\ \mathbf{B}'\mathbf{v} = \mathbf{0},\ \mathbf{v} \geqq \mathbf{0},\ \beta_1\mathbf{u} + v_1 > 0.$$

Let

$$\bar{\mathbf{v}} = \begin{bmatrix} \mathbf{v} \\ \lambda'\mathbf{v} \end{bmatrix} \geqq \mathbf{0}. \tag{5.1}$$

Then

$$\bar{\mathbf{A}}'\bar{\mathbf{v}} \ = \ \mathbf{A}'\mathbf{v} + \alpha'_{m+1}\lambda'\mathbf{v} = \mathbf{B}'\mathbf{v} = \mathbf{0}. \tag{5.2}$$

Let $\mathbf{w} = \mathbf{u} + \mu\mathbf{x}$, where

$$\mu = -\frac{\alpha_{m+1}\mathbf{u}}{\alpha_{m+1}\mathbf{v}}$$

so that $\alpha_{m+1}\mathbf{w} = 0$. Then

$$\bar{\mathbf{A}}\mathbf{w} = \begin{bmatrix} \mathbf{A}\mathbf{w} \\ 0 \end{bmatrix} = \begin{bmatrix} (\mathbf{B} - \lambda\alpha_{m+1})\mathbf{w} \\ 0 \end{bmatrix} = \begin{bmatrix} \mathbf{B}\mathbf{w} \\ 0 \end{bmatrix} = \begin{bmatrix} \mathbf{B}\mathbf{u} - \mu\mathbf{B}\mathbf{x} \\ 0 \end{bmatrix} = \begin{bmatrix} \mathbf{B}\mathbf{u} \\ 0 \end{bmatrix} \geqq \mathbf{0}. \tag{5.3}$$

Also,

$$\alpha_1\mathbf{w}+v_1=(\beta_1-\lambda_1\alpha_{m+1})\mathbf{w}+v_1$$
$$=\beta_1\mathbf{w}+v_1=\beta_1\mathbf{u}-\mu\beta_1\mathbf{x}+v_1=\beta_1\mathbf{u}+v_1>0. \tag{5.4}$$

Thus, by constructing solutions $\mathbf{u}$ and $\mathbf{v}$, the lemma is extended to $\mathbf{A}$, *i.e.*, from (5.1)-(5.4),

$$\bar{\mathbf{A}}\mathbf{w}\geqq\mathbf{0},\ \bar{\mathbf{A}}'\bar{\mathbf{v}}=\mathbf{0},\ \bar{\mathbf{v}}\geqq\mathbf{0},\ \alpha_1\mathbf{w}+v_1>0.$$

Thus the lemma holds for all rows m and induction on the same is complete. Q.E.D.

This lemma asserts that: (a) from (I), there exists a vector $\mathbf{x}$ which makes an acute angle ($\leq\pi/2$) with the rows of $\mathbf{A}$; (b) from (II), the null vector is expressible as a nonnegative linear combination of the columns of $\mathbf{A}'$; and (c) the weight y_1 must be positive if the first row of $\mathbf{A}$ is orthogonal to $\mathbf{x}$.

The importance of Tucker's lemma 5.1 is that it guarantees the existence of solutions of dual linear systems of equalities/inequalities that exhibit a special positivity property. Specifically, we have

5.2. TUCKER'S FIRST EXISTENCE THEOREM [Tucker, 1956]. For any $(m\times n)$ matrix $\mathbf{A}$, the dual systems

(I) $\mathbf{A}\mathbf{x}\geqq\mathbf{0}$ and

(II) $\mathbf{A}'\mathbf{y}=\mathbf{0},\ \mathbf{y}\geqq\mathbf{0}$

possess solutions $\mathbf{x}\in\mathbf{R}^n$, $\mathbf{y}\in\mathbf{R}^m$ satisfying $\mathbf{A}\mathbf{x}+\mathbf{y}>\mathbf{0}$.

PROOF. In the preceding lemma renumber the rows α_i, i=1,...,m, so that any row can serve as α_1. Then, by Tucker's lemma 5.1, there exist pairs $\mathbf{x}_i\in\mathbf{R}^n$, $\mathbf{y}_i\in\mathbf{R}^m$ such that

$$\left.\begin{array}{l}\mathbf{A}\mathbf{x}_i\geqq\mathbf{0}\\ \mathbf{A}'\mathbf{y}_i=\mathbf{0},\ \mathbf{y}_i\geqq\mathbf{0}\\ \alpha_i\mathbf{x}_i+y_i^i>0\end{array}\right\}\ i=1,\ldots,m, \tag{5.5}$$

where y_i^i is the ith component of $\mathbf{y}_i$. Let $\mathbf{x} = \sum_i \mathbf{x}_i$, $\mathbf{y} = \sum_i \mathbf{y}_i$. Then

$$\mathbf{Ax} = \sum_i \mathbf{Ax}_i \geqq \mathbf{0},\ \mathbf{A'y} = \sum_i \mathbf{A'y}_i = \mathbf{0},\ \mathbf{y} \geqq \mathbf{0}.$$

For y_i the ith component of $\mathbf{y}$, (5.5) implies that

$$\alpha_i\mathbf{x} + y_i = \sum_i \left(\alpha_i\mathbf{x}_i + y_i^i\right) = \alpha_i\mathbf{x}_i + y_i^i + \sum_{\substack{k=1 \\ k\neq i}}^{m} \left(\alpha_i\mathbf{x}_k + y_i^k\right) > 0$$

for *all* i or $\mathbf{Ax} + \mathbf{y} > \mathbf{0}$. Q.E.D.

The implication of this theorem is that: (a) from (I), there exists an $\mathbf{x}$ which makes an acute angle ($\leq \pi/2$) with the rows of $\mathbf{A}$; (b) from (II), $\mathbf{0}$ is expressible as a nonnegative linear combination of the columns of $\mathbf{A'}$; and (c) since $\mathbf{Ax} + \mathbf{y} > \mathbf{0}$, positive weights are associated with the rows of $\mathbf{A}$ that are orthogonal to $\mathbf{x}$. Note also that this theorem will be valid in the special circumstances where: (1) $\mathbf{Ax} > \mathbf{0}$ ($\mathbf{x}$ makes a strictly acute angle with the rows of $\mathbf{A}$) and $\mathbf{y} \geqq \mathbf{0}$; and (2) $\mathbf{Ax} \geqq \mathbf{0}$ and $\mathbf{y} > \mathbf{0}$ (the origin is expressible as a strictly positive linear combination of the columns of $\mathbf{A'}$). Moreover, the collection of one-to-one correspondences inherent within dual systems (I), (II) may be written as

$$\text{(I)} \left\{ \begin{array}{lcl} \mathbf{x} \text{ unrestricted} & \longleftrightarrow & n \text{ equations } \mathbf{A'y} = \mathbf{0} \\ \mathbf{Ax} \geqq \mathbf{0} & \longleftrightarrow & \mathbf{y} \geqq \mathbf{0} \end{array} \right\} \text{(II)}$$

wherein $y_i > 0$ if $\alpha_i\mathbf{x} = 0$, $i=1,\ldots,n$.

We next have

5.3. TUCKER'S SECOND EXISTENCE THEOREM [Tucker, 1956]. Let $\mathbf{A}, \mathbf{B}$ be $(m_1 \times n), (m_2 \times n)$ matrices respectively, with $\mathbf{A}$ nonvacuous. Then the dual systems

(I) $\mathbf{Ax} \geqq \mathbf{0}$, $\mathbf{Bx} = \mathbf{0}$ and

(II) $\mathbf{A'y}_1 + \mathbf{B'y}_2 = \mathbf{0}$, $\mathbf{y}_1 \geqq \mathbf{0}, \mathbf{y}_2$ unrestricted possess solutions $\mathbf{x} \epsilon \mathbf{R}^n$, $\mathbf{y}_1 \epsilon \mathbf{R}^{m_1}$, and $\mathbf{y}_2 \epsilon \mathbf{R}^{m_2}$ satisfying $\mathbf{Ax} + \mathbf{y}_1 > \mathbf{0}$.

PROOF. We may apply the preceding existence theorem 5.2 to the system

$$\begin{bmatrix} \mathbf{A} \\ \mathbf{B} \\ -\mathbf{B} \end{bmatrix} \mathbf{x} \geqq \mathbf{0}, \qquad (\mathbf{A}', \mathbf{B}', -\mathbf{B}') \begin{bmatrix} \mathbf{y}_1 \\ \mathbf{u}_1 \\ \mathbf{u}_2 \end{bmatrix} = 0, \qquad \begin{bmatrix} \mathbf{y}_1 \\ \mathbf{u}_1 \\ \mathbf{u}_2 \end{bmatrix} \geqq \mathbf{0} \tag{5.6}$$

so as to obtain vectors $\mathbf{x}$, $\mathbf{y}_1$ along with $\mathbf{u}_1, \mathbf{u}_2 \epsilon \mathbf{R}^{m_2}$ satisfying

$$\begin{aligned} \mathbf{Ax} + \mathbf{y}_1 &> \mathbf{0} \\ \mathbf{Bx} + \mathbf{u}_1 &> \mathbf{0} \\ -\mathbf{Bx} + \mathbf{u}_2 &> \mathbf{0}. \end{aligned}$$

For $\mathbf{y}_2 = \mathbf{u}_1 - \mathbf{u}_2$ unrestricted, we obtain solutions $\mathbf{x}$, $\mathbf{y}_1$, $\mathbf{y}_2$ satisfying (5.6), *i.e.*,

$$\begin{aligned} &\mathbf{Ax} \geqq \mathbf{0},\ \mathbf{Bx} = \mathbf{0} \\ &\mathbf{A}'\mathbf{y}_1 + \mathbf{B}'\mathbf{y}_2 = \mathbf{0},\ \mathbf{y}_1 \geqq \mathbf{0} \\ &\mathbf{Ax} + \mathbf{y}_1 > \mathbf{0}. \quad \text{Q.E.D.} \end{aligned}$$

In this particular dual system the one-to-one correspondence appears as

$$\text{(I)} \left\{ \begin{array}{lcl} \mathbf{x} \text{ unrestricted} & \longleftrightarrow & n \text{ equations } \mathbf{A}'\mathbf{y}_1 + \mathbf{B}'\mathbf{y}_2 = \mathbf{0} \\ \mathbf{Ax} \geqq \mathbf{0} & \longleftrightarrow & \mathbf{y}_1 \geqq \mathbf{0} \\ \mathbf{Bx} = \mathbf{0} & \longleftrightarrow & \mathbf{y}_2 \text{ unrestricted} \end{array} \right\} \text{(II)}$$

with $y_i^1 > \mathbf{0}$ if $\boldsymbol{\alpha}_i\mathbf{x} = 0$, i=1,…,n.

An extremely useful generalization of the preceding theorem is provided by

5.4. COROLLARY (to Tucker's Second Existence Theorem 5.3) [Mangasarian, 1969]. Let matrixes $\mathbf{A}, \mathbf{B}, \mathbf{C}$, and $\mathbf{D}$ be of order $(m_1 \times n)$, $(m_2 \times n)$, $(m_3 \times n)$, and $(m_4 \times n)$ respectively with at least one of the matrices $\mathbf{A}, \mathbf{B}, \mathbf{C}$, nonvacuous. Then the dual systems

(I) $\mathbf{Ax} \geqq 0$, $\mathbf{Bx} \geqq 0$, $\mathbf{Cx} \geqq 0$, $\mathbf{Dx} = 0$ and

(II) $\mathbf{A'y}_1 + \mathbf{B'y}_2 + \mathbf{C'y}_3 + \mathbf{D'y}_4 = 0$, $\mathbf{y}_1 \geqq 0$, $\mathbf{y}_2 \geqq 0$, $\mathbf{y}_3 \geqq 0$, $\mathbf{y}_4$ unrestricted

possess solutions $\mathbf{x} \in \mathbf{R}^n$, $\mathbf{y}_i \in \mathbf{R}^{m_i}$, $i=1,2,3,4$, satisfying $\mathbf{Ax} + \mathbf{y}_1 > 0$, $\mathbf{Bx} + \mathbf{y}_2 > 0$, and $\mathbf{Cx} + \mathbf{y}_3 > 0$.

PROOF. Mirroring the proof of the second existence theorem 5.3 let us apply the first existence theorem 5.2 to the system

$$\begin{bmatrix} \mathbf{A} \\ \mathbf{B} \\ \mathbf{C} \\ \mathbf{D} \\ -\mathbf{D} \end{bmatrix} \mathbf{x} \geqq 0, \qquad (\mathbf{A}', \mathbf{B}', \mathbf{C}', \mathbf{D}', -\mathbf{D}') \begin{bmatrix} \mathbf{y}_1 \\ \mathbf{y}_2 \\ \mathbf{y}_3 \\ \mathbf{u}_1 \\ \mathbf{u}_2 \end{bmatrix} = 0, \begin{bmatrix} \mathbf{y}_1 \\ \mathbf{y}_2 \\ \mathbf{y}_3 \\ \mathbf{u}_1 \\ \mathbf{u}_2 \end{bmatrix} \geqq 0 \qquad (5.7)$$

in order to obtain vectors $\mathbf{x}$, $\mathbf{y}_1$, $\mathbf{y}_2$, $\mathbf{y}_3$ along with $\mathbf{u}_1, \mathbf{u}_2 \in \mathbf{R}^{m_4}$ satisfying

$$\begin{aligned} \mathbf{Ax} + \mathbf{y}_1 &> 0 \\ \mathbf{Bx} + \mathbf{y}_2 &> 0 \\ \mathbf{Cx} + \mathbf{y}_3 &> 0 \\ \mathbf{Dx} + \mathbf{u}_1 &> 0 \\ -\mathbf{Dx} + \mathbf{u}_2 &> 0 \end{aligned}$$

For $\mathbf{y}_4 = \mathbf{u}_1 - \mathbf{u}_2$ unrestricted, we obtain solutions $\mathbf{x}, \mathbf{y}_1, \ldots, \mathbf{y}_4$ satisfying (5.7) or

$$\mathbf{Ax} \geqq \mathbf{0},\ \mathbf{Bx} \geqq \mathbf{0},\ \mathbf{Cx} \geqq \mathbf{0},\ \mathbf{Dx} = \mathbf{0}$$

$$\mathbf{A'y}_1 + \mathbf{B'y}_2 + \mathbf{C'y}_3 + \mathbf{D'y}_4 = \mathbf{0},\ \mathbf{y}_1 \geqq \mathbf{0},\ \mathbf{y}_2 \geqq \mathbf{0},\ \mathbf{y}_3 \geqq \mathbf{0}$$

$$\mathbf{Ax} + \mathbf{y}_1 > \mathbf{0}$$

$$\mathbf{Bx} + \mathbf{y}_2 > \mathbf{0}$$

$$\mathbf{Cx} + \mathbf{y}_3 > \mathbf{0}. \quad \text{Q.E.D.}$$

The following set of existence theorems provides the foundation for the development of the concept of complementary slackness in pairs of dual systems and in a particular self-dual system. We state first

5.5 TUCKER'S THIRD EXISTENCE THEOREM [Tucker, 1956]. For any $(m \times n)$ matrix $\mathbf{A}$, the dual systems

(I) $\mathbf{Ax} \geqq \mathbf{0},\ \mathbf{x} \geqq \mathbf{0}$ and

(II) $-\mathbf{A'y} \geqq \mathbf{0},\ \mathbf{y} \geqq \mathbf{0}$

possess solutions $\mathbf{x} \in \mathbf{R}^n$, $\mathbf{y} \in \mathbf{R}^m$ such that $\mathbf{x} - \mathbf{A'y} > \mathbf{0}$ and $\mathbf{Ax} + \mathbf{y} > \mathbf{0}$.

PROOF. Let us rewrite (I), (II) as

$$\text{(I')} \quad \begin{bmatrix} \mathbf{I}_n \\ \mathbf{A} \end{bmatrix} \mathbf{x} \geqq \mathbf{0} \qquad \text{(II')} \quad (\mathbf{I}_n, \mathbf{A'}) \begin{bmatrix} \mathbf{w} \\ \mathbf{y} \end{bmatrix} = \mathbf{0}, \quad \begin{bmatrix} \mathbf{w} \\ \mathbf{y} \end{bmatrix} \geqq \mathbf{0},$$

where $\mathbf{w} = -\mathbf{A'y} \in \mathbf{R}^n$. Then according to Tucker's first existence theorem 5.2, there exist vectors $\mathbf{x}$, $\mathbf{y}$, and $\mathbf{w}$ such that

$$\begin{bmatrix} \mathbf{I}_n \\ \mathbf{A} \end{bmatrix} \mathbf{x} \geqq \mathbf{0},\ (\mathbf{I}_n, \mathbf{A'}) \begin{bmatrix} \mathbf{w} \\ \mathbf{y} \end{bmatrix} = \mathbf{0}, \quad \begin{bmatrix} \mathbf{w} \\ \mathbf{y} \end{bmatrix} \geqq \mathbf{0}, \quad \text{and} \quad \begin{bmatrix} \mathbf{I}_n \\ \mathbf{A} \end{bmatrix} \mathbf{x} + \begin{bmatrix} \mathbf{w} \\ \mathbf{y} \end{bmatrix} > \mathbf{0} \ \text{or}$$

$\mathbf{x} \geqq \mathbf{0}$, $\mathbf{Ax} \geqq \mathbf{0}$, $\mathbf{w} = -\mathbf{A'y} \geqq \mathbf{0}$, $\mathbf{y} \geqq \mathbf{0}$, $\mathbf{x} - \mathbf{A'y} > \mathbf{0}$, and $\mathbf{Ax} + \mathbf{y} > \mathbf{0}$. Q.E.D.

An immediate consequence of this theorem is

5.6. COROLLARY (to Tucker's third existence theorem 5.5) [Tucker, 1956]. For any $(m \times n)$ matrix $\mathbf{A}$, the dual systems

(I) $\mathbf{Ax} \geqq \mathbf{0}, \mathbf{x} \geqq \mathbf{0}$ and

(II) $-\mathbf{A}'\mathbf{y} \geqq \mathbf{0}, \mathbf{y} \geqq \mathbf{0}$

possess solutions $\mathbf{x} \in \mathbf{R}^n$, $\mathbf{y} \in \mathbf{R}^m$ for which the following alternatives hold:

(a) either $\mathbf{Ax} \neq \mathbf{0}$ or $\mathbf{y} > \mathbf{0}$;

(b) either $\mathbf{Ax} > \mathbf{0}$ or $\mathbf{y} \neq \mathbf{0}$;

(c) either $\mathbf{x} > \mathbf{0}$ or $-\mathbf{A}'\mathbf{y} \neq \mathbf{0}$;

(d) either $\mathbf{x} \neq \mathbf{0}$ or $-\mathbf{A}'\mathbf{y} > \mathbf{0}$.

PROOF. Parts (a), (b) hold since we must have $\mathbf{Ax} + \mathbf{y} > \mathbf{0}$ by virtue of the preceding theorem. Parts (c), (d) hold given that the preceding theorem requires $\mathbf{x} - \mathbf{A}'\mathbf{y} > \mathbf{0}$. Q.E.D.

A generalization of theorem 5.5 is provided by

5.7. TUCKER'S FOURTH EXISTENCE THEOREM [Tucker, 1956]. Let $\mathbf{A}, \mathbf{B}, \mathbf{C}$, and $\mathbf{D}$ be given $(m \times n)$ matrices. Then the general dual systems

$$
\text{(I)}\begin{cases} \mathbf{Ax} + \mathbf{By} = \mathbf{0} \\ \mathbf{Cx} + \mathbf{Dy} \geqq \mathbf{0} \\ \mathbf{x} \geqq \mathbf{0} \\ \mathbf{y} \text{ unrestricted} \end{cases} \qquad \text{(II)}\begin{cases} \mathbf{u} \text{ unrestricted} \\ \mathbf{v} \geqq \mathbf{0} \\ -\mathbf{A}'\mathbf{u} - \mathbf{C}'\mathbf{v} \geqq \mathbf{0} \\ -\mathbf{B}'\mathbf{u} - \mathbf{D}'\mathbf{v} = \mathbf{0} \end{cases}
$$

possess solutions $\mathbf{x}, \mathbf{y} \in \mathbf{R}^n$ and $\mathbf{u}, \mathbf{v} \in \mathbf{R}^m$ such that

$\mathbf{x} - \mathbf{A}'\mathbf{u} - \mathbf{C}'\mathbf{v} > \mathbf{0}$ and $\mathbf{Cx} + \mathbf{Dy} + \mathbf{v} > \mathbf{0}$.

PROOF. Let us replace the equalities in (I), (II) by $\mathbf{Ax} + \mathbf{By} \geqq \mathbf{0}$, $-\mathbf{Ax} - \mathbf{By} \geqq \mathbf{0}$ and $-\mathbf{B}'\mathbf{u} - \mathbf{D}'\mathbf{v} \geqq \mathbf{0}$, $\mathbf{B}'\mathbf{u} + \mathbf{D}'\mathbf{v} \geqq \mathbf{0}$ respective-ly with $\mathbf{y} = \mathbf{y}_1 - \mathbf{y}_2$, $\mathbf{u} = \mathbf{u}_1 - \mathbf{u}_2$. Then the above dual systems can be rewritten in a form similar to those appearing in theorem 5.5 as

$$\text{(I}'\text{)}\qquad \begin{bmatrix} \mathbf{A} & \mathbf{B} & -\mathbf{B} \\ -\mathbf{A} & -\mathbf{B} & \mathbf{B} \\ \mathbf{C} & \mathbf{D} & -\mathbf{D} \end{bmatrix} \begin{bmatrix} \mathbf{x} \\ \mathbf{y}_1 \\ \mathbf{y}_2 \end{bmatrix} \geqq \mathbf{0}, \quad \begin{bmatrix} \mathbf{x} \\ \mathbf{y}_1 \\ \mathbf{y}_2 \end{bmatrix} \geqq \mathbf{0} \qquad \text{and}$$

$$\text{(II}'\text{)}\qquad \begin{bmatrix} -\mathbf{A}' & \mathbf{A}' & -\mathbf{C}' \\ -\mathbf{B}' & \mathbf{B}' & -\mathbf{D}' \\ \mathbf{B}' & -\mathbf{B}' & \mathbf{D}' \end{bmatrix} \begin{bmatrix} \mathbf{u}_1 \\ \mathbf{u}_2 \\ \mathbf{v} \end{bmatrix} \geqq \mathbf{0}, \quad \begin{bmatrix} \mathbf{u}_1 \\ \mathbf{u}_2 \\ \mathbf{v} \end{bmatrix} \geqq \mathbf{0},$$

where the coefficient matrix in (II′) is the negative transpose of its counterpart in (I′).

By applying Tucker's third existence theorem 5.5 to (I′), (II′) we are assured that there exist vectors $\mathbf{x} \geqq \mathbf{0}$, $\mathbf{y}_1 \geqq \mathbf{0}$, $\mathbf{y}_2 \geqq \mathbf{0}$, $\mathbf{u}_1 \geqq \mathbf{0}$, $\mathbf{u}_2 \geqq \mathbf{0}$ and $\mathbf{v} \geqq \mathbf{0}$ such that (I′), (II′) (and thus (I), (II)) hold with

$$\begin{bmatrix} \mathbf{x} \\ \mathbf{y}_1 \\ \mathbf{y}_2 \end{bmatrix} + \begin{bmatrix} -\mathbf{A}' & \mathbf{A}' & -\mathbf{C}' \\ -\mathbf{B}' & \mathbf{B}' & -\mathbf{D}' \\ \mathbf{B}' & -\mathbf{B}' & \mathbf{D}' \end{bmatrix} \begin{bmatrix} \mathbf{u}_1 \\ \mathbf{u}_2 \\ \mathbf{v} \end{bmatrix} > \mathbf{0}, \quad \begin{bmatrix} \mathbf{A} & \mathbf{B} & -\mathbf{B} \\ -\mathbf{A} & -\mathbf{B} & \mathbf{B} \\ \mathbf{C} & \mathbf{D} & -\mathbf{D} \end{bmatrix} \begin{bmatrix} \mathbf{x} \\ \mathbf{y}_1 \\ \mathbf{y}_2 \end{bmatrix} + \begin{bmatrix} \mathbf{u}_1 \\ \mathbf{u}_2 \\ \mathbf{v} \end{bmatrix} > \mathbf{0}.$$

These latter inequalities ultimately yield $\mathbf{x} - \mathbf{A}'\mathbf{u} - \mathbf{C}'\mathbf{v} > \mathbf{0}$, $\mathbf{C}\mathbf{x} + \mathbf{D}\mathbf{y} + \mathbf{v} > \mathbf{0}$ respectively. Q.E.D.

We next state

5.8. TUCKER'S FIFTH EXISTENCE THEOREM [Tucker, 1956]. Let $\mathbf{K}$ be an $(n \times n)$ skew-symmetric matrix $(\mathbf{K} = -\mathbf{K}')$. Then the self-dual system $\mathbf{K}\mathbf{w} \geqq \mathbf{0}$, $\mathbf{w} \geqq \mathbf{0}$ has a solution $\mathbf{w} \in R^n$ such that $\mathbf{K}\mathbf{w} + \mathbf{w} > \mathbf{0}$.

PROOF. For $\mathbf{A} = \mathbf{K} = -\mathbf{K}'$ we may apply theorem 5.5 by constructing the dual systems

(I) $\mathbf{K}\mathbf{x} \geqq \mathbf{0}$, $\mathbf{x} \geqq \mathbf{0}$ and

(II) $\mathbf{K}\mathbf{y} \geqq \mathbf{0}$, $\mathbf{y} \geqq \mathbf{0}$.

Then there exist vectors $\mathbf{x}, \mathbf{y} \in \mathbf{R}^n$ such that

$$\mathbf{x} \geqq \mathbf{0},\ \mathbf{Kx} \geqq \mathbf{0},\ \mathbf{Ky} \geqq \mathbf{0},\ \mathbf{y} \geqq \mathbf{0},\ \mathbf{x} + \mathbf{Ky} > \mathbf{0}, \text{ and } \mathbf{Kx} + \mathbf{y} > \mathbf{0}.$$

Hence

$$\mathbf{Kx} + \mathbf{Ky} = \mathbf{K}(\mathbf{x}+\mathbf{y}) \geqq \mathbf{0},\ \mathbf{x} + \mathbf{y} \geqq \mathbf{0}, \text{ and}$$
$$(\mathbf{x}+\mathbf{Ky}) + (\mathbf{Kx}+\mathbf{y}) = \mathbf{K}(\mathbf{x}+\mathbf{y}) + (\mathbf{x}+\mathbf{y}) > \mathbf{0}.$$

For $\mathbf{w} = \mathbf{x} + \mathbf{y}$, we obtain $\mathbf{Kw} \geqq \mathbf{0}$, $\mathbf{w} \geqq \mathbf{0}$, and $\mathbf{Kw} + \mathbf{w} > \mathbf{0}$. Q.E.D.

5.3. Exercises

1. Let system (I) of section 5.1 appear as

$$\begin{array}{llll} 3x_1 - & x_2 + & 2y_1 + 4y_2 & = 0 \\ 8x_1 + & 2x_2 + & y_1 & \geq 0 \\ x_1 & & & \geq 0 \\ & x_2 & & \geq 0 \\ & & y_1 \text{ unrestricted} & \\ & & y_2 \text{ unrestricted.} & \end{array}$$

Determine the corresponding (II).

2. Suppose (II′) in section 5.1 has the form

$$\begin{array}{rcl} v_1 & & \geq 0 \\ & v_2 & \geq 0 \\ v_1 + & v_2 & \geq 0 \\ -3v_1 - & v_2 & \geq 0. \end{array}$$

Find the appropriate (I′).

3. In theorem 5.2 let $\mathbf{A} = \begin{bmatrix} 1 & -1 \\ 1 & 0 \\ -1 & 1 \end{bmatrix}, \ \mathbf{x} = \begin{bmatrix} 1 \\ 1 \end{bmatrix}.$

Find a $\mathbf{y}$ satisfying (II) and $\mathbf{Ax} + \mathbf{y} > \mathbf{0}$.

4. In corollary 5.6 let $\mathbf{A} = \begin{bmatrix} 3 & -1 \\ 1 & 2 \\ -1 & 1 \end{bmatrix}$

For $\mathbf{x}' = (1,2)$, can (II) possess the trivial solution $\mathbf{y} = \mathbf{0}$?

5. Let $\mathbf{C}'\mathbf{x} \leq \mathbf{b}'\mathbf{u}$ when $\mathbf{Ax} \leq \mathbf{b}$, $\mathbf{A}'\mathbf{u} \geq \mathbf{C}$ for $\mathbf{x} \in \mathbf{R}^n$, $\mathbf{u} \in \mathbf{R}^m$.
Use these $m+n+1$ inequalities to form the inequality $\mathbf{Ky} \geqq \mathbf{0}$, where $\mathbf{y}' = (\mathbf{u}', \mathbf{x}', 1)$. Is $\mathbf{K}$ skew-symmetric? Does $\mathbf{Ky} \geq \mathbf{0}$ constitute a homogeneous inequality system? If not, how may it be converted to one?

CHAPTER 6
THEOREMS OF THE ALTERNATIVE FOR LINEAR SYSTEMS

6.1 The Structure of a Theorem of the Alternative

We now turn to an assortment of theorems which are generally characterized as "theorems of the alternative" or, in particular instances, so-called "transposition theorems." Specifically, a theorem of the alternative involves two mutually exclusive systems of linear inequalities and/or equalities denoted simply as (I), (II). Moreover, it asserts that either system (I) has a solution or system (II) has a solution, but never both. A transposition theorem, which is a special type of theorem of the alternative, asserts for systems (I), (II) the disjoint alternatives of solvability or contradiction given that in one system a particular vector (usually the null vector) is a linear combination of vectors from the other.

Theorems of the alternative possess the following structure:

"(I) if and only if $(\overline{\text{II}})$" or equivalently

"$(\overline{\text{I}})$ if and only if (II),"

where "–" indicates the nonoccurrence of a solution to a given system. In this regard, a typical proof of a theorem of the alternative is executed in two steps and is structured as:

1. (I) implies $(\overline{\text{II}})$ (or equivalently, (II) implies $(\overline{\text{I}})$); and
2. $(\overline{\text{I}})$ implies (II) (or equivalently, $(\overline{\text{II}})$ implies (I)).

To carry out step 1 we assume that (I) has a solution and show that if (II) also has a solution, then a contradiction occurs. Step 2 typically employs the preceding existence theorems and exploits the special assumptions made about the coefficient matrices of the given systems of linear inequalities and/or equalities.

6.2. THEOREMS OF THE ALTERNATIVE.

Let us initiate our study of this class of theorems with

6.1. SLATER'S THEOREM OF THE ALTERNATIVE [Slater, 1951]. Let matrices $\mathbf{A}, \mathbf{B}, \mathbf{C}$, and $\mathbf{D}$ be of order $(m_1 \times n), (m_2 \times n), (m_3 \times n)$, and $(m_4 \times n)$ respectively with $\mathbf{A}, \mathbf{B}$ nonvacuous. Then either

(I) $\mathbf{Ax} > \mathbf{0},\ \mathbf{Bx} \geq \mathbf{0},\ \mathbf{Cx} \geqq \mathbf{0},\ \mathbf{Dx} = \mathbf{0}$ has a solution $\mathbf{x} \in \mathbf{R}^n$ or

(II) $\mathbf{A}'\mathbf{y}_1 + \mathbf{B}'\mathbf{y}_2 + \mathbf{C}'\mathbf{y}_3 + \mathbf{D}'\mathbf{y}_4 = \mathbf{0}$ with

$\mathbf{y}_1 \geq \mathbf{0},\ \mathbf{y}_2 \geqq \mathbf{0},\ \mathbf{y}_3 \geqq \mathbf{0},\ \mathbf{y}_4$ unrestricted or

$\mathbf{y}_1 \geqq \mathbf{0},\ \mathbf{y}_2 > \mathbf{0},\ \mathbf{y}_3 \geqq \mathbf{0},\ \mathbf{y}_4$ unrestricted

has a solution $\mathbf{y}_i \in \mathbf{R}^{m_i}, i=1,\ldots,4$,

but never both.

PROOF. ((I) implies ($\overline{\text{II}}$)). Let (I) possess a solution. If (II) also has a solution, then there exist vectors $\mathbf{x}, \mathbf{y}_1, \ldots, \mathbf{y}_4$ such that

$$\underbrace{\mathbf{x}'\mathbf{A}'\mathbf{y}_1 + \mathbf{x}'\mathbf{B}'\mathbf{y}_2}_{\substack{\text{with } \mathbf{A},\mathbf{B} \text{ nonvacuous} \\ \text{either } \mathbf{x}'\mathbf{A}'\mathbf{y}_1 > 0 \\ \text{and } \mathbf{x}'\mathbf{B}'\mathbf{y}_2 \geq 0 \\ \text{or} \\ \mathbf{x}'\mathbf{A}'\mathbf{y}_1 \geq 0 \text{ and} \\ \mathbf{x}'\mathbf{B}'\mathbf{y}_2 > 0}} + \underbrace{\mathbf{x}'\mathbf{C}'\mathbf{y}_3}_{\geq 0} + \underbrace{\mathbf{x}'\mathbf{D}'\mathbf{y}_4}_{=0} > 0.$$

Clearly this expression must be strictly positive and consequently contradicts the requisite equality in (II) so that (II) has no solution.

(($\overline{\text{I}}$) implies (II)). If (I) does not possess a solution $\mathbf{x}$, then it can be modified to read $\mathbf{Ax} \geqq \mathbf{0},\ \mathbf{Bx} \geqq \mathbf{0},\ \mathbf{Cx} \geqq \mathbf{0},\ \mathbf{Dx} = \mathbf{0}$. Hence it is possible that either $\mathbf{Ax} \not> \mathbf{0}$ or $\mathbf{Bx} = \mathbf{0}$. Then from corollary 5.4 to theorem 5.3, systems

(1) $\mathbf{Ax} \geqq \mathbf{0},\ \mathbf{Bx} \geqq \mathbf{0},\ \mathbf{Cx} \geqq \mathbf{0},\ \mathbf{Dx} = \mathbf{0}$ and

(2) $\mathbf{A}'\mathbf{y}_1 + \mathbf{B}'\mathbf{y}_2 + \mathbf{C}'\mathbf{y}_3 + \mathbf{D}'\mathbf{y}_4 = \mathbf{0}$

$\mathbf{y}_1 \geqq \mathbf{0},\ \mathbf{y}_2 \geqq \mathbf{0},\ \mathbf{y}_3 \geqq \mathbf{0}$

possess solutions $\mathbf{x}, \mathbf{y}_1, \ldots, \mathbf{y}_4$ satisfying $\mathbf{Ax} + \mathbf{y}_1 > \mathbf{0}$, $\mathbf{Bx} + \mathbf{y}_2 > \mathbf{0}$, $\mathbf{Cx} + \mathbf{y}_3 > \mathbf{0}$. With $\mathbf{Ax} \not> \mathbf{0}$ or $\mathbf{Bx} = \mathbf{0}$, we have $\mathbf{y}_1 \geq \mathbf{0}$ or $\mathbf{y}_2 > \mathbf{0}$ respectively. Hence (II) has a solution. Q.E.D.

If in the preceding theorem we require that ***only* A *is nonvacuous***, then we obtain

6.2. MOTZKIN'S THEOREM OF THE ALTERNATIVE [Motzkin, 1936]. Let matrices $\mathbf{A}, \mathbf{C}$, and $\mathbf{D}$ be of order $(m_1 \times n), (m_3 \times n)$, and $(m_4 \times n)$ respectively with $\mathbf{A}$ nonvacuous. Then either

(I) $\mathbf{Ax} > \mathbf{0}$, $\mathbf{Cx} \geqq \mathbf{0}$, $\mathbf{Dx} = \mathbf{0}$ has a solution $\mathbf{x} \in \mathbf{R}^n$ or

(II) $\mathbf{A'y}_1 + \mathbf{C'y}_2 + \mathbf{D'y}_4 = \mathbf{0}$ with $\mathbf{y}_1 \geq \mathbf{0}$, $\mathbf{y}_3 \geqq \mathbf{0}$, $\mathbf{y}_4$ unrestricted has a solution $\mathbf{y}_1 \in \mathbf{R}^{m_1}$, $\mathbf{y}_3 \in \mathbf{R}^{m_3}$, and $\mathbf{y}_4 \in \mathbf{R}^{m_4}$,

but never both.

PROOF. ((I) implies $(\overline{\text{II}})$). Let (I) possess a solution. If (II) also has a solution, then there exist vectors $\mathbf{x}$, $\mathbf{y}_1$, $\mathbf{y}_3$ and $\mathbf{y}_4$ such that

$$\underbrace{\mathbf{x'A'y}_1}_{>0} + \underbrace{\mathbf{x'C'y}_3}_{\geq 0} + \underbrace{\mathbf{x'D'y}_4}_{=0} > 0.$$

Since the positivity of this expression contradicts the equality in (II), it is evident that (II) has no solution.

($(\overline{\text{I}})$ implies (II)). If (I) does not possess a solution, then it can be rewritten as $\mathbf{Ax} \geqq \mathbf{0}$, $\mathbf{Cx} \geqq \mathbf{0}$, $\mathbf{Dx} = \mathbf{0}$ and thus it is possible that $\mathbf{Ax} \not> \mathbf{0}$. By virtue of corollary 5.4 to theorem 5.3, systems

(1) $\mathbf{Ax} \geqq \mathbf{0}$, $\mathbf{Cx} \geqq \mathbf{0}$, $\mathbf{Dx} = \mathbf{0}$ and

(2) $\mathbf{A'y}_1 + \mathbf{C'y}_3 + \mathbf{D'y}_4 = \mathbf{0}$

$\mathbf{y}_1 \geqq \mathbf{0}$, $\mathbf{y}_2 \geqq \mathbf{0}$

possess solutions $\mathbf{x}$, $\mathbf{y}_1$, $\mathbf{y}_3$ and $\mathbf{y}_4$ satisfying $\mathbf{Ax} + \mathbf{y}_1 > \mathbf{0}$ and $\mathbf{Cx} + \mathbf{y}_3 > \mathbf{0}$. Since $\mathbf{Ax} \not> \mathbf{0}$, we must have $\mathbf{y}_1 \geqq \mathbf{0}$. Thus (II) has a solution. Q.E.D.

Next, if in Slater's theorem of the alternative 6.1 ***only* B *is nonvacuous***, then this restriction leads to

6.3. TUCKER'S THEOREM OF THE ALTERNATIVE [Tucker, 1956]. Let the matrices $\mathbf{B}, \mathbf{C},$ and $\mathbf{D}$ be of order $(m_2 \times n), (m_3 \times n),$ and $(m_4 \times n)$ respectively with $\mathbf{B}$ nonvacuous. Then either

(I) $\mathbf{Bx} \geq \mathbf{0},\ \mathbf{Cx} \geqq \mathbf{0},\ \mathbf{Dx} = \mathbf{0}$ has a solution $\mathbf{x} \epsilon \mathbf{R}^n$ or

(II) $\mathbf{B'Y}_2 + \mathbf{C'y}_3 + \mathbf{D'y}_4 = \mathbf{0}$ with $\mathbf{y}_2 > \mathbf{0},\ \mathbf{y}_3 \geqq \mathbf{0},\ \mathbf{y}_4$ unrestricted has a solution $\mathbf{y}_2 \epsilon \mathbf{R}^{m_2}$, $\mathbf{y}_3 \epsilon \mathbf{R}^{m_3}$ and $\mathbf{y}_4 \epsilon \mathbf{R}^{m_4}$,

but never both.

PROOF. ((I) implies $(\overline{\text{II}})$). Given that (I) has a solution, if (II) also has a solution, then there exist vectors $\mathbf{x}, \mathbf{y}_2, \mathbf{y}_3,$ and $\mathbf{y}_4$ such that

$$\underbrace{\mathbf{x'B'y}_2}_{>0} + \underbrace{\mathbf{x'C'y}_3}_{\geq 0} + \underbrace{\mathbf{x'D'y}_4}_{=0} > 0.$$

Thus (II) has no solution since the positive sign of the preceding expression contradicts the equality in (II).

($(\overline{\text{I}})$ implies (II)). If $(\overline{\text{I}})$ holds, then system (I) can be written as $\mathbf{Bx} \geqq \mathbf{0},\ \mathbf{Cx} \geqq \mathbf{0},\ \mathbf{Dx} = \mathbf{0}$ and thus we admit the possibility that $\mathbf{Bx} = \mathbf{0}$. If we invoke corollary 5.4 to theorem 5.3, the systems

(1) $\mathbf{Bx} \geqq \mathbf{0},\ \mathbf{Cx} \geqq \mathbf{0},\ \mathbf{Dx} = \mathbf{0}$ and

(2) $\mathbf{B'y}_2 + \mathbf{C'y}_3 + \mathbf{D'y}_4 = \mathbf{0}$

$\mathbf{y}_2 \geq \mathbf{0},\ \mathbf{y}_3 \geqq \mathbf{0}$

possess solutions $\mathbf{x}, \mathbf{y}_2, \mathbf{y}_3,$ and $\mathbf{y}_4$ satisfying $\mathbf{Bx} + \mathbf{y}_2 > \mathbf{0}$ and $\mathbf{Cx} + \mathbf{y}_3 > \mathbf{0}$. Under $\mathbf{Bx} = \mathbf{0}$, it must be true that $\mathbf{y}_2 > \mathbf{0}$ so that (II) has a solution. Q.E.D.

One of the principal uses of Slater's theorem 6.1, Motzkin's theorem 6.2, and Tucker's theorem 6.3, is that they are used to derive other (typically structurally simpler) theorems of the alternative.

If in Slater's theorem 6.1 the matrices **B**, **C**, and **D** are vacuous (or equivalently, if in Motzkin's theorem 6.2 **C**, **D** are vacuous), then we obtain

6.4. GORDAN'S THEOREM OF THE ALTERNATIVE FOR SEMIPOSITIVE SOLUTIONS OF HOMOGENEOUS EQUALITIES [Gordan, 1873]. For any $(m \times n)$ matrix **A**, either

(I) $\mathbf{Ax} > \mathbf{0}$ has a solution $\mathbf{x} \in \mathbf{R}^n$ or

(II) $\mathbf{A}'\mathbf{y} = \mathbf{0}$, $\mathbf{y} \geq \mathbf{0}$ has a solution $\mathbf{y} \in \mathbf{R}^m$,

but never both.

PROOF. From Motzkin's theorem 6.2 (with **C**, **D** vacuous), either (I) has a solution or

$$\text{(II}') \quad \mathbf{A}'\mathbf{y}_1 = \mathbf{0},\ \mathbf{y}_1 \geq \mathbf{0}$$

has a solution $\mathbf{y}_1 \in \mathbf{R}^m$, but never both. For $\mathbf{y} = \mathbf{y}_1$, (II′) is equivalent to (II). Q.E.D.

Geometrically, Gordan's theorem 6.4 states that either: (a) there exists a vector **x** which forms a strictly acute angle ($<\pi/2$) with the rows $\boldsymbol{\alpha}_i$ of **A**, $i=1,\ldots,m$ (Figure 6.1.a); or (b) the null vector is expressible as a nonnegative nontrivial linear combination of the columns $\boldsymbol{\alpha}_i'$ of $\mathbf{A}'$, $i=1,\ldots,m$ (Figure 6.1.b).

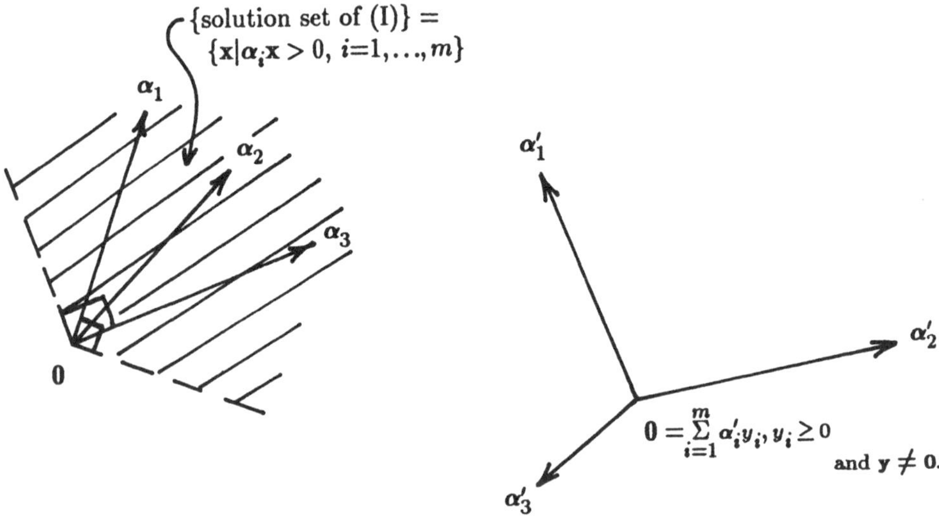

(I) has a solution;
(II) has no solution.

a.

(II) has a solution;
(I) has no solution.

b.

Figure 6.1

Proceeding in this same vein we have, for $\mathbf{A}, \mathbf{C}$, and $\mathbf{D}$ vacuous in Slater's theorem 6.1 (or for $\mathbf{C}, \mathbf{D}$ vacuous in Tucker's theorem 6.3),

6.5. STIEMKE'S THEOREM OF THE ALTERNATIVE FOR POSITIVE SOLUTIONS OF HOMOGENEOUS EQUATIONS [Stiemke, 1915]. For any $(m \times n)$ matrix $\mathbf{B}$, either

(I) $\mathbf{Bx} \geq \mathbf{0}$ has a solution $\mathbf{x} \in \mathbf{R}^n$ or

(II) $\mathbf{B'y} = \mathbf{0}$, $\mathbf{y} > \mathbf{0}$ has a solution $\mathbf{y} \in \mathbf{R}^m$,

but never both.

PROOF. From Tucker's theorem 6.3 (with $\mathbf{C}, \mathbf{D}$ vacuous), either (I) has a solution or

$$\text{(II}') \quad \mathbf{B'y}_2 = \mathbf{0}, \ \mathbf{y}_2 > \mathbf{0}$$

has a solution $\mathbf{y}_2 \in \mathbf{R}^m$, but never both. For $\mathbf{y} = \mathbf{y}_2$, (II′) is equivalent to (II). Q.E.D.

It is apparent that Stiemke's theorem 6.5 requires that either: (a) there exists a vector $\mathbf{x}$ which forms a non-obtuse angle ($\leq \pi/2$) with the rows β_i of B, $i=1,\ldots,m$ (Figure 6.2.a); or (b) the null vector is expressible as a strictly positive linear combination of the columns β_i' of B′, $i=1,\ldots,m$ (Figure 6.2.b).

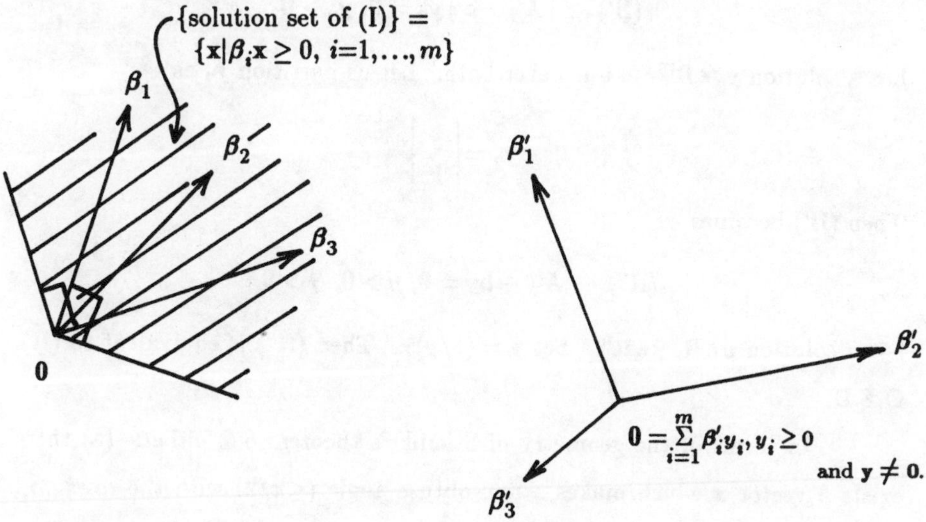

(I) has a solution;
(II) does not.

a.

(II) has a solution;
(I) does not.

b.

Figure 6.2

6.6. STIEMKE'S THEOREM OF THE ALTERNATIVE FOR POSITIVE SOLUTIONS OF NONHOMOGENEOUS EQUATIONS [Stiemke, 1915]. For any $(m \times n)$ matrix $\mathbf{A}$ and a given vector $\mathbf{b} \in \mathbf{R}^n$, either

(I) $\mathbf{A}\mathbf{x} \geq \mathbf{0},\ -\mathbf{b}'\mathbf{x} \geq 0$ has a solution $\mathbf{x} \in \mathbf{R}^n$ or

(II) $\mathbf{A}'\mathbf{y} = \mathbf{b},\ \mathbf{y} > \mathbf{0}$ has a solution $\mathbf{y} \in \mathbf{R}^m$,

but never both.

PROOF. Let us rewrite system (I) as

$$(\text{I}') \qquad \begin{bmatrix} \mathbf{A} \\ -\mathbf{b}' \end{bmatrix} \mathbf{x} \geq \mathbf{0}$$

has a solution $\mathbf{x} \in \mathbf{R}^n$. Then by Tucker's theorem 6.3 (with $\mathbf{C}, \mathbf{D}$ vacuous), either (I′) has a solution or

$$(\text{II}') \qquad [\,\mathbf{A}', \ -\mathbf{b}\,]\,\mathbf{y}_2 = \mathbf{0},\ \mathbf{y}_2 > \mathbf{0}$$

has a solution $\mathbf{y}_2 \in \mathbf{R}^{m+1}$, but never both. Let us partition $\mathbf{y}_2$ as

$$\mathbf{y}_2 = \begin{bmatrix} \bar{\mathbf{y}} \\ y \end{bmatrix}.$$

Then (II′) becomes

$$(\text{II}'') \qquad \mathbf{A}'\bar{\mathbf{y}} - \mathbf{b}y = \mathbf{0},\ y > 0,\ \bar{\mathbf{y}} > \mathbf{0}$$

has a solution $y \in \mathbf{R}$, $\bar{\mathbf{y}} \in \mathbf{R}^m$. Set $\mathbf{y} = (1/y)\bar{\mathbf{y}}$. Then (II″) is equivalent to (II). Q.E.D.

In terms of the geometry of Stiemke's theorem 6.6, either: (a) there exists a vector $\mathbf{x}$ which makes a non-obtuse angle ($\leq \pi/2$) with the rows $\boldsymbol{\alpha}_i$, $i=1,\ldots,m$, of $\mathbf{A}$ and a strictly obtuse angle ($>\pi/2$) with the vector $\mathbf{b}$ (*i.e.*, the open half-space formed by $\mathbf{b}'\mathbf{x} \leq 0$ does not have any points in common with the polyhedral convex cone $\mathcal{C}$ formed by $\mathbf{Ax} \geq \mathbf{0}$)(Figure 6.3.a); or (b) the vector $\mathbf{b}$ is expressible as a strictly positive linear combination of the columns $\boldsymbol{\alpha}_i'$, $i=1,\ldots,m$, of $\mathbf{A}'$ (Figure 6.3.b).

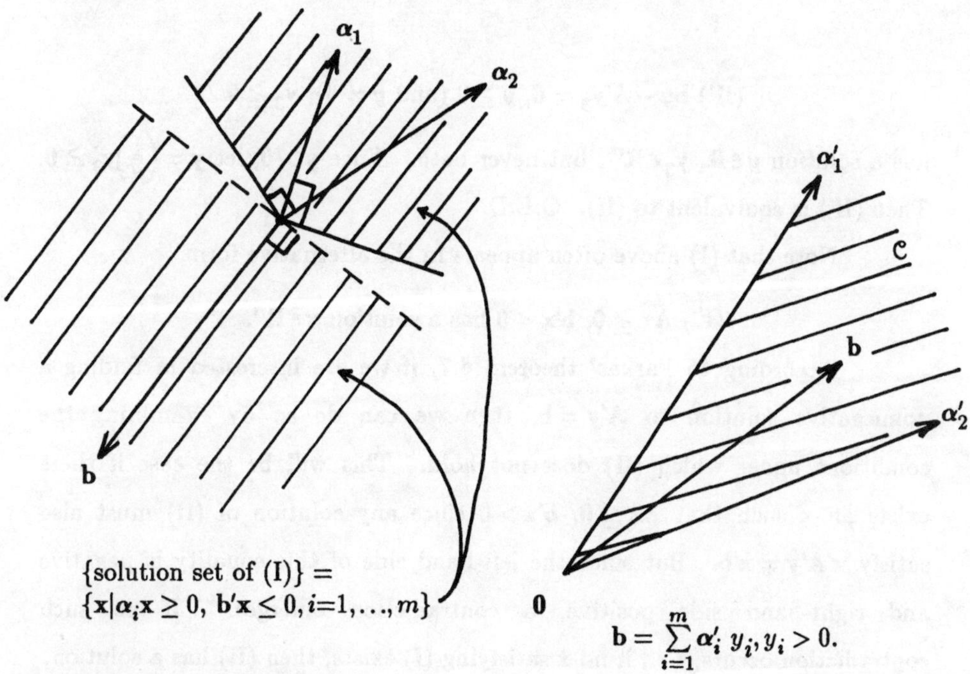

(I) has a solution;
(II) does not.

a.

(II) has a solution;
(I) does not.

b.

Figure 6.3

We next examine

6.7. FARKAS' THEOREM OF THE ALTERNATIVE FOR NONNEGATIVE SOLUTIONS OF NONHOMOGENEOUS EQUATIONS [Farkas, 1902]. For any $(m \times n)$ matrix $\mathbf{A}$ and vector $\mathbf{b} \in \mathbf{R}^n$, either

(I) $\mathbf{A}\mathbf{x} \leqq \mathbf{0}$, $\mathbf{b}'\mathbf{x} > 0$ has a solution $\mathbf{x} \in \mathbf{R}^n$ or

(II) $\mathbf{A}'\mathbf{y} = \mathbf{b}$, $\mathbf{y} \geqq \mathbf{0}$ has a solution $\mathbf{y} \in \mathbf{R}^m$,

but never both.

PROOF. System (I) is equivalent to

$$\text{(I}'\text{)} \qquad \mathbf{b}'\mathbf{x} > 0,\ -\mathbf{A}\mathbf{x} \geqq \mathbf{0}$$

has a solution $\mathbf{x} \in \mathbf{R}^n$. Then by Motzkin's theorem 6.2 (with $\mathbf{D}$ vacuous), either (I′) holds or

$$\text{(II}'\text{)}\ \mathbf{b}y-\mathbf{A}'\mathbf{y}_3=\mathbf{0},\ y\geq 0\ (\text{and } y\neq 0),\ \mathbf{y}_3\geqq\mathbf{0}$$

has a solution $y\in\mathbf{R}$, $\mathbf{y}_3\in\mathbf{R}^m$, but never both. Since $y>0$, let $\mathbf{y}=\left(\frac{1}{y}\right)\mathbf{y}_3\geqq\mathbf{0}$. Then (II′) is equivalent to (II). Q.E.D.

Note that (I) above often appears in the alternative form

(I″) $\mathbf{Ax}\geqq\mathbf{0}$, $\mathbf{b}'\mathbf{x}<0$ has a solution $\mathbf{x}\in\mathbf{R}^n$.

According to Farkas' theorem 6.7, if we are interested in finding a nonnegative solution to $\mathbf{A}'\mathbf{y}=\mathbf{b}$, then we can do so by examining the conditions under which (II) does not hold. This will be the case if there exists an $\mathbf{x}$ such that $\mathbf{Ax}\leqq\mathbf{0}$, $\mathbf{b}'\mathbf{x}>0$ since any solution of (II) must also satisfy $\mathbf{x}'\mathbf{A}'\mathbf{y}=\mathbf{x}'\mathbf{b}$. But since the left-hand side of this equality is negative and right-hand side positive, a contradiction emerges. If no such contradiction occurs, *i.e.*, if no $\mathbf{x}$ satisfying (I) exists, then (II) has a solution.

Looking to the geometry of Farkas' theorem 6.7, either: (a) there exists a vector $\mathbf{x}$ that makes an obtuse angle ($\geq\pi/2$) with the rows $\boldsymbol{\alpha}_i, i=1,\ldots,m$, of $\mathbf{A}$ and a strictly acute angle ($<\pi/2$) with $\mathbf{b}$ (Figure 6.4.a); or (b) $\mathbf{b}$ is expressible as a nonnegative linear combination of the columns $\boldsymbol{\alpha}_i', i=1,\ldots,m$, of $\mathbf{A}'$, *i.e.*, $\mathbf{b}$ is an element of the polyhedral convex cone $\mathscr{C}$ spanned by the columns of $\mathbf{A}'$ (Figure 6.4.b). Stated in another fashion, if (II) has no solution (*i.e.*, $\mathbf{b}\notin\mathscr{C}$), then according to (I″), there exists

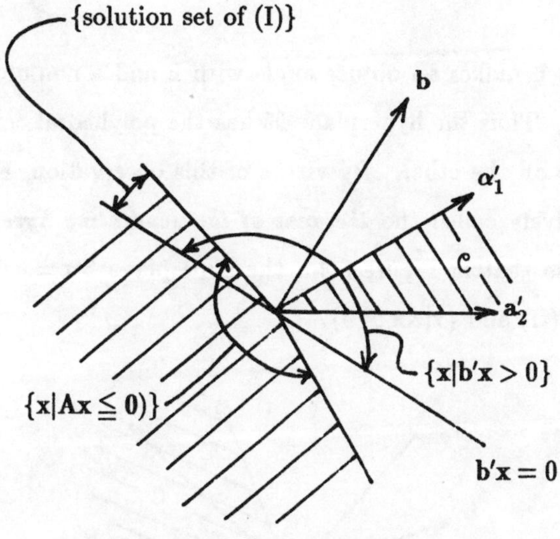

$\{\mathbf{x}|\mathbf{Ax} \leqq \mathbf{0}\} \cap \{\mathbf{x}|\mathbf{b}'\mathbf{x} > 0\} \neq \phi$, *i.e.*,
(I) has a solution; (II) does not.

a.

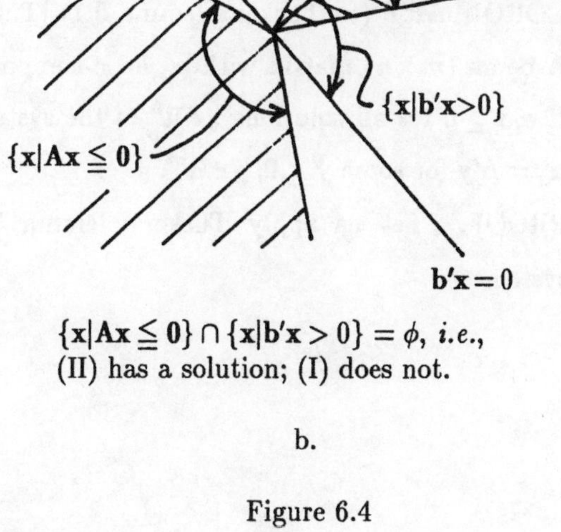

$\{\mathbf{x}|\mathbf{Ax} \leqq \mathbf{0}\} \cap \{\mathbf{x}|\mathbf{b}'\mathbf{x} > 0\} = \phi$, *i.e.*,
(II) has a solution; (I) does not.

b.

Figure 6.4

a vector $\mathbf{x}$ which makes an obtuse angle with $\mathbf{b}$ and a non-obtuse angle with the rows of $\mathbf{A}$. Thus the hyperplane $\mathcal{H}$ has the polyhedral convex cone $\mathcal{C}$ on one side and $\mathbf{b}$ on the other. By virtue of this observation, Farkas' theorem 6.7 is alternatively called the ***theorem of the separating hyperplane*** (Figure 6.5). Note also that in Figure 6.4.b the hyperplane $\mathbf{b}'\mathbf{x} = 0$ "separates" the solution set of (II) and $\{\mathbf{x}|\mathbf{A}\mathbf{x} \leqq \mathbf{0}\}$.

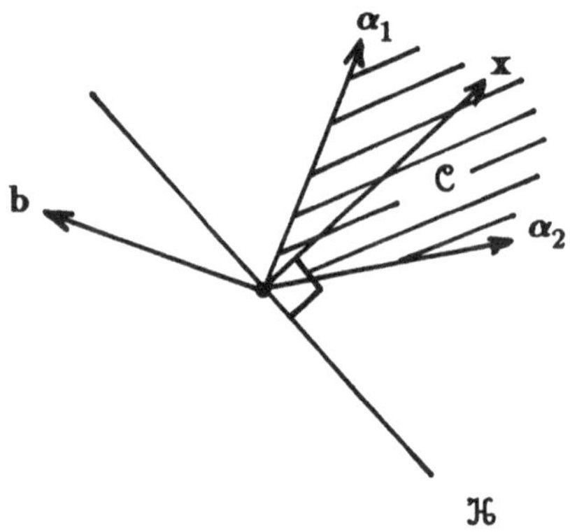

Figure 6.5

It is interesting to note that Farkas' theorem 6.7 can be viewed as a direct consequence of Tucker's lemma 5.1. To see this we state the following

6.8. COROLLARY (to Tucker's lemma 5.1) [Tucker, 1956]. Let $\mathbf{A}$ be an $(m \times n)$ matrix with $\boldsymbol{\alpha}_0$ an n-component row vector. If $\boldsymbol{\alpha}_0\mathbf{x} \geq 0$ for all solutions $\mathbf{x} \in \mathbf{R}^n$ of the system $\mathbf{A}\mathbf{x} \geqq \mathbf{0}$, then $\boldsymbol{\alpha}_0' = \mathbf{A}'\mathbf{y}$ for some $\mathbf{y} \geqq \mathbf{0}$, $\mathbf{y} \in \mathbf{R}^m$.

PROOF. Let us apply Tucker's lemma 5.1 to the dual systems

$$\text{(I)} \quad \begin{bmatrix} -\boldsymbol{\alpha}_0 \\ \mathbf{A} \end{bmatrix} \mathbf{x} \geqq \mathbf{0} \quad \text{and}$$

$$\text{(II)} \quad (-\boldsymbol{\alpha}_0', \mathbf{A}') \begin{bmatrix} y_0 \\ \mathbf{y} \end{bmatrix} = \mathbf{0}, \begin{bmatrix} y_0 \\ \mathbf{y} \end{bmatrix} \geqq \mathbf{0}$$

where $-\boldsymbol{\alpha}_0$, y_0 assume the leading positions held by $\boldsymbol{\alpha}_1, y_1$ in the original statement of the lemma. Hence there exist solutions $\mathbf{x}$ and y_0, $\mathbf{y}_0$ such that

$$\begin{aligned} -\boldsymbol{\alpha}_0\,\mathbf{x} \geq 0,\ \mathbf{A}\mathbf{x} \geqq \mathbf{0} \quad & \text{and} \\ -\boldsymbol{\alpha}_0'\,y_0 + \mathbf{A}'\mathbf{y}_0 = \mathbf{0},\ y_0 \geq 0,\ \mathbf{y}_0 \geqq \mathbf{0} & \end{aligned} \tag{6.1}$$

satisfying

$$-\boldsymbol{\alpha}_o\,\mathbf{x} + y_0 > 0. \tag{6.2}$$

By hypothesis, $\boldsymbol{\alpha}_0\mathbf{x} \geq 0$ since $\mathbf{A}\mathbf{x} \geqq \mathbf{0}$. Hence $y_0 > \boldsymbol{\alpha}_0\mathbf{x} \geq 0$ (from (6.2)) and thus, from (6.1), $-\boldsymbol{\alpha}_0' + \left(\frac{1}{y_0}\right)\mathbf{A}'\mathbf{y}_0 = \mathbf{0}$ or $\boldsymbol{\alpha}_0' = \mathbf{A}'\mathbf{y}$, where $\mathbf{y} = \left(\frac{1}{y_0}\right)\mathbf{y}_0$. Q.E.D.

This corollary tends to reinforce the notion that Farkas' theorem 6.7 addresses the existence of a nonnegative solution to a nonhomogeneous system of linear equations, *i.e.*, a necessary and sufficient condition for $\boldsymbol{\alpha}_0' = \mathbf{A}'\mathbf{y}$ to have a solution $\mathbf{y} \geqq \mathbf{0}$ is that $\boldsymbol{\alpha}_0\,\mathbf{x} \geq 0$ for all $\mathbf{x}$ satisfying $\mathbf{A}\mathbf{x} \geqq \mathbf{0}$.

6.9. FARKAS' THEOREM OF THE ALTERNATIVE FOR NONHOMOGENEOUS SYSTEMS [Duffin, 1956].

Given any $(m \times n)$ matrix $\mathbf{A}$, vectors $\mathbf{b} \epsilon \mathbf{R}^n$ and $\mathbf{C} \epsilon \mathbf{R}^n$, and a scalar $\beta \epsilon \mathbf{R}$, either

(I) $\mathbf{b}'\mathbf{x} > \beta$, $\mathbf{A}\mathbf{x} \leqq \mathbf{C}$ has a solution $\mathbf{x} \epsilon \mathbf{R}^n$ or

(II) $\mathbf{A}'\mathbf{y} = \mathbf{b}$, $\mathbf{C}'\mathbf{y} \leq \beta$, $\mathbf{y} \geqq \mathbf{0}$ or

$\mathbf{A}'\mathbf{y} = \mathbf{0}$, $\mathbf{C}'\mathbf{y} < 0$, $\mathbf{y} \geqq \mathbf{0}$ has a solution $\mathbf{y} \epsilon \mathbf{R}^m$,

but never both.

PROOF. Let us rewrite (I) in the equivalent form

$$\text{(I)}\qquad \left.\begin{aligned} \mathbf{b}'\mathbf{x}-\beta\eta &> 0 \\ \eta &> 0 \\ -\mathbf{A}\mathbf{x}+\mathbf{C}\eta &\geqq \mathbf{0} \end{aligned}\right\}\text{ or }\left\{\begin{aligned} &\begin{bmatrix} \mathbf{b}' & -\beta \\ \mathbf{0}' & 1 \end{bmatrix}\begin{bmatrix} \mathbf{x} \\ \eta \end{bmatrix} > \mathbf{0} \\ &(-\mathbf{A},\mathbf{C})\begin{bmatrix} \mathbf{x} \\ \eta \end{bmatrix} \geqq \mathbf{0} \end{aligned}\right.$$

has a solution $\mathbf{x}\in\mathbf{R}^n$, $\eta\in\mathbf{R}$. Then, by Motzkin's theorem 6.2 (with $\mathbf{D}$ vacuous), either (I') holds or

$$\text{(II')}\qquad \left.\begin{aligned} &\begin{bmatrix} \mathbf{b} & \mathbf{0} \\ -\beta & 1 \end{bmatrix}\begin{bmatrix} y_1 \\ y_2 \end{bmatrix}+\begin{bmatrix} -\mathbf{A}' \\ \mathbf{C}' \end{bmatrix}\mathbf{y}_3=\mathbf{0} \\ &\begin{bmatrix} y_1 \\ y_2 \end{bmatrix} \geq \mathbf{0},\ \mathbf{y}_3 \geqq \mathbf{0} \end{aligned}\right\}\text{ or }\left\{\begin{aligned} &\mathbf{b}y_1 \qquad -\mathbf{A}'\mathbf{y}_3 = \mathbf{0} \\ &-\beta y_1+y_2+\mathbf{C}'\mathbf{y}_3 = 0 \\ &y_1 \geq 0,\ y_2 \geq 0,\ \mathbf{y}_3 \geqq \mathbf{0} \end{aligned}\right.$$

has a solution $y_1, y_2\in\mathbf{R}$ and $\mathbf{y}_3\in\mathbf{R}^m$, but never both. In this regard, either $y_1 > 0$ or $y_1 = 0$ (in which case $y_2 > 0$). If $y_1 > 0$, let $\mathbf{y} = \left(\frac{1}{y_1}\right)\mathbf{y}_3$; if $y_1 = 0$, let $\mathbf{y} = \mathbf{y}_3$. These two cases taken separately reveal that (II') is equivalent to

$$\text{(II'')}\qquad \begin{aligned} &\mathbf{b}-\mathbf{A}'\mathbf{y}=\mathbf{0},\ -\beta+\mathbf{C}'\mathbf{y} = -\frac{y_2}{y_1} \leq 0,\ \mathbf{y} \geqq \mathbf{0}, \quad \text{or} \\ &-\mathbf{A}'\mathbf{y}=\mathbf{0},\ \mathbf{C}'\mathbf{y} = -y_2 < 0,\ \mathbf{y} \geqq \mathbf{0} \end{aligned}$$

which in turn is simply (II). Q.E.D.

Farkas' theorem 6.7 of the alternative for nonnegative solutions of nonhomogeneous linear equations, in which we have a necessary and sufficient condition for a vector to be expressed as a "nonnegative linear combination" of a set of vectors, will now be extended to the case where a given vector is expressible as a "convex combination" of a set of vectors.

To this end we have

6.10. FARKAS' THEOREM OF THE ALTERNATIVE UNDER CONVEX COMBINATION [Panik, 1990]. For an $(m \times n)$ matrix $\mathbf{A}$ and a vector $\mathbf{b} \in R^m$, either

(I) $\mathbf{A}\mathbf{x} = \mathbf{b},\ \mathbf{x} \geqq \mathbf{0},\ \mathbf{1}'\mathbf{x} = 1$ has a solution $\mathbf{x} \in R^n$ or

(II) $\mathbf{A}'\mathbf{y} + \mathbf{1}y \geqq \mathbf{0},\ (\mathbf{b}'\mathbf{y}, y) \leq \mathbf{0}'$ has a solution $y \in R$, $\mathbf{y} \in R^m$,

but never both.

PROOF. System (I) is equivalent to

(I′) $\eta_1 > 0,\ \eta_2 > 0,\ \mathbf{x} \geqq \mathbf{0},\ \mathbf{b}\eta_1 - \mathbf{A}\mathbf{x} = \mathbf{0},\ \eta_2 - \mathbf{1}'\mathbf{x} = 0$ or

$$\begin{bmatrix} \mathbf{0}' & 1 & 0 \\ \mathbf{0}' & 0 & 1 \end{bmatrix}\begin{bmatrix} \mathbf{x} \\ \eta_1 \\ \eta_2 \end{bmatrix} > \mathbf{0},\ (\mathbf{I}_n, \mathbf{0}, \mathbf{0})\begin{bmatrix} \mathbf{x} \\ \eta_1 \\ \eta_2 \end{bmatrix} \geqq \mathbf{0},\ \begin{bmatrix} -\mathbf{A} & \mathbf{b} & \mathbf{0} \\ -\mathbf{1}' & 0 & 1 \end{bmatrix}\begin{bmatrix} \mathbf{x} \\ \eta_1 \\ \eta_2 \end{bmatrix} = \mathbf{0}$$

has a solution $\eta_1, \eta_2 \in R$ and $\mathbf{x} \in R^n$. By Motzkin's theorem 6.2, either (I′) has a solution or

$$\text{(II}'\text{)} \quad \begin{bmatrix} \mathbf{0} & \mathbf{0} \\ 1 & 0 \\ 0 & 1 \end{bmatrix}\mathbf{y}_1 + \begin{bmatrix} \mathbf{I}_n \\ \mathbf{0}' \\ \mathbf{0}' \end{bmatrix}\mathbf{y}_3 + \begin{bmatrix} -\mathbf{A}' & -\mathbf{1} \\ \mathbf{b}' & 0 \\ \mathbf{0}' & 1 \end{bmatrix}\mathbf{y}_4 = \mathbf{0}$$

$$\mathbf{y}_1 \geq \mathbf{0},\ \mathbf{y}_3 \geqq \mathbf{0},\ \mathbf{y}_4 \text{ unrestricted}$$

has a solution $\mathbf{y}_1 \in R^2$, $\mathbf{y}_3 \in R^n$, and $\mathbf{y}_4 \in R^{m+1}$, but never both.

For $\mathbf{y}_1' = (y_{11}, y_{21})$ and $\mathbf{y}_4' = (\mathbf{y}', y)$, where $\mathbf{y} \in R^m$, (II′) becomes

$$\text{(II}''\text{)} \quad \begin{aligned} \mathbf{y}_3 - \mathbf{A}'\mathbf{y} - \mathbf{1}y &= \mathbf{0} \\ y_{11} \qquad + \mathbf{b}'\mathbf{y} &= 0 \\ y_{21} \qquad\qquad + y &= 0. \end{aligned}$$

Since $\mathbf{y}_1 \geq \mathbf{0}$ (*i.e.*, $\mathbf{y}_1 \geqq \mathbf{0}$ and $\mathbf{y}_1 \neq \mathbf{0}$), it follows that $y = -y_{21} \leq 0$, $\mathbf{b}'\mathbf{y} = -y_{11} \leq 0$ (where y_{11}, y_{21} cannot both equal zero), and $\mathbf{A}'\mathbf{y} + \mathbf{1}y = \mathbf{y}_3 \geqq \mathbf{0}$. Hence, (II″) is equivalent to (II). Q.E.D.

Note in system (II) that: (a) if $y < 0$, then

$$\mathbf{A}'\mathbf{y} \geqq -\mathbf{1}y > \mathbf{0},\ \ \mathbf{b}'\mathbf{y} \leq 0;$$

(b) if $y = 0$, then

$$\mathbf{A}'\mathbf{y} \geqq \mathbf{0},\ \ \mathbf{b}'\mathbf{y} < 0.$$

Clearly this latter set of homogeneous linear inequalities constitutes a system (II) for Farkas' theorem proper in the absence of the system (I) restriction $\mathbf{1}'\mathbf{x} = 1$.

Geometrically, the above theorem asserts that: either (a) **b** lies within the convex polyhedron generated by **A** (*i.e.,* the convex hull of the vectors in **A**); or (b) there exists a vector **y** that lies within the intersection of the polytope formed from $\mathbf{A}'\mathbf{y} \geqq -\mathbf{1}y$ and the closed half-space $\mathbf{b}'\mathbf{y} \leq 0$.

Two variations of theorem 6.10, one which admits a semi-positive solution and the other which requires that the variables be strictly positive, will now be offered. First,

6.10.1. FARKAS' THEOREM UNDER CONVEX COMBINATION AND SEMIPOSITIVE SOLUTION. For an $(m \times n)$ matrix **A** and a vector $\mathbf{b} \in \mathbf{R}^m$, either

(I) $\mathbf{A}\mathbf{x} = \mathbf{b},\ \mathbf{x} \geq \mathbf{0},\ \mathbf{1}'\mathbf{x} = 1$ has a solution $\mathbf{x} \in \mathbf{R}^n$ or

$$\text{(II)} \begin{cases} \text{(i)}\ \ \mathbf{A}'\mathbf{y} + \mathbf{1}y \geqq \mathbf{0},\ (\mathbf{b}'\mathbf{y}, y) \leq \mathbf{0}' \text{ or} \\ \text{(ii)}\ \mathbf{A}'\mathbf{y} + \mathbf{1}y > \mathbf{0},\ (\mathbf{b}'\mathbf{y}, y) \leqq \mathbf{0}' \end{cases}$$

has a solution $y \in \mathbf{R}$, $\mathbf{y} \in \mathbf{R}^m$,

but never both.

PROOF. System (I) is equivalent to

$$\text{(I}'\text{)} \quad \eta_1 > 0,\ \eta_2 > 0,\ \mathbf{x} \geq \mathbf{0},\ \mathbf{b}\eta_1 - \mathbf{A}\mathbf{x} = \mathbf{0},\ \eta_2 - \mathbf{1}'\mathbf{x} = 0 \quad \text{or}$$

$$\begin{bmatrix} \mathbf{0}' & 1 & 0 \\ \mathbf{0}' & 0 & 1 \end{bmatrix} \begin{bmatrix} \mathbf{x} \\ \eta_1 \\ \eta_2 \end{bmatrix} > \mathbf{0},\ (\mathbf{I}_n, \mathbf{0}, \mathbf{0}) \begin{bmatrix} \mathbf{x} \\ \eta_1 \\ \eta_2 \end{bmatrix} \geq \mathbf{0},\ \begin{bmatrix} -\mathbf{A} & \mathbf{b} & \mathbf{0} \\ -\mathbf{1}' & 0 & 1 \end{bmatrix} \begin{bmatrix} \mathbf{x} \\ \eta_1 \\ \eta_2 \end{bmatrix} = \mathbf{0}$$

has a solution $\eta_1, \eta_2 \in \mathbf{R}$ and $\mathbf{x} \in \mathbf{R}^n$. By Slater's theorem 6.1, either (I′) has a solution or

$$
\text{(II}'\text{)}\quad
\begin{array}{ll}
\text{(a)} & \begin{bmatrix} 0 & 0 \\ 1 & 0 \\ 0 & 1 \end{bmatrix} \mathbf{y}_1 + \begin{bmatrix} \mathbf{I}_n \\ \mathbf{0}' \\ \mathbf{0}' \end{bmatrix} \mathbf{y}_2 + \begin{bmatrix} -\mathbf{A}' & -\mathbf{1} \\ \mathbf{b}' & 0 \\ \mathbf{0}' & 1 \end{bmatrix} \mathbf{y}_3 = \mathbf{0} \\
\text{(b)} & \mathbf{y}_1 \geq \mathbf{0},\ \mathbf{y}_2 \geqq \mathbf{0},\ \mathbf{y}_3 \text{ unrestricted} \quad \text{or} \\
\text{(c)} & \mathbf{y}_1 \geqq \mathbf{0},\ \mathbf{y}_2 > \mathbf{0},\ \mathbf{y}_3 \text{ unrestricted}
\end{array}
$$

has a solution $\mathbf{y}_1 \in \mathbf{R}^2$, $\mathbf{y}_2 \in \mathbf{R}^2$, and $\mathbf{y}_3 \in \mathbf{R}^{m+1}$, but never both.

For $\mathbf{y}_1' = (y_{11}, y_{21})$ and $\mathbf{y}_3' = (\mathbf{y}', y)$, where $\mathbf{y} \in \mathbf{R}^m$ and $y \in \mathbf{R}$, (II′. a,b) yield (II. i) (given that y_{11}, y_{21} cannot both be zero) while (II. ii) follows from (II′. a,c). Q.E.D.

We next have

6.10.2. FARKAS' THEOREM UNDER CONVEX COMBINATION AND STRICTLY POSITIVE SOLUTION. For an $(m \times n)$ matrix $\mathbf{A}$ and a vector $\mathbf{b} \in \mathbf{R}^m$, either

(I) $\mathbf{A}\mathbf{x} = \mathbf{b},\ \mathbf{x} > \mathbf{0},\ \mathbf{1}'\mathbf{x} = 1$ has a solution $\mathbf{x} \in \mathbf{R}^n$ or

$$
\text{(II)} \begin{cases} \text{(i)} & \mathbf{A}'\mathbf{y} + \mathbf{1}y \geq \mathbf{0},\ (\mathbf{b}'\mathbf{y}, y) \leqq \mathbf{0}' \quad \text{or} \\ \text{(ii)} & \mathbf{A}'\mathbf{y} + \mathbf{1}y \geqq \mathbf{0},\ (\mathbf{b}'\mathbf{y}, y) \leq \mathbf{0}' \end{cases}
$$

has a solution $y \in \mathbf{R}$, $\mathbf{y} \in \mathbf{R}^m$,

but never both.

PROOF. System (I) is equivalent to

$$
\text{(I}'\text{)}\quad \eta_1 > 0, \eta_2 > 0, \mathbf{x} > \mathbf{0}, \mathbf{b}\eta_1 - \mathbf{A}\mathbf{X} = \mathbf{0}, \eta_2 - \mathbf{1}'\mathbf{x} = 0 \quad \text{or}
$$

$$
\begin{bmatrix} \mathbf{I}_n & \mathbf{0} & \mathbf{0} \\ \mathbf{0}' & 1 & 0 \\ \mathbf{0}' & 0 & 1 \end{bmatrix} \begin{bmatrix} \mathbf{x} \\ \eta_1 \\ \eta_2 \end{bmatrix} > \mathbf{0}, \quad \begin{bmatrix} -\mathbf{A} & \mathbf{b} & \mathbf{0} \\ -\mathbf{1}' & 0 & 1 \end{bmatrix} \begin{bmatrix} \mathbf{x} \\ \eta_1 \\ \eta_2 \end{bmatrix} = \mathbf{0}
$$

has a solution $\eta_1, \eta_2 \in \mathbf{R}$ and $\mathbf{x} \in \mathbf{R}^n$. By Motzkin's theorem 6.2, either (I′) has a solution or

$$
\text{(II}') \quad \begin{bmatrix} \mathbf{I}_n & \mathbf{0} & \mathbf{0} \\ \mathbf{0}' & 1 & 0 \\ \mathbf{0}' & 0 & 1 \end{bmatrix} \mathbf{y}_1 + \begin{bmatrix} -\mathbf{A}' & -\mathbf{1} \\ \mathbf{b}' & 0 \\ \mathbf{0}' & 1 \end{bmatrix} \mathbf{y}_2 = \mathbf{0}
$$

$$\mathbf{y}_1 \geq \mathbf{0}, \ \mathbf{y}_2 \text{ unrestricted}$$

has a solution $\mathbf{y}_1 \in \mathbf{R}^{n+2}$, $\mathbf{y}_2 \in \mathbf{R}^{m+1}$, but never both.

Let $\mathbf{y}_1' = (\bar{\mathbf{y}}', y_{11}, y_{21})$, where $\bar{\mathbf{y}} \in \mathbf{R}^n$ and $y_{11}, y_{21} \in \mathbf{R}$, and $\mathbf{y}_2' = (\mathbf{y}', y)$, where $\mathbf{y} \in \mathbf{R}^m$ and $y \in \mathbf{R}$. If $\bar{\mathbf{y}} \geq \mathbf{0}$, then $y_{11}, y_{21} \geq 0$ and thus (II′) simplifies to (II. i); and if $\bar{\mathbf{y}} \geqq \mathbf{0}$, then $y_{11} > 0, y_{21} \geq 0$ or $y_{11} \geq 0$, $y_{21} > 0$ so that (II′) now becomes (II. ii). Q.E.D.

Clearly theorem 6.10 holds as a special case of theorems 6.10.1, 6.10.2.

6.11. VON NEUMANN'S THEOREM OF THE ALTERNATIVE FOR SEMIPOSITIVE SOLUTIONS OF HOMOGENEOUS INEQUALITIES [von Neumann, 1944]. For any $(m \times n)$ matrix $\mathbf{A}$, either

(I) $\mathbf{A}\mathbf{x} > \mathbf{0}$, $\mathbf{x} \geqq \mathbf{0}$ has a solution $\mathbf{x} \in \mathbf{R}^n$ or

(II) $\mathbf{A}'\mathbf{y} \leqq \mathbf{0}$, $\mathbf{y} \geq \mathbf{0}$ has a solution $\mathbf{y} \in \mathbf{R}^m$,

but never both.

PROOF. Motzkin's theorem 6.2 (with D vacuous) indicates that either (I) has a solution or

$$\text{(II}') \quad \mathbf{A}'\mathbf{y}_1 + \mathbf{y}_3 = \mathbf{0}, \ \mathbf{y}_1 \geq \mathbf{0}, \ \mathbf{y}_3 \geqq \mathbf{0}$$

has a solution $\mathbf{y}_1 \in \mathbf{R}^m$, $\mathbf{y}_3 \in \mathbf{R}^m$, but never both. Let $\mathbf{y} = \mathbf{y}_1$. Then with $\mathbf{y}_3 \geqq \mathbf{0}$, we have $\mathbf{A}'\mathbf{y} = -\mathbf{y}_3 \leqq \mathbf{0}$ so that (II′) is equivalent to (II). Q.E.D.

6.12. GALE'S THEOREM FOR THE SOLVABILITY OF LINEAR EQUALITIES [Gale,1960]. For any $(m\times n)$ matrix $\mathbf{A}$ and a given vector $\mathbf{b}\in\mathbf{R}^m$, either

(I) $\mathbf{Ax}=\mathbf{b}$ has a solution $\mathbf{x}\in\mathbf{R}^n$ or

(II) $\mathbf{A'y}=\mathbf{0}$, $\mathbf{b'y}=1$ has a solution $\mathbf{y}\in\mathbf{R}^m$,

but never both.

PROOF. If (I) is rewritten as

$$(\text{I}')\ \eta>0,\ \ \mathbf{Ax}-\eta\mathbf{b}=\mathbf{0}\ \text{ or }\ \begin{cases}(\mathbf{0}',1)\begin{bmatrix}\mathbf{x}\\ \eta\end{bmatrix}>0\\[2ex] (\mathbf{A},-\mathbf{b})\begin{bmatrix}\mathbf{x}\\ \eta\end{bmatrix}=\mathbf{0}\end{cases}$$

has a solution $\eta\in\mathbf{R}$, $\mathbf{x}\in\mathbf{R}^n$, then by Motzkin's theorem 6.2 (with $\mathbf{C}$ vacuous), either (I$'$) holds or

$$(\text{II}')\ \left.\begin{matrix}\begin{bmatrix}0\\ 1\end{bmatrix}y+\begin{bmatrix}\mathbf{A}'\\ -\mathbf{b}'\end{bmatrix}\mathbf{y}_4=\mathbf{0}\\[2ex] y>0,\ \mathrm{Y}_4\ \text{unrestricted}\end{matrix}\right\}\ \text{or}\ \begin{cases}\mathbf{A'y}_4=\mathbf{0}\\ y-\mathbf{b'y}_4=0\\ y\geq 0\end{cases}$$

has a solution $y\in\mathbf{R}$, $\mathbf{y}_4\in\mathbf{R}^m$, but never both. Since $y>0$, let $\mathbf{y}=\left(\frac{1}{y}\right)\mathbf{y}_4$. Hence (II$'$) is equivalent to (II). Q.E.D.

Following Gale (1960), we note that this theorem provides us with a positive criterion for determining when a nonhomogeneous system of linear equations has ***no solution***. Specifically, to demonstrate that (I) does not hold, we need only produce a solution to (II), *i.e.*, if $\mathbf{x}$ satisfies (I), then given $\mathbf{y}$, $\mathbf{x}$ must also satisfy $\mathbf{x'A'y}=\mathbf{b'y}$. But if $\mathbf{y}$ satisfies (II), then the preceding equality yields a contradiction so that (I) has no solution.

In terms of the geometry of theorem 6.12, either: (a) $\mathbf{b}$ is expressible as a linear combination of the columns of $\mathbf{A}$; or (b) there exists a vector $\mathbf{y}$ which is orthogonal to the rows of $\mathbf{A}'$ and making an acute angle $(<\pi/2)$ with $\mathbf{b}$.

6.13. GALE'S THEOREM FOR THE SOLVABILITY OF LINEAR INEQUALITIES [Gale, 1960]. For any $m \times n)$ matrix $\mathbf{A}$ and a given vector $\mathbf{b} \in \mathbf{R}^m$, either

(I) $\mathbf{Ax} \leqq \mathbf{b}$ has a solution $\mathbf{x} \in \mathbf{R}^n$, or

(II) $\mathbf{A}'\mathbf{y} = \mathbf{0}$, $\mathbf{b}'\mathbf{y} = -1$, $\mathbf{y} \geqq \mathbf{0}$ has a solution $\mathbf{y} \in \mathbf{R}^m$,

but never both.

PROOF. Since (I) is equivalent to

$$(\mathrm{I}') \quad \eta > 0, \ \ \eta\mathbf{b} - \mathbf{Ax} \geqq \mathbf{0} \quad \text{or} \quad \begin{cases} (\mathbf{0}', 1)\begin{bmatrix} \mathbf{x} \\ \eta \end{bmatrix} > 0 \\ (-\mathbf{A}, \mathbf{b})\begin{bmatrix} \mathbf{x} \\ \eta \end{bmatrix} \geqq \mathbf{0} \end{cases}$$

has a solution $\eta \in \mathbf{R}$, $\mathbf{x} \in \mathbf{R}^n$, Motzkin's theorem 6.2 (with $\mathbf{D}$ vacuous) indicates that either (I') has a solution or

$$(\mathrm{II}') \quad \left.\begin{matrix} \begin{bmatrix} \mathbf{0} \\ 1 \end{bmatrix} y + \begin{bmatrix} -\mathbf{A}' \\ \mathbf{b}' \end{bmatrix} \mathbf{y}_3 = \mathbf{0} \\ y > 0, \ \mathbf{y}_3 \geqq \mathbf{0} \end{matrix}\right\} \quad \text{or} \quad \begin{cases} -\mathbf{A}'\mathbf{y}_3 = \mathbf{0} \\ y + \mathbf{b}'\mathbf{y}_3 = 0 \\ y \geq 0, \ \mathbf{y}_3 \geqq \mathbf{0} \end{cases}$$

has a solution $y \in \mathbf{R}$, $\mathbf{y}_3 \in \mathbf{R}^m$, but never both. With $y > 0$, let $\mathbf{y} = \left(\frac{1}{y}\right)\mathbf{y}_3$. Hence (II') is equivalent to (II). Q.E.D.

Note that if in system (I) of the preceding theorem $\mathbf{Ax} \geqq \mathbf{b}$ appears, then in (II) we must have $\mathbf{b}'\mathbf{y} = 1$.

In terms of the geometry of this theorem, either: (a) $\mathbf{x}$ lies within the intersection of the m closed half-spaces $\boldsymbol{\alpha}_i\mathbf{x} = \mathrm{b}_i, i=1,\ldots,m$, where $\boldsymbol{\alpha}_i$ is the ith row of $\mathbf{A}$; or (b) there exists a vector $\mathbf{y}$ which facilitates the expression of the null vector as a linear combination of the columns of $\mathbf{A}'$ and which makes a strictly obtuse angle $(>\pi/2)$ with $\mathbf{b}$.

6.14. **GALE'S THEOREM FOR NONNEGATIVE SOLUTIONS OF LINEAR INEQUALITIES [Gale, 1960]. For any $(m \times n)$ matrix A and a given vector $\mathbf{b} \in \mathbf{R}^m$, either**

(I) $\mathbf{Ax} \leqq \mathbf{b}, \mathbf{x} \geqq \mathbf{0}$ has a solution $\mathbf{x} \in \mathbf{R}^n$ or

(II) $\mathbf{A'y} \geqq \mathbf{0}, \mathbf{b'y} < 0, \mathbf{y} \geqq \mathbf{0}$ has a solution $\mathbf{y} \in \mathbf{R}^m$,

but never both.

PROOF. System (I) is equivalent to

$$(\text{I}') \quad \eta > 0,\ \eta\mathbf{b} - \mathbf{Ax} \geqq \mathbf{0},\ \mathbf{x} \geqq \mathbf{0} \quad \text{or} \quad \begin{cases} (\mathbf{0}', 1)\begin{bmatrix}\mathbf{x}\\ \eta\end{bmatrix} > 0 \\ \begin{bmatrix}-\mathbf{A} & \mathbf{b}\\ \mathbf{I}_n & \mathbf{0}\end{bmatrix}\begin{bmatrix}\mathbf{x}\\ \eta\end{bmatrix} \geqq \mathbf{0}\end{cases}$$

has a solution $\eta \in \mathbf{R}$, $\mathbf{x} \in \mathbf{R}^n$. Then by Motzkin's theorem 6.2 (with D vacuous), either (I′) holds or

(II′)

$$\left.\begin{matrix}\begin{bmatrix}0\\1\end{bmatrix} y + \begin{bmatrix}-\mathbf{A}'\mathbf{I}_n\\ \mathbf{b}'\mathbf{0}'\end{bmatrix}\mathbf{y}_3 = \begin{bmatrix}0\\1\end{bmatrix} y + \begin{bmatrix}-\mathbf{A}'\mathbf{I}_n\\ \mathbf{b}'\mathbf{0}'\end{bmatrix}\begin{bmatrix}\mathbf{y}_{13}\\ \mathbf{y}_{23}\end{bmatrix} = \mathbf{0} \\ y > 0,\ \mathbf{y}_3 \geqq \mathbf{0}\end{matrix}\right\} \text{or} \begin{cases} -\mathbf{A}'\mathbf{y}_{13} + \mathbf{y}_{23} = \mathbf{0} \\ y + \mathbf{b}'\mathbf{y}_{13} = 0 \\ y \geqq 0, \mathbf{y}_{13} \geqq \mathbf{0}, \mathbf{y}_{23} \geqq \mathbf{0}\end{cases}$$

has a solution $y \in \mathbf{R}$, $\mathbf{y}_{13} \in \mathbf{R}^m$, and $\mathbf{y}_{23} \in \mathbf{R}^n$, but never both. With $y > 0$, $\mathbf{y} = \mathbf{y}_{13} \geqq \mathbf{0}$, and $\mathbf{y}_{23} \geqq \mathbf{0}$, we obtain $\mathbf{A'y} = \mathbf{y}_{23} \geqq \mathbf{0}$ and $\mathbf{b'y} = -y < 0$. Hence (II′) is equivalent to (II). Q.E.D.

6.15. **GALE'S THEOREM FOR SEMIPOSITIVE SOLUTIONS OF HOMOGENEOUS INEQUALITIES [Gale, 1960]. For any $(m \times n)$ matrix A, either**

(I) $\mathbf{Ax} \leqq \mathbf{0}, \mathbf{x} \geq \mathbf{0}$ has a solution $\mathbf{x} \in \mathbf{R}^n$ or

(II) $\mathbf{A'y} > \mathbf{0}, \mathbf{y} \geqq \mathbf{0}$ has a solution $\mathbf{y} \in \mathbf{R}^m$,

but never both.

PROOF. Rewriting the inequalities in (I) as

$$(\text{I}') \qquad -\mathbf{Ax} \geqq \mathbf{0},\ \mathbf{x} \geq \mathbf{0}$$

we find, from Tucker's theorem 6.3 (with **D** vacuous), that either (I′) holds or

$$\text{(II}')\quad \mathbf{y}_2 - \mathbf{A}'\mathbf{y}_3 = \mathbf{0},\ \mathbf{y}_2 > \mathbf{0},\ \mathbf{y}_3 \geqq \mathbf{0}$$

has a solution $\mathbf{y}_2 \in \mathbf{R}^n$, $\mathbf{y}_3 \in \mathbf{R}^m$, but never both. Set $\mathbf{y} = \mathbf{y}_3 \geqq \mathbf{0}$. Then $\mathbf{A}'\mathbf{y} = \mathbf{y}_2 > \mathbf{0}$ so that (II′) is equivalent to (II). Q.E.D.

6.16. GALE'S THEOREM FOR NONNEGATIVE SOLUTIONS OF HOMOGENEOUS INEQUALITIES [Gale, 1960]. For any $(m \times n)$ matrix **A**, either

(I) $\mathbf{Ax} \leq \mathbf{0},\ \mathbf{x} \geqq \mathbf{0}$ has a solution $\mathbf{x} \in \mathbf{R}^n$ or

(II) $\mathbf{A}'\mathbf{y} \geqq \mathbf{0},\ \mathbf{y} > \mathbf{0}$ has a solution $\mathbf{y} \in \mathbf{R}^m$,

but never both.

PROOF. Since (I) is equivalent to

$$\text{(I}')\quad -\mathbf{AX} \geq \mathbf{0},\ \mathbf{x} \geqq \mathbf{0},$$

Tucker's theorem 6.3 (with **D** vacuous) informs us that either (I′) has a solution or

$$\text{(II}')\quad -\mathbf{A}'\mathbf{y}_2 + \mathbf{y}_3 = \mathbf{0},\ \mathbf{y}_2 > \mathbf{0},\ \mathbf{y}_3 \geqq \mathbf{0}$$

has a solution $\mathbf{y}_2 \in \mathbf{R}^m$, $\mathbf{y}_3 \in \mathbf{R}^n$, but never both. Let $\mathbf{y} = \mathbf{y}_2 > \mathbf{0}$. Then $\mathbf{y}_3 = \mathbf{A}'\mathbf{y} \geqq \mathbf{0}$ and thus (II′) is equivalent to (II). Q.E.D.

6.17 MANGASARIAN'S THEOREM FOR THE SOLVABILITY OF LINEAR INEQUALITIES [Mangasarian, 1969]. For any $(m \times n)$ matrix **A** and a given vector $\mathbf{b} \in \mathbf{R}^m$, either

(I) $\mathbf{Ax} \leq \mathbf{b}$ has a solution $\mathbf{x} \in \mathbf{R}^n$ or

(II) $\mathbf{A}'\mathbf{y} = \mathbf{0},\ \mathbf{b}'\mathbf{y} = -1,\ \mathbf{y} \geqq \mathbf{0}$ or

$\mathbf{A}'\mathbf{y} = \mathbf{0},\ \mathbf{b}'\mathbf{y} \leq 0,\ \mathbf{y} > \mathbf{0}$ has a solution $\mathbf{y} \in \mathbf{R}^m$,

but never both

PROOF. If (I) is rewritten as

$$\text{(I}'\text{)} \qquad \eta > 0, \eta \mathbf{b} - \mathbf{AX} \geq \mathbf{0} \ \text{ or } \begin{cases} (\mathbf{0}', 1)\begin{bmatrix} \mathbf{x} \\ \eta \end{bmatrix} > 0 \\ (-\mathbf{A}, \mathbf{b})\begin{bmatrix} \mathbf{x} \\ \eta \end{bmatrix} \geqq \mathbf{0} \end{cases}$$

has a solution $\eta \in R$, $\mathbf{x} \in R^n$, then by Slater's theorem 6.1 (with C, D vacuous), either (I′) holds or

$$\begin{bmatrix} 0 \\ 1 \end{bmatrix} y + \begin{bmatrix} -\mathbf{A}' \\ \mathbf{b}' \end{bmatrix} \mathbf{y}_2 = \mathbf{0} \ \text{ with } \qquad \begin{matrix} -\mathbf{A}'\mathbf{y}_2 = \mathbf{0} \end{matrix}$$

$$\text{(II}'\text{)} \quad \left.\begin{matrix} y > 0, \ \mathbf{y}_2 \geqq \mathbf{0} \quad \text{or} \\ y \geq 0, \ \mathbf{y}_2 > 0 \end{matrix}\right\} \text{ or } \begin{cases} y + \mathbf{b}'\mathbf{y}_2 = 0 \\ y \geq 0, \mathbf{y}_2 \geqq \mathbf{0} \ \text{ or} \\ y \geqq 0, \mathbf{y}_2 > 0 \end{cases}$$

has a solution $y \in R$, $\mathbf{y}_2 \in R^m$, but never both. For $y > 0$ and $\mathbf{y}_2 \geqq \mathbf{0}$, let $\mathbf{y} = \left(\frac{1}{y}\right)\mathbf{y}_2$; for $y \geq 0$ and $\mathbf{y}_2 > \mathbf{0}$, let $\mathbf{y} = \mathbf{y}_2$. In either case (II′) is equivalent to (II). Q.E.D.

For convenience, the entire collection of theorems of the alternative considered in this section is summarized in Table 6.1.

Table 6.1 Theorems of the Alternative

	(I)	(II)
1.	(I) $\mathbf{Ax} > \mathbf{0}$, $\mathbf{Bx} \geq \mathbf{0}$, $\mathbf{Cx} \geqq \mathbf{0}$, $\mathbf{Dx}=\mathbf{0}$ (A, B nonvacuous) [Slater, 1951]	(II) $\mathbf{A}'\mathbf{y}_1+\mathbf{B}'\mathbf{y}_2+\mathbf{C}'\mathbf{y}_3+\mathbf{D}'\mathbf{y}_4=\mathbf{0}$, $\mathbf{y}_1 \geq \mathbf{0}$, $\mathbf{y}_2 \geqq \mathbf{0}$, $\mathbf{y}_3 \geqq \mathbf{0}$, $\mathbf{y}_4$ unrestricted or $\mathbf{y}_1 \geqq \mathbf{0}$, $\mathbf{y}_2 > \mathbf{0}$, $\mathbf{y}_3 \geqq \mathbf{0}$. $\mathbf{y}_4$ unrestricted
2.	(I) $\mathbf{Ax} > \mathbf{0}$, $\mathbf{Cx} \geqq \mathbf{0}$, $\mathbf{Dx}=\mathbf{0}$ (A nonvacuous) [Motzkin, 1936]	(II) $\mathbf{A}'\mathbf{y}_1+\mathbf{C}'\mathbf{y}_3+\mathbf{D}'\mathbf{y}_4=\mathbf{0}$, $\mathbf{y}_1 \geq \mathbf{0}$, $\mathbf{y}_3 \geqq \mathbf{0}$, $\mathbf{y}_4$ unrestricted
3.	(I) $\mathbf{Bx} \geq \mathbf{0}$, $\mathbf{Cx} \geqq \mathbf{0}$, $\mathbf{Dx}=\mathbf{0}$ (B nonvacuous) [Tucker, 1956]	(II) $\mathbf{B}'\mathbf{y}_2+\mathbf{C}'\mathbf{y}_3+\mathbf{D}'\mathbf{y}_4=\mathbf{0}$, $\mathbf{y}_2 > \mathbf{0}$, $\mathbf{y}_3 \geqq \mathbf{0}$, $\mathbf{y}_4$ unrestricted
4.	(I) $\mathbf{Ax} > \mathbf{0}$ [Gordan, 1873]	(II) $\mathbf{A}'\mathbf{y}=\mathbf{0}$, $\mathbf{y} \geq \mathbf{0}$
5.	(I) $\mathbf{Bx} \geq \mathbf{0}$, [Stiemke, 1915]	(II) $\mathbf{B}'\mathbf{y}=\mathbf{0}$, $\mathbf{y} > \mathbf{0}$
6.	(I) $\mathbf{Ax} \geq \mathbf{0}$, $-\mathbf{b}'\mathbf{x} \geq 0$ [Stiemke, 1915]	(II) $\mathbf{A}'\mathbf{y}=\mathbf{b}$, $\mathbf{y} > \mathbf{0}$
7.	(I) $\mathbf{Ax} \leqq \mathbf{0}$, $\mathbf{b}'\mathbf{x} > 0$ [Farkas, 1902]	(II) $\mathbf{A}'\mathbf{y}=\mathbf{b}$, $\mathbf{y} \geqq \mathbf{0}$
8.	(I) $\mathbf{b}'\mathbf{x} > \beta$, $\mathbf{Ax} \leqq \mathbf{C}$ [Duffin, 1956]	(II) $\mathbf{A}'\mathbf{y}=\mathbf{b}$, $\mathbf{C}'\mathbf{y} \leq \beta$, $\mathbf{y} \geqq \mathbf{0}$ or $\mathbf{A}'\mathbf{y}=\mathbf{0}$, $\mathbf{C}'\mathbf{y} < 0$, $\mathbf{y} \geqq \mathbf{0}$
9.	(I) $\mathbf{Ax}=\mathbf{b}$, $\mathbf{x} \geqq \mathbf{0}$, $\mathbf{1}'\mathbf{x}=1$ [Panik, 1990]	(II) $\mathbf{A}'\mathbf{y}+\mathbf{1}y \geqq \mathbf{0}$, $(\mathbf{b}'\mathbf{y}, y) \leq \mathbf{0}'$

Table 6.1 Theorems of the Alternative (continued)

	(I)	(II)
10.	(I) $\mathbf{Ax}=\mathbf{b},\ \mathbf{x} \geq \mathbf{0},\ \mathbf{1}'\mathbf{x}=1$	(II) $\mathbf{A}'\mathbf{y}+\mathbf{1}y \geqq \mathbf{0},\ (\mathbf{b}'\mathbf{y},\ y) \leq \mathbf{0}'$ or $\mathbf{A}'\mathbf{y}+\mathbf{1}y > \mathbf{0},\ (\mathbf{b}'\mathbf{y},y) \leqq \mathbf{0}'$
11.	(I) $\mathbf{Ax}=\mathbf{b},\ \mathbf{x} > \mathbf{0},\ \mathbf{1}'\mathbf{x}=1$	(II) $\mathbf{A}'\mathbf{y}+\mathbf{1}y \geq \mathbf{0},\ (\mathbf{b}'\mathbf{y},\ y) \leqq \mathbf{0}'$ or $\mathbf{A}'\mathbf{y}+\mathbf{1}y \geqq \mathbf{0},\ (\mathbf{b}'\mathbf{y},y) \leq \mathbf{0}'$
12.	(I) $\mathbf{Ax} > \mathbf{0},\ \mathbf{x} \geqq \mathbf{0}$ [von Neumann, 1944]	(II) $\mathbf{A}'\mathbf{y} \leqq \mathbf{0},\ \mathbf{y} \geq \mathbf{0}$
13.	(I) $\mathbf{Ax}=\mathbf{b}$ [Gale, 1960]	(II) $\mathbf{A}'\mathbf{y}=\mathbf{0},\ \mathbf{b}'\mathbf{y}=1$
14.	(I) $\mathbf{Ax} \leqq \mathbf{b}$ [Gale, 1960]	(II) $\mathbf{A}'\mathbf{y}=\mathbf{0},\ \mathbf{b}'\mathbf{y}=-1,\ \mathbf{y} \geqq \mathbf{0}$
15.	(I) $\mathbf{Ax} \leqq \mathbf{b},\ \mathbf{x} \geqq \mathbf{0}$ [Gale, 1960]	(II) $\mathbf{A}'\mathbf{y} \geqq \mathbf{0},\ \mathbf{b}'\mathbf{y} < 0,\ \mathbf{y} \geqq \mathbf{0}$
16.	(I) $\mathbf{Ax} \leqq \mathbf{0},\ \mathbf{x} \geq \mathbf{0}$ [Gale, 1960]	(II) $\mathbf{A}'\mathbf{y} > \mathbf{0},\ \mathbf{y} \geqq \mathbf{0}$
17.	(I) $\mathbf{Ax} \leq \mathbf{0},\ \mathbf{x} \geqq \mathbf{0}$ [Gale, 1960]	(II) $\mathbf{A}'\mathbf{y} \geqq \mathbf{0},\ \mathbf{y} > \mathbf{0}$
18.	(I) $\mathbf{Ax} \leq \mathbf{b}$ [Mangasarian, 1969]	(II) $\mathbf{A}'\mathbf{y}=\mathbf{0},\ \mathbf{b}'\mathbf{y}=-1,\ \mathbf{y} \geqq \mathbf{0}$ or $\mathbf{A}'\mathbf{y}=\mathbf{0},\ \mathbf{b}'\mathbf{y} \leq 0,\ \mathbf{y} > \mathbf{0}$

6.3 Homogeneous Inequalities/Equalities Under Convex Combination

This section endeavors to extend certain theorems of the alternative to the case where the variables in one of the systems of homogeneous inequalities/equalities are subject to the additional requirement that they must satisfy a "convex combination." The structural systems, as well as the variables under consideration, are taken to be either nonnegative, semipositive, or strictly positive. We first have

6.18. THEOREM. For any $(m \times n)$ matrix $\mathbf{A}$, either

(I) $\mathbf{Ax} > \mathbf{0},\ \mathbf{x} \geq \mathbf{0},\ \mathbf{1}'\mathbf{x} = 1$ has a solution $\mathbf{x} \in \mathbf{R}^n$ or

$$
\text{(II)} \begin{cases} \text{(i)} \ \mathbf{A}'\mathbf{y} + \mathbf{1}y \leqq \mathbf{0},\ y \geq 0,\ \mathbf{y} \geq \mathbf{0} \ \text{ or} \\ \text{(ii)} \ \mathbf{A}'\mathbf{y} + \mathbf{1}y \leqq \mathbf{0},\ y > 0,\ \mathbf{y} \geqq \mathbf{0} \end{cases}
$$

has a solution $y \in \mathbf{R}$, $\mathbf{y} \in \mathbf{R}^m$, but not both.

PROOF. System (I) is equivalent to

$$
\eta > 0, \mathbf{Ax} > \mathbf{0},\ \mathbf{x} \geq \mathbf{0},\ \eta - \mathbf{1}'\mathbf{x} = 0 \quad \text{or} \tag{I$'$}
$$

$$
\begin{bmatrix} \mathbf{0}' & 1 \\ \mathbf{A} & \mathbf{0} \end{bmatrix} \begin{bmatrix} \mathbf{x} \\ \eta \end{bmatrix} > \mathbf{0},\ (\mathbf{I}_n, \mathbf{0}) \begin{bmatrix} \mathbf{x} \\ \eta \end{bmatrix} \geq \mathbf{0},\ (-\mathbf{1}', 1) \begin{bmatrix} \mathbf{x} \\ \eta \end{bmatrix} = 0
$$

has a solution $\eta \epsilon \mathbf{R}$, $\mathbf{x} \epsilon \mathbf{R}^n$. By Slater's theorem 6.1, either (I′) has a solution or

$$\text{(a)} \begin{bmatrix} \mathbf{0} & \mathbf{A}' \\ 1 & \mathbf{0}' \end{bmatrix} \mathbf{y}_1 + \begin{bmatrix} \mathbf{I}_n \\ \mathbf{0}' \end{bmatrix} \mathbf{y}_2 + \begin{bmatrix} -\mathbf{1} \\ 1 \end{bmatrix} y_3 = \mathbf{0}$$

$$\text{(b) } \mathbf{y}_1 \geq \mathbf{0},\ \mathbf{y}_2 \geqq \mathbf{0},\ y_3 \text{ unrestricted or} \qquad \text{(II}')$$

$$\text{(c) } \mathbf{y}_1 \geqq \mathbf{0},\ \mathbf{y}_2 > \mathbf{0},\ y_3 \text{ unrestricted}$$

has a solution $\mathbf{y}_1 \epsilon \mathbf{R}^{m+1}$, $\mathbf{y}_2 \epsilon \mathbf{R}^n$, and $y_3 \epsilon \mathbf{R}$, but never both.

For $\mathbf{y}_1' = (y, \mathbf{y}')$, where $y \epsilon \mathbf{R}$ and $\mathbf{y} \epsilon \mathbf{R}^m$, (II′. a, b) yield (II. i, ii) above while (II′. a, c) render

$$\mathbf{A}'\mathbf{y} + \mathbf{1}y < \mathbf{0}, y \geq 0, \mathbf{y} \geq \mathbf{0}. \qquad \text{(II. iii)}.$$

But if (II.iii) holds, then (II.ii) holds for some $y > 0$. Q.E.D.

Note that (II. i, ii) serve as a system (II) for a system (I) of the form $\mathbf{Ax} > \mathbf{0}$, $\mathbf{x} \geqq \mathbf{0}$, $\mathbf{1}'\mathbf{x} = 1$;[1] and (II. ii, iii) serve as a (II) when (I) appears as $\mathbf{Ax} \geqq \mathbf{0}$, $\mathbf{x} \geq \mathbf{0}$, $\mathbf{1}'\mathbf{x} = 1$.[2] Note also that (II. ii) assumes the role of a system (II) when the system (I) alternative is $\mathbf{Ax} \geqq \mathbf{0}$, $\mathbf{x} \geqq \mathbf{0}$, $\mathbf{1}'\mathbf{x} = 1$.

We next have

6.19. THEOREM. For any $(m \times n)$ matrix $\mathbf{A}$, either

(I) $\mathbf{Ax} > \mathbf{0}$, $\mathbf{x} > \mathbf{0}$, $\mathbf{1}'\mathbf{x} = 1$ has a solution $\mathbf{x} \epsilon \mathbf{R}^n$ or

$$\text{(II)} \begin{cases} \text{(i) } \mathbf{A}'\mathbf{y} + \mathbf{1}y \leqq \mathbf{0},\ y \geq 0,\ \mathbf{y} \geq \mathbf{0} \text{ or} \\ \text{(ii) } \mathbf{A}'\mathbf{y} + \mathbf{1}y \leq \mathbf{0},\ y \geq 0,\ \mathbf{y} \geqq \mathbf{0} \text{ or} \\ \text{(iii) } \mathbf{A}'\mathbf{y} + \mathbf{1}y \leqq \mathbf{0},\ y > 0,\ \mathbf{y} \geqq \mathbf{0} \end{cases}$$

has a solution $y \epsilon \mathbf{R}$, $\mathbf{y} \epsilon \mathbf{R}^m$,

but not both.

PROOF. System (I) is equivalent to

[1]This particular system one represents an extension of von Neumann's theorem to the case where the variables satisfy a convex combination.

[2]This system one extends a theorem by Gale (see 6.15 above) to the case where $\mathbf{1}'\mathbf{x} = 1$ holds.

$$\eta > 0, \mathbf{Ax} > \mathbf{0}, \ \mathbf{x} > \mathbf{0}, \ \eta - \mathbf{1}'\mathbf{x} = 0 \quad \text{or} \tag{I'}$$

$$\begin{bmatrix} \mathbf{0}' & 1 \\ \mathbf{A} & \mathbf{0} \\ \mathbf{I}_n & \mathbf{0} \end{bmatrix} \begin{bmatrix} \mathbf{x} \\ \eta \end{bmatrix} > \mathbf{0}, \ (-\mathbf{1}', 1) \begin{bmatrix} \mathbf{x} \\ \eta \end{bmatrix} = 0$$

has a solution $\eta \in \mathbf{R}$, $\mathbf{x} \in \mathbf{R}^n$. From Motzkin's theorem 6.2, either (I') has a solution or

$$\begin{bmatrix} \mathbf{0} & \mathbf{A}' & \mathbf{I}_n \\ 1 & \mathbf{0}' & \mathbf{0}' \end{bmatrix} \mathbf{y}_1 + \begin{bmatrix} -\mathbf{1} \\ 1 \end{bmatrix} y_2 = \mathbf{0} \tag{II'}$$

$$\mathbf{y}_1 \geq \mathbf{0}, \ y_2 \text{ unrestricted}$$

has a solution $\mathbf{y}_1 \in \mathbf{R}^{m+n+1}$, $y_2 \in \mathbf{R}$, but never both.

For $\mathbf{y}_1' = (y, \mathbf{y}', \bar{\mathbf{y}}') \geq \mathbf{0}'$, where $y \in \mathbf{R}$, $\mathbf{y} \in \mathbf{R}^m$, and $\bar{\mathbf{y}} \in \mathbf{R}^n$, let $y \geq 0$. Then from (II'), either $\mathbf{y} \geq \mathbf{0}$, $\bar{\mathbf{y}} \geqq \mathbf{0}$, in which case we obtain (II. i); or $\mathbf{y} \geqq \mathbf{0}$, $\bar{\mathbf{y}} \geq \mathbf{0}$, so that (II. ii) holds. And if $y > 0$, then both $\mathbf{y}, \bar{\mathbf{y}} \geqq \mathbf{0}$ and (II. iii) results. Q.E.D.

A theorem which considers a homogeneous semipositive inequality system under $\mathbf{1}'\mathbf{x} = 1$ is

6.20. THEOREM. For any $(m \times n)$ matrix $\mathbf{A}$, either

(I) $\mathbf{Ax} \geq \mathbf{0}$, $\mathbf{x} \geq \mathbf{0}$, $\mathbf{1}'\mathbf{x} = 1$ has a solution $\mathbf{x} \in \mathbf{R}^n$ or

(II) (i) $\mathbf{A}'\mathbf{y} + \mathbf{1}y \leqq \mathbf{0}$, $y > 0$, $\mathbf{y} \geqq \mathbf{0}$ or

(ii) $\mathbf{A}'\mathbf{y} + \mathbf{1}y \leqq \mathbf{0}$, $y \geq 0$, $\mathbf{y} > \mathbf{0}$

has a solution $y \in \mathbf{R}$, $\mathbf{y} \in \mathbf{R}^m$,

but not both.

PROOF. Rewrite (I) as

$$\eta > 0, \mathbf{Ax} \geq \mathbf{0}, \ \mathbf{x} \geq \mathbf{0}, \ \eta - \mathbf{1}'\mathbf{x} = 0 \quad \text{or} \tag{I'}$$

$$(\mathbf{0}', 1) \begin{bmatrix} \mathbf{x} \\ \eta \end{bmatrix} > 0, \ \begin{bmatrix} \mathbf{A} & \mathbf{0} \\ \mathbf{I}_n & \mathbf{0} \end{bmatrix} \begin{bmatrix} \mathbf{x} \\ \eta \end{bmatrix} \geq \mathbf{0}, \ (-\mathbf{1}', 1) \begin{bmatrix} \mathbf{x} \\ \eta \end{bmatrix} = 0$$

has a solution $\eta \in \mathbf{R}$, $\mathbf{x} \in \mathbf{R}^n$. Using Slater's theorem 6.1, either (I') has a solution or

$$\text{(a)} \begin{bmatrix} \mathbf{0} \\ 1 \end{bmatrix} y + \begin{bmatrix} \mathbf{A}' & \mathbf{I}_n \\ \mathbf{0}' & \mathbf{0}' \end{bmatrix} \mathbf{y}_1 + \begin{bmatrix} -\mathbf{1} \\ 1 \end{bmatrix} y_3 = \mathbf{0}$$

$$\text{(b) } y > 0,\ \mathbf{y}_1 \geqq \mathbf{0},\ y_3 \text{ unrestricted or} \qquad \text{(II}'\text{)}$$

$$\text{(c) } y \geq 0,\ \mathbf{y}_1 > \mathbf{0},\ y_3 \text{ unrestricted}$$

has a solution y, $y_3 \in \mathbf{R}$, and $\mathbf{y}_2 \in \mathbf{R}^{m+n}$, but never both.

For $\mathbf{y}_1' = (\mathbf{y}', \overline{\mathbf{y}}')$, where $\mathbf{y} \in \mathbf{R}^m$ and $\overline{\mathbf{y}} \in \mathbf{R}^n$, (II′. a, b) simplify to (II.i) while (II′. a, c) yield (II. ii). Q.E.D.

The final theorem involving homogeneous inequalities is

6.21. THEOREM. For any $(m \times n)$ matrix $\mathbf{A}$, either

(I) $\mathbf{A}\mathbf{x} \geq \mathbf{0},\ \mathbf{x} > \mathbf{0},\ \mathbf{1}'\mathbf{x} = 1$ has a solution $\mathbf{x} \in \mathbf{R}^n$ or

$$\text{(II)} \begin{cases} \text{(i) } \mathbf{A}'\mathbf{y} + \mathbf{1}y \leq \mathbf{0},\ y \geq 0,\ \mathbf{y} \geqq \mathbf{0} \text{ or} \\ \text{(ii) } \mathbf{A}'\mathbf{y} + \mathbf{1}y \leqq \mathbf{0},\ y > 0,\ \mathbf{y} \geqq \mathbf{0} \text{ or} \\ \text{(iii) } \mathbf{A}'\mathbf{y} + \mathbf{1}y \leqq \mathbf{0},\ y \geq 0,\ \mathbf{y} > \mathbf{0} \end{cases}$$

has a solution $y \in \mathbf{R}$, $\mathbf{y} \in \mathbf{R}^m$,

but not both.

PROOF. Alternative (I) is equivalent to

$$\eta > 0, \mathbf{x} > \mathbf{0},\ \mathbf{A}\mathbf{x} \geq \mathbf{0},\ \eta - \mathbf{1}'\mathbf{x} = 0 \text{ or} \qquad \text{(I}'\text{)}$$

$$\begin{bmatrix} \mathbf{I}_n & \mathbf{0} \\ \mathbf{0}' & 1 \end{bmatrix} \begin{bmatrix} \mathbf{x} \\ \eta \end{bmatrix} > \mathbf{0},\ (\mathbf{A}, \mathbf{0}) \begin{bmatrix} \mathbf{x} \\ \eta \end{bmatrix} \geq \mathbf{0},\ (-\mathbf{1}', 1) \begin{bmatrix} \mathbf{x} \\ \eta \end{bmatrix} = 0$$

has a solution $\eta \in \mathbf{R}$, $\mathbf{x} \in \mathbf{R}^n$. By Slater's theorem 6.1, either (I′) has a solution or

$$\text{(a)} \begin{bmatrix} \mathbf{I}_n & \mathbf{0} \\ \mathbf{0}' & 1 \end{bmatrix} \mathbf{y}_1 + \begin{bmatrix} \mathbf{A}' \\ \mathbf{0}' \end{bmatrix} \mathbf{y} + \begin{bmatrix} -\mathbf{1} \\ 1 \end{bmatrix} y_3 = \mathbf{0}$$

$$\text{(b) } \mathbf{y}_1 \geq \mathbf{0},\ \mathbf{y} \geqq \mathbf{0},\ y_3 \text{ unrestricted or} \qquad \text{(II}'\text{)}$$

$$\text{(c) } \mathbf{y}_1 \geqq \mathbf{0},\ \mathbf{y} > \mathbf{0},\ y_3 \text{ unrestricted}$$

has a solution $\mathbf{y}_1 \in \mathbf{R}^{n+1}$, $\mathbf{y}_2 \in \mathbf{R}^m$, and $y_3 \in \mathbf{R}$, but never both.

For $\mathbf{y}_1' = (\overline{\mathbf{y}}', y)$, where $\overline{\mathbf{y}} \in \mathbf{R}^n$ and $y \in \mathbf{R}$, (II′. a, b) simplify to (II.i, ii); and when (II′. a, c) is considered, we obtain (II. iii). Q.E.D.

If (II. i, iii) are taken together, then they constitute a (II) for a (I) of the form $\mathbf{A}\mathbf{x} \geq \mathbf{0}$, $\mathbf{x} \geqq \mathbf{0}$, $\mathbf{1}'\mathbf{x} = 1$.

Two well-known theorems by Gale (1960) -- one pertaining to a semi-positive and the other a positive solution to a homogeneous equation system -- will now be modified to include $\mathbf{1}'\mathbf{x} = 1$ in alternative (I). First,

6.22. THEOREM. For any $(m \times n)$ matrix $\mathbf{A}$, either

(I) $\mathbf{Ax} = \mathbf{0}$, $\mathbf{x} \geq \mathbf{0}$, $\mathbf{1}'\mathbf{x} = 1$ has a solution $\mathbf{x} \in \mathbf{R}^n$ or

(II) $\mathbf{A}'\mathbf{y} + \mathbf{1}y \leqq \mathbf{0}$, $y > 0$, $\mathbf{y}$ unrestricted

has a solution $y \in \mathbf{R}$, $\mathbf{y} \in \mathbf{R}^m$,

but not both.

PROOF. System (I) is equivalent to

$$\eta > 0, \mathbf{x} \geq \mathbf{0},\ \mathbf{Ax} = \mathbf{0},\ \eta - \mathbf{1}'\mathbf{x} = 0 \quad \text{or} \tag{I'}$$

$$(\mathbf{0}', 1)\begin{bmatrix} \mathbf{x} \\ \eta \end{bmatrix} > 0,\ (\mathbf{I}_n, \mathbf{0})\begin{bmatrix} \mathbf{x} \\ \eta \end{bmatrix} \geq \mathbf{0},\ \begin{bmatrix} \mathbf{A} & \mathbf{0} \\ -\mathbf{1}' & 1 \end{bmatrix}\begin{bmatrix} \mathbf{x} \\ \eta \end{bmatrix} = 0$$

has a solution $\eta \in \mathbf{R}$, $\mathbf{x} \in \mathbf{R}^n$. Invoking Slater's theorem 6.1 leads us to conclude that either (I′) has a solution or

$$\text{(a)} \begin{bmatrix} \mathbf{0} \\ 1 \end{bmatrix} y + \begin{bmatrix} \mathbf{I}_n \\ \mathbf{0}' \end{bmatrix} \mathbf{y}_2 + \begin{bmatrix} \mathbf{A}' & -\mathbf{1} \\ \mathbf{0}' & 1 \end{bmatrix} \mathbf{y}_3 = \mathbf{0}$$

$$\text{(b)}\ y > 0,\ \mathbf{y}_2 \geqq \mathbf{0},\ \mathbf{y}_3 \text{ unrestricted or} \tag{II'}$$

$$\text{(c)}\ y \geq 0,\ \mathbf{y}_2 > \mathbf{0},\ \mathbf{y}_3 \text{ unrestricted}$$

has a solution $y \in \mathbf{R}$, $\mathbf{y}_2 \in \mathbf{R}^n$, and $\mathbf{y}_3 \in \mathbf{R}^{m+1}$, but never both.

For $\mathbf{y}_3' = (\mathbf{y}', y_3)$, where $\mathbf{y} \in \mathbf{R}^m$ and $y_3 \in \mathbf{R}$, (II′. a, b) simplify to (II) while (II′. a, c) transform to

$$\mathbf{A}'\mathbf{y} + \mathbf{1}y < \mathbf{0}, y \geq 0, \mathbf{y} \text{ unrestricted.} \tag{II.1}$$

But it (II.1) holds, then (II) does likewise for some $y > 0$. Q.E.D.

Moreover, (II.i) can be thought of as the appropriate (II) for the (I) alternative $\mathbf{Ax} = \mathbf{0}$, $\mathbf{x} \geqq \mathbf{0}$, $\mathbf{1}'\mathbf{x} = 1$.

Next in this subgroup of theorems appears

6.23. THEOREM. For any $(m \times n)$ matrix $\mathbf{A}$, either

(I) $\mathbf{Ax} = \mathbf{0}$, $\mathbf{x} > \mathbf{0}$, $\mathbf{1}'\mathbf{x} = 1$ has a solution $\mathbf{x} \in \mathbf{R}^n$ or

(II) $\mathbf{A}'\mathbf{y} + \mathbf{1}y \leq \mathbf{0}$, $y \geq 0$, $\mathbf{y}$ unrestricted

has a solution $y \in \mathbf{R}$, $\mathbf{y} \in \mathbf{R}^m$,

but not both.

PROOF. System (I) can also be written as

$$\eta > 0, \mathbf{x} > \mathbf{0}, \mathbf{Ax} = \mathbf{0}, \eta - \mathbf{1}'\mathbf{x} = 0 \quad \text{or} \tag{I'}$$

$$\begin{bmatrix} \mathbf{I}_n & \mathbf{0} \\ \mathbf{0}' & 1 \end{bmatrix} \begin{bmatrix} \mathbf{x} \\ \eta \end{bmatrix} > \mathbf{0}, \begin{bmatrix} \mathbf{A} & \mathbf{0} \\ -\mathbf{1}' & 1 \end{bmatrix} \begin{bmatrix} \mathbf{x} \\ \eta \end{bmatrix} = \mathbf{0}$$

has a solution $\eta \in \mathbf{R}$, $\mathbf{x} \in \mathbf{R}^n$. By Motzkin's theorem 6.2, either (I′) has a solution or

$$\begin{bmatrix} \mathbf{I}_n & \mathbf{0} \\ \mathbf{0}' & 1 \end{bmatrix} \mathbf{y}_1 + \begin{bmatrix} \mathbf{A}' & -\mathbf{1} \\ \mathbf{0}' & 1 \end{bmatrix} \mathbf{y}_2 = \mathbf{0} \tag{II'}$$

$$\mathbf{y}_1 \geq \mathbf{0}, \ \mathbf{y}_2 \text{ unrestricted}$$

has a solution $\mathbf{y}_1 \in \mathbf{R}^{n+1}$, $\mathbf{y}_2 \in \mathbf{R}^{m+1}$, but never both.

Let $\mathbf{y}_1' = (\bar{\mathbf{y}}_1', y)$, $\mathbf{y}_2' = (\mathbf{y}', y_2)$, where both y, $y_2 \in \mathbf{R}$ and $\bar{\mathbf{y}}_1 \in \mathbf{R}^n$, $\mathbf{y} \in \mathbf{R}^m$. For $y \geq 0$, (II′) simplifies to (II) Q.E.D.

The results of this section are summarized in Tables 6.2, 6.3 which follow.

Table 6.2 Theorems of the Alternative

Homogeneous Inequalities under Convex Combination	
1. (I) $\mathbf{Ax} > \mathbf{0}$, $\mathbf{x} \geqq \mathbf{0}$, $\mathbf{1}'\mathbf{x}=1$	(II) $\mathbf{A}'\mathbf{y}+\mathbf{1}y \leqq \mathbf{0}$, $y \geq 0$, $\mathbf{y} \geq \mathbf{0}$ or $\mathbf{A}'\mathbf{y}+\mathbf{1}y \leqq \mathbf{0}$, $y > 0$, $\mathbf{y} \geqq \mathbf{0}$
2. (I) $\mathbf{Ax} > \mathbf{0}$, $\mathbf{x} \geq \mathbf{0}$, $\mathbf{1}'\mathbf{x}=1$	(II) $\mathbf{A}'\mathbf{y}+\mathbf{1}y \leqq \mathbf{0}$, $y \geq 0$, $\mathbf{y} \geq \mathbf{0}$ or $\mathbf{A}'\mathbf{y}+\mathbf{1}y \leqq \mathbf{0}$, $y > 0$, $\mathbf{y} \geqq \mathbf{0}$
3. (I) $\mathbf{Ax} > \mathbf{0}$, $\mathbf{x} > \mathbf{0}$, $\mathbf{1}'\mathbf{x}=1$	(II) $\mathbf{A}'\mathbf{y}+\mathbf{1}y \leqq \mathbf{0}$, $y \geq 0$, $\mathbf{y} \geq \mathbf{0}$ or $\mathbf{A}'\mathbf{y}+\mathbf{1}y \leq \mathbf{0}$, $y \geq 0$, $\mathbf{y} \geqq \mathbf{0}$ or $\mathbf{A}'\mathbf{y}+\mathbf{1}y \leqq \mathbf{0}$, $y > 0$, $\mathbf{y} \geqq \mathbf{0}$
4. (I) $\mathbf{Ax} \geq \mathbf{0}$, $\mathbf{x} \geqq \mathbf{0}$, $\mathbf{1}'\mathbf{x}=1$	(II) $\mathbf{A}'\mathbf{y}+\mathbf{1}y \leq \mathbf{0}$, $y \geq 0$, $\mathbf{y} \geqq \mathbf{0}$ or $\mathbf{A}'\mathbf{y}+\mathbf{1}y \leqq \mathbf{0}$, $y \geq 0$, $\mathbf{y} > \mathbf{0}$

Table 6.2 Theorems of the Alternative (continued)

	(I)	(II)
5.	(I) $\mathbf{Ax} \geq \mathbf{0},\ \mathbf{x} \geq \mathbf{0},\ \mathbf{1}'\mathbf{x}=1$	(II) $\mathbf{A}'\mathbf{y}+\mathbf{1}y \leqq \mathbf{0},\ y > 0,\ \mathbf{y} \geqq \mathbf{0}$ or $\mathbf{A}'\mathbf{y}+\mathbf{1}y \leqq \mathbf{0},\ y \geq 0,\ \mathbf{y} > \mathbf{0}$
6.	(I) $\mathbf{Ax} \geq \mathbf{0},\ \mathbf{x} > \mathbf{0},\ \mathbf{1}'\mathbf{x}=1$	(II) $\mathbf{A}'\mathbf{y}+\mathbf{1}y \leqq \mathbf{0},\ y \geq 0,\ \mathbf{y} > \mathbf{0}$ or $\mathbf{A}'\mathbf{y}+\mathbf{1}y \leq \mathbf{0},\ y \geq 0,\ \mathbf{y} \geqq \mathbf{0}$ or $\mathbf{A}'\mathbf{y}+\mathbf{1}y \leqq \mathbf{0},\ y > 0,\ \mathbf{y} \geqq \mathbf{0}$
7.	(I) $\mathbf{Ax} \geqq \mathbf{0},\ \mathbf{x} \geqq \mathbf{0},\ \mathbf{1}'\mathbf{x}=1$	(II) $\mathbf{A}'\mathbf{y}+\mathbf{1}y \leqq \mathbf{0},\ y > 0,\ \mathbf{y} \geqq \mathbf{0}$
8.	(I) $\mathbf{Ax} \geqq \mathbf{0},\ \mathbf{x} \geq \mathbf{0},\ \mathbf{1}'\mathbf{x}=1$	(II) $\mathbf{A}'\mathbf{y}+\mathbf{1}y \leqq \mathbf{0},\ y > 0,\ \mathbf{y} \geqq \mathbf{0}$ or $\mathbf{A}'\mathbf{y}+\mathbf{1}y < \mathbf{0},\ y \geq 0,\ \mathbf{y} \geqq \mathbf{0}$
9.	(I) $\mathbf{Ax} \geqq \mathbf{0},\ \mathbf{x} > \mathbf{0},\ \mathbf{1}'\mathbf{x}=1$	(II) $\mathbf{A}'\mathbf{y}+\mathbf{1}y \leq \mathbf{0},\ y \geq 0,\ \mathbf{y} \geqq \mathbf{0}$ or $\mathbf{A}'\mathbf{y}+\mathbf{1}y \leqq \mathbf{0},\ y > 0,\ \mathbf{y} \geqq \mathbf{0}$

Table 6.3 Theorems of the Alternative

Homogeneous Equalities under Convex Combination

	(I)	(II)
1.	(I) $\mathbf{Ax}=\mathbf{0},\ \mathbf{x} \geqq \mathbf{0},\ \mathbf{1}'\mathbf{x}=1$	(II) $\mathbf{A}'\mathbf{y}+\mathbf{1}y \leqq \mathbf{0},\ y > 0,\ \mathbf{y}$ unrestricted
2.	(I) $\mathbf{Ax}=\mathbf{0},\ \mathbf{x} \geq \mathbf{0},\ \mathbf{1}'\mathbf{x}=1$	(II) $\mathbf{A}'\mathbf{y}+\mathbf{1}y \leqq \mathbf{0},\ y > 0,\ \mathbf{y}$ unrestricted
3.	(I) $\mathbf{Ax}=\mathbf{0},\ \mathbf{x} > \mathbf{0},\ \mathbf{1}'\mathbf{x}=1$	(II) $\mathbf{A}'\mathbf{y}+\mathbf{1}y \leq \mathbf{0},\ y \geq 0,\ \mathbf{y}$ unrestricted

6.4. Exercises

1. Use theorem 6.12 to verify that the system

$$\begin{aligned} x_1 - 3x_2 &= 1 \\ 2x_1 + 3x_2 &= 1 \\ -x_1 + \ x_2 &= 0 \end{aligned}$$

has no solution.

2. Use theorem 6.14 to demonstrate that the system

$$\begin{aligned} 5x_1 - 4x_2 &\leq \ 7 \\ -3x_1 + 3x_2 &\leq -5 \end{aligned}$$

has no nonnegative solution.

3. Use Farkas' theorem to demonstrate that the system

$$\begin{aligned} x_1 + 3x_2 - 5x_3 &= 2 \\ x_1 - 4x_2 - 7x_3 &= 3 \end{aligned}$$

has no nonnegative solution.

4. Using theorem 6.13 show that the following inequality system has no solution.

$$\begin{aligned} -4x_1 + 5x_2 &\leq -2 \\ 2x_1 + 7x_2 &\leq -2 \\ x_1 + 3x_2 &\leq 1 \end{aligned}$$

5. Use Gordan's theorem to determine if the following system has a semipositive solution.

$$\begin{aligned} -3x_1 + 5x_2 - 2x_3 &= 0 \\ -2x_1 + 4x_2 - x_3 &= 0 \end{aligned}$$

6. Does Stiemke's II hold in theorem 6.6 when

$$\mathbf{A} = \begin{bmatrix} 1 & 2 & 5 \\ 4 & 2 & 1 \end{bmatrix}, \mathbf{b}' = (3, 2, 2)?$$

CHAPTER 7

BASIC SOLUTIONS AND COMPLEMENTARY SLACKNESS IN PAIRS OF DUAL SYSTEMS

7.1 Basic Solutions to Linear Equalities

Consider the simultaneous linear system $\mathbf{Ax} = \mathbf{b}$, where $\mathbf{A}$ is of order $(m \times n)$ and $\rho(\mathbf{A}) = \rho[\mathbf{A}, \mathbf{b}]$ (the system is consistent). If $m < n$, then we have an ***underdetermined equation system*** which, given that it is consistent, possesses an infinite number of particular solutions. Furthermore, let $\rho(\mathbf{A}) = m$ so that $\mathbf{Ax} = \mathbf{b}$ is consistent for every $\mathbf{b}$ and none of the linear equations is ***redundant***, *i.e.*, expressible as a linear combination of one or more other equations in the system.

Let us now select m linearly independent columns from $\mathbf{A}$ and express the variables associated with these columns in terms of the remaining $n - m$ variables. If arbitrary values are then assigned to the latter, it is evident that we obtain an infinity of particular solutions. Specifically, let us rearrange the columns of $\mathbf{A}$ so that the first m are linearly independent and constitute the columns of the mth order basis matrix $\mathbf{B} = [\mathbf{b}_1, \ldots, \mathbf{b}_m]$. The remaining $n - m$ columns of $\mathbf{A}$ then form the $(m \times n - m)$ matrix $\mathbf{R} = [\mathbf{r}_1, \ldots, \mathbf{r}_{n-m}]$. Hence

$$\mathbf{Ax} = [\mathbf{B}, \mathbf{R}] \begin{bmatrix} \mathbf{x}_B \\ \mathbf{x}_R \end{bmatrix} = \mathbf{B}\mathbf{x}_B + \mathbf{R}\mathbf{x}_R = \mathbf{b},$$

where $\mathbf{x}$ has been partitioned into an $(m \times 1)$ vector $\mathbf{x}_B$, $\mathbf{x}_B' = (x_{B1}, \ldots, x_{Bm})$, containing the m variables corresponding to the m linearly independent columns of $\mathbf{A}$ and an $(n - m \times 1)$ vector $\mathbf{x}_R$, $\mathbf{x}_R' = (x_{R1}, \ldots, x_{R, n-m})$, containing the $n - m$ remaining variables. Since $\rho(\mathbf{B}) = m$, $\mathbf{B}$ is nonsingular. Hence

$$\mathbf{x}_B = \mathbf{B}^{-1}\mathbf{b} - \mathbf{B}^{-1}\mathbf{R}\mathbf{x}_R \tag{7.1}$$

and all solutions to $\mathbf{Ax} = \mathbf{b}$ can be generated by assigning arbitrary values to $\mathbf{x}_R$. In particular, if $\mathbf{x}_R = \mathbf{0}$, $\mathbf{x}_B = \mathbf{B}^{-1}\mathbf{b}$ and we obtain what will be called a unique ***basic solution*** to $\mathbf{Ax} = \mathbf{b}$, namely[1]

$$\mathbf{x} = \begin{bmatrix} \mathbf{x}_B \\ \mathbf{0} \end{bmatrix} = \begin{bmatrix} \mathbf{B}^{-1}\mathbf{b} \\ \mathbf{0} \end{bmatrix} \tag{7.2}$$

This type of solution is so named because the columns of $\mathbf{B}$ constitute a basis for $\mathbf{R}^m$, *i.e.*, each column of $\mathbf{R}$ and the vector $\mathbf{b}$ are expressible as a linear combination of the linearly independent columns of $\mathbf{B}$. Hence the variables in $\mathbf{x}_B$ are said to be ***basic variables*** while those within $\mathbf{x}_R$ are termed ***nonbasic variables.*** Moreover, a basic solution to $\mathbf{Ax} = \mathbf{b}$ is said to be ***degenerate*** if one or more of the basic variables within $\mathbf{x}_B$ vanishes, *i.e.,* $\mathbf{x}$ contains fewer than m positive variables or, equivalently, $\mathbf{x}$ contains more than $n - m$ zeros. In sum, given $\mathbf{Ax} = \mathbf{b}$ with $\rho(\mathbf{A}) = \rho[\mathbf{A},\mathbf{b}] = m$, if any m linearly independent columns are chosen from $\mathbf{A}$ and if all the remaining $n - m$ variables not associated with these columns are set equal to zero, then the resulting solution contains no more than m nonzero variables, and is a basic solution to $\mathbf{Ax} = \mathbf{b}$. Clearly the number of basic solutions must be finite since it cannot exceed the number of ways of selecting subsets of m columns form $\mathbf{A}$, *i.e.*, the number of basic solutions cannot exceed $n!/m!(n-m)!$.

[1]We assumed above that for $\mathbf{Ax} = \mathbf{b}$ we have $m < n$. In general, if $\mathbf{A}$ is $(m \times n)$ and $m \gtrless n$, then a solution $\mathbf{x}$ of $\mathbf{Ax} = \mathbf{b}$ is a basic solution if the columns of $\mathbf{A}$ corresponding to nonzero components of $\mathbf{x}$ are linearly independent. In this regard, if $\rho(\mathbf{A}) = r$ and $\mathbf{x}$ is a basic solution to $\mathbf{Ax} = \mathbf{b}$, then $\mathbf{A}$ has at least one $(m \times r)$ submatrix $\mathbf{N}$ such that

(1) $\rho(\mathbf{N}) = r$; and

(2) the components of $\mathbf{x}$ corresponding to columns of $\mathbf{A}$ not in $\mathbf{N}$ are zero.

Conversely, associated with each $(m \times r)$ submatrix $\mathbf{N}$ of $\mathbf{A}$ having rank r there exists a unique basic solution.

In practice, a basic solution $\mathbf{x}$ can be found by the following procedure:

1. Choose a set of m linearly independent columns of $\mathbf{A}$ to form a basis matrix $\mathbf{B}$.
2. Set all components of $\mathbf{x}$ corresponding to columns of $\mathbf{A}$ not in $\mathbf{B}$ equal to zero.
3. Solve the m resulting equations to determine the remaining components of $\mathbf{x}$. These are the desired basic variables.

We also observe that:

(a) We may determine an upper bound on the absolute value of the components of any basic solution to $\mathbf{Ax}=\mathbf{b}$ by invoking the assumption that the entries in $\mathbf{A}, \mathbf{b}$ are integers. In this regard, for $\mathbf{x}' = (\mathbf{x}'_B, \mathbf{0}')$ a basic solution to $\mathbf{Ax}=\mathbf{b}$, we have

$$|x_{Bi}| \leqq m! a^{m-1} b,$$

where

$$a = \max_{i,j}\{|a_{ij}|\},\ b = \max_{i}\{|b_i|\},$$

$i=1,\ldots,m;\ j=1,\ldots,n.$

(b) If two distinct bases correspond to the same basic solution $\mathbf{X}$, then $\mathbf{x}$ is degenerate. (To see this suppose that the distinct basis matrices $\mathbf{B}$, $\bar{\mathbf{B}}$ determine the same basic solution $\mathbf{x}$. Since the components of $\mathbf{x}$ corresponding to the $n-m$ columns of $\mathbf{A}$ not in $\mathbf{B}$ are all zeros, it follows that $\mathbf{x}$ must also have zeros for its components corresponding to the columns of $\mathbf{A}$ common to $\mathbf{B}$ and $\bar{\mathbf{B}}$. Hence $\mathbf{x}$ is degenerate.)

One of the important aspects of the notion of a basic solution to a linear equation system is its applicability in solving a problem of the form: given an $(m \times n)$ matrix $\mathbf{A}$ and vector $\mathbf{b} \in \mathbf{R}^m$, find an $\mathbf{x} \in \mathbf{R}^n$ such that $\mathbf{Ax}=\mathbf{b}$

with $\mathbf{x} \geqq \mathbf{0}$. A problem such as this can serve as system (I) of the following version of Farkas' theorem of the alternative: either

(I) $\mathbf{Ax} = \mathbf{b}$, $\mathbf{x} \geqq \mathbf{0}$ has a solution $\mathbf{x} \epsilon \mathbf{R}^n$ or

(II) $\mathbf{A}'\mathbf{y} \geqq \mathbf{0}$, $\mathbf{b}'\mathbf{y} < \mathbf{0}$ has a solution $\mathbf{y} \epsilon \mathbf{R}^m$,

but never both. For system (I), if $\mathbf{x}_B = \mathbf{B}^{-1}\mathbf{b} \geqq \mathbf{0}$, then

$$\mathbf{x} = \begin{bmatrix} \mathbf{B}^{-1}\mathbf{b} \\ \mathbf{0} \end{bmatrix} \geqq \mathbf{0}$$

is termed a ***basic feasible solution*** to $\mathbf{Ax} = \mathbf{b}$. Moreover, a basic feasible solution in ***nondegenerate*** if each basic variable is strictly positive.

Looking to Farkas' system (I) of the preceding paragraph, given any feasible solution $\mathbf{x} \geqq \mathbf{0}$ to $\mathbf{Ax} = \mathbf{b}$, under what circumstances can we find at least one basic feasible solution? The answer is provided by the following

7.1.1. EXISTENCE THEOREM FOR BASIC FEASIBLE SOLUTIONS TO NONHOMOGENEOUS LINEAR EQUALITIES. For an $(m \times n)$ matrix $\mathbf{A}$ with $\rho(\mathbf{A}) = m < n$ and a vector $\mathbf{b} \epsilon \mathbf{R}^m$, if $\mathbf{Ax} = \mathbf{b}$, $\mathbf{x} \geqq \mathbf{0}$ has a feasible solution $\mathbf{x} \epsilon \mathbf{R}^n$, then it also has a basic feasible solution.

PROOF.[2] Assume that there exists a feasible solution to $\mathbf{Ax} = \mathbf{b}$ in $k \leq m$ positive variables. (Without loss of generality, let the first k components of $\mathbf{x}$ be strictly positive.) Then $\mathbf{Ax} = \mathbf{b}$ becomes

$$\sum_{j=1}^{k} x_j \mathbf{a}_j = \mathbf{b},\ x_j > 0,\ j = 1, \ldots, k \leq m; \qquad x_j = 0, j = k+1, \ldots, n. \tag{7.3}$$

What sort of solution does (7.3) represent? Since the vectors $\mathbf{a}_j, j = 1, \ldots, k \leq m$, may or may not be linearly independent, two cases present themselves.

[2]The proof of this theorem is "constructive" in that is spells out a computational procedure which actually allows us to obtain a basic feasible solution from any feasible solution to $\mathbf{Ax} = \mathbf{b}$. The example which follows outlines the essential features of the process.

First, let the vectors $\mathbf{a}_j, j=1,\ldots,k \leq m$, be linearly independent. If $k=m$, then associated with this set of vectors is an mth order nonsingular basis matrix $\mathbf{B}$ which provides us with a unique nondegenerate basic feasible solution. If $k < m$, we may add to the vectors $\mathbf{a}_j, j=1,\ldots,k < m$, any $m-k$ additional vectors selected from those remaining in $\mathbf{A}$ to form

$$\sum_{j=1}^{k} x_j\mathbf{a}_j + \sum_{j=k+1}^{m} 0\mathbf{a}_j = \mathbf{b}$$

and thus a basis for $\mathbf{R}^m$. Since the vectors $\mathbf{a}_j, j=k+1,\ldots,m$, are admitted to the basis at a zero level, the resulting basic feasible solution is degenerate.

Next, let the vectors $\mathbf{a}_j, j=1,\ldots,k \leq m$, be linearly dependent. Since our goal is to ultimately attain a basic feasible solution, we shall introduce a procedure which allows us to systematically reduce the number of positive variables step by step until the vectors associated with those that remain are linearly independent. By hypothesis, there exist coefficients y_j not all zero such that

$$\sum_{j=1}^{k} y_j\mathbf{a}_j = \mathbf{0}. \tag{7.4}$$

A multiplication of both sides of this equation by a nonzero scalar θ yields

$$\sum_{j=1}^{k} \theta y_j\mathbf{a}_j = \mathbf{0}. \tag{7.4.1}$$

Upon subtracting this last equation from (7.3) we obtain

$$\sum_{j=1}^{k} (x_j-\theta y_j)\mathbf{a}_j = \sum_{j=1}^{k} \hat{x}_j\mathbf{a}_j = \mathbf{b}. \tag{7.5}$$

If (7.5) is to yield a feasible solution to $\mathbf{Ax} = \mathbf{b}$, then we require that

$$\hat{x}_j = x_j - \theta y_j \geq 0, j=1,\ldots,k. \tag{7.6}$$

Under what conditions will this inequality hold? If $\theta > 0$, then $\hat{x}_j \geq 0, j=1,\ldots,k$, for all $y_j \leq 0$. Hence we need only consider those y_j which are positive. (If all $y_j \neq 0$ are negative, we may multiply (7.4) by -1 so as to insure that there exists at least one $y_j > 0$.) Rearranging (7.6) yields $x_j/y_j \geq \theta$ so that any θ which satisfies

$$\theta = \hat{\theta} = \min_j \left\{ \frac{x_j}{y_j}, \ y_j > 0 \right\} \tag{7.7}$$

guarantees that a feasible solution is attained. Suppose the minimum is assumed for $\hat{\theta} = x_h/y_h$. Then if this ratio is substituted into (7.5), the hth coefficient in this equation turns to zero with the result that $\mathbf{a}_h$ is dropped from this linear combination. Hence

$$\sum_{\substack{j=1 \\ j \neq h}}^{k} \left[x_j - \left(\frac{x_h}{y_h} \right) y_j \right] \mathbf{a}_j = \mathbf{b} \tag{7.5.1}$$

represents a new feasible solution with no more than $k-1$ positive variables. If the vectors associated with these $k-1$ positive variables are linearly independent, we have actually attained a basic feasible solution; if they are linearly dependent, we may repeat the above procedure and reduce one of the $k-1$ positive variables to zero. If necessary, this elimination technique may be applied continually until the remaining vectors are linearly independent and thus provide us with a basic feasible solution. Q.E.D.

EXAMPLE 7.1 Let $\mathbf{Ax} = \sum_{j=1}^{3} x_j \mathbf{a}_j = \mathbf{b}$ assume the form

$$x_1 \begin{bmatrix} 2 \\ 3 \end{bmatrix} + x_2 \begin{bmatrix} 2 \\ 1 \end{bmatrix} + x_3 \begin{bmatrix} 3 \\ 5 \end{bmatrix} = \begin{bmatrix} 5 \\ 6 \end{bmatrix} \quad \text{or} \quad \begin{cases} 2x_1 + 2x_2 + 3x_3 = 5 \\ 3x_1 + \ \ x_2 + 5x_3 = 6 \end{cases}$$

where $\mathbf{a}_1' = (2,3)$, $\mathbf{a}_2' = (2,1)$, $\mathbf{a}_3' = (3,5)$, and $\mathbf{b}' = (5,6)$. It is easily verified that a particular feasible solution is $\mathbf{x}' = \left(\frac{7}{8}, \frac{7}{8}, \frac{1}{2}\right)$. Since $\rho(\mathbf{A}) = \rho[\mathbf{a}_1, \mathbf{a}_2, \mathbf{a}_3] = 2$, the vectors $\mathbf{a}_j$, j=1,2,3, are linearly dependent, *i.e.*, there exists an alternative feasible solution to $\mathbf{Ax} = \mathbf{b}$ in at most two positive variables. In this regard, a set of y_j, j=1,2,3, which satisfy (7.4) are $y_1 = -\frac{7}{8}$, $y_2 = \frac{1}{8}$, and $y_3 = \frac{1}{2}$ so that

$$\sum_{j=1}^{3} y_j \mathbf{a}_j = -\tfrac{7}{8}\mathbf{a}_1 + \tfrac{1}{8}\mathbf{a}_2 + \tfrac{1}{2}\mathbf{a}_3 = \mathbf{0}.$$

To obtain a new feasible solution to $\mathbf{Ax} = \mathbf{b}$, one of the three positive variables ($x_1 = \frac{7}{8}$ or $x_2 = \frac{7}{8}$ or $x_3 = \frac{1}{2}$) must be driven to zero. Form (7.7) we find that

$$\hat{\theta} = min\left\{\frac{x_2}{y_2}, \frac{x_3}{y_3}\right\} = min\{7, 1\} = 1.$$

Hence we may eliminate $\mathbf{a}_3$ since $\frac{x_3}{y_3} = \hat{\theta}$. Thus, for $\theta = \hat{\theta}$ in (7.6), x_3 reduces to zero and we obtain an alternative feasible solution in only two positive variables, namely

$$\hat{x}_1 = x_1 - \hat{\theta} y_1 = \frac{7}{8} - \left(-\frac{7}{8}\right) = \frac{7}{4}$$

$$\hat{x}_2 = x_2 - \hat{\theta} y_2 = \frac{7}{8} - \frac{1}{8} = \frac{3}{4}$$

$$\hat{x}_3 = x_3 - \hat{\theta} y_3 = \frac{1}{2} - \frac{1}{2} = 0.$$

Thus our new solution to $\mathbf{Ax} = \mathbf{b}$ is given by $X' = \left(\frac{7}{4}, \frac{3}{4}, 0\right)$ or

$$\hat{x}_1 \mathbf{a}_1 + \hat{x}_2 \mathbf{a}_2 = \frac{7}{4}\mathbf{a}_1 + \frac{3}{4}\mathbf{a}_2 = \mathbf{b}.$$

Moreover, since $\mathbf{a}_1, \mathbf{a}_2$ are linearly independent and $\hat{x}_1, \hat{x}_2 > 0$, we have actually attained a nondegenerate basic feasible solution.

In chapter eight we shall encounter an extremely important geometric interpretation of this theorem. Specifically, it will be shown that basic feasible solutions to the system $\mathbf{Ax} = \mathbf{b}$ are always found at the vertices of the convex polyhedron $\mathcal{S} = \{\mathbf{x} \mid \mathbf{Ax} = \mathbf{b}, \mathbf{x} \geq \mathbf{0}\ \mathbf{x} \in \mathbf{R}^n\}$.

7.2 Moving from One Basic (Feasible) Solution to Another

As in the preceding section, let us assume that **A** is of order $(m \times n)$ with $\rho(\mathbf{A}) = m < n$ and $\mathbf{b} \epsilon \mathbf{R}^m$. Then a basic solution to $\mathbf{Ax} = \mathbf{b}$ is provided by

$$\mathbf{x} = \begin{bmatrix} B^{-1}b \\ 0 \end{bmatrix} \geqq \mathbf{0}.$$

Our aim is to develop a procedure for replacing a given vector in the basis by another so that the new set of vectors is linearly independent. Let us again partition **A** as $\mathbf{A} = [\mathbf{B}, \mathbf{R}]$. In this regard, in generating a new basic solution to $\mathbf{Ax} = \mathbf{b}$, some vector which currently constitutes a column of the basis matrix **B** must be replaced by one of the vectors comprising the columns of **R**.

Under what conditions can a given nonbasis vector $\mathbf{r}_j, j=1,\ldots,n-m$, within **R** replace one of the basis vectors in the set $\mathfrak{B} = \{\mathbf{b}_1,\ldots,\mathbf{b}_m\}$ so that the new set of vectors $\bar{\mathfrak{B}}$ forms a basis for $\mathbf{R}^m$? The answer is provided by

7.2.1. THEOREM. Let the set of vectors $\{\mathbf{b}_1,\ldots,\mathbf{b}_m\}$ constitute a basis for $\mathbf{R}^m$. If a vector $\mathbf{r} \epsilon \mathbf{R}^m$ is expressible as a linear combination of the $\mathbf{b}_i, \mathbf{r} = \Sigma_{i=1}^m \lambda_i \mathbf{b}_i$, then any $\mathbf{b}_i$ for which $\lambda_i \neq 0$ can be removed from the basis and replaced by $\mathbf{r}$ with the result that the new collection of vectors is also a basis for $\mathbf{R}^m$.

PROOF. Without loss generality, let us assume that $\lambda_m \neq 0$ so that $\mathbf{r}$ replaces $\mathbf{b}_m$ in the basis. If the new set of vectors $\bar{\mathfrak{B}} = \{\mathbf{b}_1,\ldots,\mathbf{b}_{m-1},\mathbf{r}\}$ constitutes a basis for $\mathbf{R}^m$, it must be linearly independent and span $\mathbf{R}^m$. Let the null vector be expressible as a linear combination of the elements in $\bar{\mathfrak{B}}$, *i.e.*,

$$\sum_{i=1}^{m-1} \mu_i \boldsymbol{b}_i + \mu_m \boldsymbol{r} = \mathbf{0}.$$

A substitution of $\mathbf{r} = \Sigma_{i=1}^{m} \lambda_i \boldsymbol{b}_i$ into this expression yields

$$\sum_{i=1}^{m-1} \mu_i \boldsymbol{b}_i + \sum_{i=1}^{m} \mu_m \lambda_i \boldsymbol{b}_i = \sum_{i=1}^{m-1} (\mu_i + \mu_m \lambda_i) \boldsymbol{b}_i + \mu_m \lambda_m \boldsymbol{b}_m = \mathbf{0}.$$

Since the original set $\mathfrak{B}$ is linearly independent, this last expression equals the null vector if and only if $\mu_i + \mu_m \lambda_i = 0$, $i=1,\ldots,m-1$, $\mu_m = 0$. With $\mu_m = 0$, $\mu_i = -\mu_m \lambda_i = 0$, $i=1,\ldots,m-1$. Hence $\mu_i = 0$, $i=1,\ldots,m$, so that $\bar{\mathfrak{B}}$ is linearly independent and thus forms a basis for $\mathbf{R}^m$. Note that $\lambda_m \neq 0$ is a sufficient condition for $\bar{\mathfrak{B}}$ to be linearly independent; it is also necessary since if $\lambda_m = 0$, then $\mathbf{r} = \Sigma_{i=1}^{m-1} \lambda_i \boldsymbol{b}_i = \mathbf{0}$ and thus $\bar{\mathfrak{B}}$ would be linearly dependent.

We know that $\mathfrak{B}$ spans $\mathbf{R}^m$. If it can be demonstrated that $\mathfrak{B}$ is expressible in terms of the new set $\bar{\mathfrak{B}}$, then the latter must also span $\mathbf{R}^m$. Since $\lambda_m \neq 0$,

$$\boldsymbol{b}_m = \lambda_m^{-1}\, \boldsymbol{r} - \sum_{i=1}^{m-1} \lambda_i \lambda_m^{-1} \boldsymbol{b}_i$$

and thus

$$\mathfrak{B} = \{\boldsymbol{b}_1, \ldots, \boldsymbol{b}_{m-1}, \lambda_m^{-1} \boldsymbol{r} - \Sigma_{i=1}^{m-1} \lambda_i \lambda_m^{-1} \boldsymbol{b}_i\}.$$

And since $\bar{\mathfrak{B}}$ was also found to be linearly independent, it thus represents a basis for $\mathbf{R}^m$. Q.E.D.

Based upon the preceding theorem, it should be evident that since the vectors in **B** form a basis for $\mathbf{R}^m$, any vector in **R** may be written as a linear combination of the columns of **B**. If we denote the jth column of **R** by $\mathbf{r}_j$, $j=1,\ldots,n-m$, then for all j,

$$\mathbf{r}_j = \sum_{i=1}^{m} \lambda_{ij} \mathbf{b}_i = \mathbf{B} \boldsymbol{\lambda}_j\,. \tag{7.8}$$

Let us assume that we have chosen some vector $\mathbf{r}_j$ to enter the basis and that the $(m \times 1)$ vector λ_j in (7.8) associated with $\mathbf{r}_j$ has at least one positive component λ_{ij}. At the present basic solution $\mathbf{B}\mathbf{x}_B = \mathbf{b}$. Upon multiplying both sides of (4.8) by some number θ and subtracting the result from $\mathbf{B}\mathbf{x}_B = \mathbf{b}$, we obtain

$$\mathbf{B}\mathbf{x}_B - \mathbf{B}(\theta\boldsymbol{\lambda}_j) = \mathbf{b} - \theta\mathbf{r}_j \quad \text{or}$$

$$\mathbf{x}_B - \theta\boldsymbol{\lambda}_j = \mathbf{B}^{-1}(\mathbf{b} - \theta\mathbf{r}_j).$$

Here $(\mathbf{x}_B - \theta\boldsymbol{\lambda}_j, \mathbf{0})'$ is a basic solution to $\mathbf{A}\mathbf{x} = \mathbf{b}$ and, if

$$\mathbf{x}_B - \theta\boldsymbol{\lambda}_j \geqq \mathbf{0} \quad \text{or} \quad x_{Bi} - \theta\lambda_{ij} \geq 0,\ i=1,\ldots,m;\ j=1,\ldots,n-m, \tag{7.9}$$

then $(\mathbf{x}_B - \theta\boldsymbol{\lambda}_j, \mathbf{0})'$ represents a basic feasible solution to $\mathbf{A}\mathbf{x} = \mathbf{b}$.

Under what circumstances will (7.9) be satisfied? If $\theta > 0$, then (7.9) always holds for all $\lambda_{ij} \leq 0$. Hence we need only concern ourselves with those components of $\boldsymbol{\lambda}_j$ which are positive. In this regard, we need to find a $\theta > 0$ such that (7.9) holds for all $\lambda_{ij} > 0$. Rearranging (7.9) yields $\frac{x_{Bi}}{\lambda_{ij}} \geq \theta$ so that any θ for which

$$0 < \theta \leq \min_i \left\{ \frac{x_{Bi}}{\lambda_{ij}},\ \lambda_{ij} > 0 \right\} \tag{7.10}$$

will do, *i.e.*, will ensure that $(\mathbf{x}_B - \theta\boldsymbol{\lambda}_j, \mathbf{0})'$ is a basic feasible solution to $\mathbf{A}\mathbf{x} = \mathbf{b}$. In fact, this latter inequality enables us to select the $\mathbf{b}_i$ to be removed from the basis and the $\mathbf{r}_j$ to be entered into the basis. If

$$\theta = \hat{\theta} = \min_i \left\{ \frac{x_{Bi}}{\lambda_{ij}},\ \lambda_{ij} > 0 \right\} \tag{7.10.1}$$

is substituted into $\mathbf{x}_B - \theta\boldsymbol{\lambda}_j$, then the component of this vector for which the minimum is attained reduces to zero. Hence the column of $\mathbf{B}$ to be replaced by $\mathbf{r}_j$ can always be determined by (7.10.1). In this regard, (7.10.1) is termed an ***exit criterion*** for moving from one basic feasible solution to another.

If $\hat{\mathfrak{B}} = \{\hat{\mathbf{b}}_1, \ldots, \hat{\mathbf{b}}_m\}$ denotes the set of linearly independent vectors obtained by replacing the rth column $\mathbf{b}_r$ within $\mathbf{B}$ by $\mathbf{r}_j$, then the new basic feasible solution can be represented as

$$\hat{\mathbf{x}} = \begin{bmatrix} \mathbf{x}_{\hat{B}} \\ \mathbf{0} \end{bmatrix} \begin{bmatrix} \mathbf{x}_B - \hat{\theta}\boldsymbol{\lambda}_j \\ \mathbf{0} \end{bmatrix} = \begin{bmatrix} \hat{\mathbf{B}}^{-1}\mathbf{b} \\ \mathbf{0} \end{bmatrix}, \tag{7.11}$$

where $\hat{\mathbf{B}} = \left[\hat{b}_1, \ldots, \hat{b}_m\right]$ with

$$\hat{\mathbf{b}}_i = \mathbf{b}_i,\ i \neq r;\ \hat{\mathbf{b}}_r = \mathbf{r}_j,$$

and

$$x_{\hat{B}i} = x_{Bi} - \hat{\theta}\lambda_{ij},\ i \neq r;\ x_{\hat{B}r} = \hat{\theta} = \frac{x_{Br}}{\lambda_{rj}}. \tag{7.11.1}$$

In sum, (7.10.1):

(1) determines the vector $\mathbf{b}_r$ to be removed from the present basis $\mathfrak{B}$;

(2) determines the value of the variable $x_{\hat{B}r}$ corresponding to the incoming vector $\mathbf{r}_j$ for the new basis $\hat{\mathfrak{B}}$; and

(3) ensures that $\hat{\mathbf{x}}$, the new basic solution to $\mathbf{Ax} = \mathbf{b}$, is feasible, *i.e.*, $\hat{\mathbf{x}} \geqq \mathbf{0}$.

EXAMPLE 7.2 To see exactly how this procedure works in practice, let us assume that from some given $\mathbf{Ax} = \mathbf{b}$, $\mathbf{x} \geqq \mathbf{0}$ system we have obtained the following basic feasible solution, *i.e.*, given

$$\mathbf{x}_B = \mathbf{B}^{-1}\mathbf{b} - \mathbf{B}^{-1}\mathbf{R}\mathbf{x}_R \quad \text{or}$$

$$\begin{bmatrix} x_2 \\ x_3 \\ x_5 \end{bmatrix} = \begin{bmatrix} 10 - x_1 - x_4 + 2x_6 \\ 20 + 2x_1 - x_4 - \frac{1}{2}x_6 \\ 5 + x_1 - 3x_4 - x_6 \end{bmatrix},$$

set $\mathbf{x}'_R = (x_{R1}, x_{R2}, x_{R3}) = \mathbf{0}'$ so that $\mathbf{x}'_B = (x_{B1}, x_{B2}, x_{B3}) = (10, 20, 5)$. Hence $\mathbf{x}' = (\mathbf{x}'_B, \mathbf{x}_R') = (0, 10, 20, 0, 5, 0)$. In addition, assume that the nonbasic variable x_4 is to become basic. Which basic variable should now become nonbasic? In addition, how large of an increase in x_4 is consistent with the feasibility requirement? Since we introduce only one nonbasic variable into the basis at a time, x_1 and x_6 are still held fixed at their zero level. Hence the above system becomes

$$\begin{aligned}
\mathbf{x}_B &= \mathbf{B}^{-1}\mathbf{b} - x_{R2}\boldsymbol{\lambda}_2 \qquad \text{or} \\
x_2 &= 10 - x_4 \\
x_3 &= 20 - x_4 \\
x_5 &= 5 - 3x_4
\end{aligned}$$

where $\boldsymbol{\lambda}_2$ is the second column of $\mathbf{B}^{-1}\mathbf{R}$ associated with nonbasic variable $x_{R2} = x_4$. Now, if x_4 increases in value, it is obvious that each basic variable decreases. How far may x_4 increase without violating the nonnegativity conditions? To determine this let us set each basic variable equal to zero and solve for the value of x_4 which makes it vanish. To this end,

$$\left.\begin{aligned}
x_2 &= x_{B1} - \lambda_{12}x_{R2} = 0 \\
x_3 &= x_{B2} - \lambda_{22}x_{R2} = 0 \\
x_5 &= x_{B3} - \lambda_{32}x_{R2} = 0
\end{aligned}\right\} \quad \text{or} \quad \left\{\begin{aligned}
&10 - x_4 = 0, \quad x_4 = 10 \\
&20 - x_4 = 0, \quad x_4 = 20 \\
&5 - 3x_4 = 0, \quad x_4 = \tfrac{5}{3}
\end{aligned}\right.$$

From (7.10.1),

$$\hat{\theta} = \frac{5}{3} = min\left\{\frac{x_{B1}}{\lambda_{12}} = 10, \frac{x_{B2}}{\lambda_{22}} = 20, \frac{x_{B3}}{\lambda_{32}} = \frac{5}{3}\right\}.$$

Hence x_5 becomes nonbasic and $\mathbf{r}_2$ replaces $\mathbf{b}_3$ in the basis. Finally, an application of (7.11.1) yields the values of the variables corresponding to the columns of the new basis matrix $\hat{\mathbf{B}} = [\mathbf{b}_1, \mathbf{b}_2, \mathbf{r}_2]$, namely

$$x_{\hat{B}1} = x_2 = 10 - \frac{5}{3} = \frac{25}{3}$$

$$x_{\hat{B}2} = x_3 = 20 - \frac{5}{3} = \frac{55}{3}$$

$$x_{\hat{B}3} = x_4 = \frac{5}{3}.$$

Armed with the concept of a basic feasible solution to $\mathbf{Ax} = \mathbf{b}$, let us now address the specific structure of the solution set $\mathcal{S} = \{\mathbf{x} \mid \mathbf{AX} = \mathbf{b}, \mathbf{x} \geqq \mathbf{0}, \mathbf{x} \in R^n\}$. As we shall now see, some of the results provided in Section 1.2 may be extended to the case where a simultaneous linear equation system comes under a nonnegativity restriction $\mathbf{x} \geqq \mathbf{0}$.

Given $\mathcal{S}$, the ***homogeneous system associated with*** $\mathbf{Ax} = \mathbf{b}, \mathbf{x} \geqq \mathbf{0}$ (*i.e.*, associated with $\mathcal{S}$) assumes the form $\mathbf{Ay} = \mathbf{0}$, $\mathbf{y} \geqq \mathbf{0}$. Let $\mathcal{T} = \{\mathbf{y} \mid \mathbf{Ay} = \mathbf{0}, \mathbf{y} \geqq \mathbf{0}, \mathbf{y} \in R^n\}$. As indicated in Chapter 4, this set of all $\mathbf{y}$ yielding homogeneous solutions is a finite or polyhedral convex cone. Furthermore, if $\bar{\mathbf{x}} \in \mathcal{S}$ and $\bar{\mathbf{y}} \in \mathcal{T}$, then $\bar{\mathbf{x}} + \theta\bar{\mathbf{y}} \in \mathcal{S}$, $0 \leq \theta \in R$. Let us now look to the following

7.2.2 RESOLUTION THEOREM. Let $\{\mathbf{x}_1, \ldots, \mathbf{x}_k\}$ be the set of basic feasible solutions to $\mathbf{Ax} = \mathbf{b}$ with $\hat{\mathbf{x}} \in co\{\mathbf{x}_1, \ldots, \mathbf{x}_k\}$ and $\hat{\mathbf{y}} \in \mathcal{T} = \{\mathbf{y} \mid \mathbf{Ay} = \mathbf{0}, \mathbf{y} \geqq \mathbf{0}, \mathbf{y} \in R^n\}$. If $\mathbf{x} \in \mathcal{S} = \{\mathbf{x} \mid \mathbf{Ax} = \mathbf{b}, \mathbf{x} \geqq \mathbf{0}, \mathbf{x} \in R^n\} \neq \phi$, then $\mathbf{x} = \hat{\mathbf{x}} + \hat{\mathbf{y}}$.

PROOF. Suppose $\mathbf{x}$ is a feasible solution of $\mathbf{Ax} = \mathbf{b}$ with

$$x_j > 0, \ j \in \mathcal{J};$$
$$x_j = 0 \text{ otherwise},$$

where $\mathcal{J}$ is the index set of subscripts j on the vectors from $\mathbf{A}$ actually used to obtain $\mathbf{x}$. To execute the proof, let us proceed by induction on the number p of subscripts in $\mathcal{J}$.

Let $p = 0$. Then $\mathbf{x} = \hat{\mathbf{x}} = \hat{\mathbf{y}} = \mathbf{0}$ is a basic feasible solution to $\mathbf{Ax} = \mathbf{b}$ as well as a homogeneous solution and thus the theorem holds.

For $p > 1$, let the theorem be true for feasible solutions with fewer than p vectors (*i.e.*, with fewer than p subscripts in $\mathfrak{J}$). Let this number be $p-1$ and assume that $\mathbf{x}$ is obtained using p vectors from $\mathbf{A}$ so that $\mathfrak{J}$ contains p distinct subscripts. If the associated set of vectors is linearly independent, then $\mathbf{x} \neq \mathbf{0}$ is a basic feasible solution of $\mathbf{Ax} = \mathbf{b}$ and thus $\mathbf{x} = \hat{\mathbf{x}} + \hat{\mathbf{y}} = \hat{\mathbf{x}} + \mathbf{0}$. Clearly the theorem holds for $\mathbf{x} = \hat{\mathbf{x}}$. But if the p vectors from $\mathbf{A}$ are linearly dependent, then by virtue of an argument similar to the one underlying (7.4) – (7.7), let

$$\theta_1 = \begin{cases} \min_r\{x_{jr}/\lambda_{jr}, \lambda_{jr} > 0, j \in \mathfrak{J}\}, \\ +\infty \text{ if } \lambda_{jr} \leq 0 \text{ for all } r,\ j \in \mathfrak{J}; \end{cases}$$

$$\theta_2 = \begin{cases} \max_r\{x_{jr}/\lambda_{jr}, \lambda_{jr} < 0, j \in \mathfrak{J}\}, \\ -\infty \text{ if } \lambda_{jr} \geq 0 \text{ for all } r,\ j \in \mathfrak{J}. \end{cases}$$

Since at least one λ_{jr} value must be different from zero, at least one of the values θ_1, θ_2 must be finite. If θ_s is finite, let

$$\bar{x}_{jr} = \begin{cases} x_{jr} - \theta_s \lambda_{jr}, r=1,\ldots,p; \\ 0\ ,\quad j \notin \mathfrak{J}. \end{cases} \tag{7.12}$$

Then $\bar{\mathbf{x}}' = (\bar{x}_1,\ldots,\bar{x}_n)$ is a feasible solution to $\mathbf{Ax} = \mathbf{b}$ that employs at most $p-1$ vectors in $\mathbf{A}$. At this point two possible subcases emerge.

First, assume that all of the λ_{jr}'s are of the same sign, *i.e.*, if all $\lambda_{jr} \leq 0$ for $r=1,\ldots,p$, then $\theta_1 = +\infty$ and $\theta_2 < 0$ is finite; if $\lambda_{jr} \geq 0$ for all $r=1,\ldots,p$, then $\theta_1 > 0$ is finite and $\theta_2 = -\infty$. Hence if

$$\bar{y}_{jr} = \begin{cases} |\theta_s \lambda_{jr}|, r=1,\ldots,p; \\ 0,\quad j \notin \mathfrak{J} \end{cases}$$

then $\bar{\mathbf{x}} = \mathbf{x} - \bar{\mathbf{y}}$, where $\bar{\mathbf{y}}' = (\bar{y}_1,\ldots,\bar{y}_n) \in \mathfrak{T}$. Given that $\bar{\mathbf{x}}$ is a feasible solution of $\mathcal{S}$ that employs a set of no more than $p-1$ columns of $\mathbf{A}$, the theorem holds for this solution (by the induction hypothesis) and thus $\bar{\mathbf{x}} = \hat{\mathbf{x}} + \mathbf{y}_1$, where $\hat{\mathbf{x}} \in \text{co}\{\mathbf{x}_1, \ldots, \mathbf{x}_k\}$ and $\mathbf{y}_1 \in \mathcal{T}$. Thus $\hat{\mathbf{x}} + \mathbf{y}_1 = \mathbf{x} - \bar{\mathbf{y}}$ or $\mathbf{x} = \hat{\mathbf{x}} + (\mathbf{y}_1 + \bar{\mathbf{y}}) = \hat{\mathbf{x}} + \hat{\mathbf{y}}$. Hence the theorem holds for $\mathbf{x}$.

Second, suppose the λ_{jr}'s are not all of the same sign. In this instance $0 < \theta_1 < +\infty$, $-\infty < \theta_2 < 0$ and thus, when $\theta_s = \theta_1$, (7.12) yields $\bar{\mathbf{x}}(\theta_1)$; and when $\theta_s = \theta_2$, (7.12) renders $\bar{\mathbf{x}}(\theta_2)$. Then

$$\begin{aligned} \mathbf{x} &= \frac{(-\theta_2)\bar{\mathbf{x}}(\theta_1) + \theta_1 \bar{\mathbf{x}}(\theta_2)}{\theta_1 - \theta_2} \\ &= \beta \bar{\mathbf{x}}(\theta_1) + (1-\beta)\bar{\mathbf{x}}(\theta_2), \end{aligned}$$

where $\beta = \theta_2/(\theta_2 - \theta_1)$, $0 < \beta < 1$. Given that both $\bar{\mathbf{x}}(\theta_1)$, $\bar{\mathbf{x}}(\theta_2)$ are feasible solutions for $\mathbf{Ax} = \mathbf{b}$ using $p-1$ or fewer column vectors for $\mathbf{A}$, then, by the induction hypothesis, the theorem holds for these solutions and thus

$$\begin{aligned} \bar{\mathbf{x}}(\theta_1) &= \hat{\mathbf{x}}_1 + \mathbf{y}_1, \\ \bar{\mathbf{x}}(\theta_2) &= \hat{\mathbf{x}}_2 + \mathbf{y}_2, \end{aligned}$$

where both $\hat{\mathbf{x}}_1$, $\hat{\mathbf{x}}_2 \in \text{co}\{\mathbf{x}_1, \ldots, \mathbf{x}_k\}$ and $\mathbf{y}_1, \mathbf{y}_2 \in \mathcal{T}$. Hence

$$\begin{aligned} \mathbf{x} &= \beta(\hat{\mathbf{x}}_1 + \mathbf{y}_1) + (1-\beta)(\hat{\mathbf{x}}_2 + \mathbf{y}_2) \\ &= \beta \hat{\mathbf{x}}_1 + (1-\beta)\hat{\mathbf{x}}_2 + \beta \mathbf{y}_1 + (1-\beta)\mathbf{y}_2 \\ &= \hat{\mathbf{x}} + \hat{\mathbf{y}}. \end{aligned}$$

Thus the theorem holds for this $\mathbf{x}$ and, by induction, holds for p as well so that the theorem is true in general. Q.E.D.

As this theorem indicates, $\mathbf{x} \in \mathcal{S}$ is expressible as a convex combination of the set of basic feasible solutions to $\mathbf{Ax} = \mathbf{b}$ plus a homogeneous solution associated with $\mathbf{Ax} = \mathbf{b}$, $\mathbf{x} \geqq \mathbf{0}$. We observe further that $\mathcal{S}$ is bounded if and only if $\mathcal{T}$ admits a unique solution $\mathbf{y} = \mathbf{0}$, *i.e.*, $\mathcal{T} = \{\mathbf{0}\}$. (In the next chapter a bounded $\mathcal{S} = \{\mathbf{x} | \mathbf{Ax} = \mathbf{b}, \mathbf{x} \geqq \mathbf{0}, \mathbf{x} \in R^n\}$ will be referred to as a convex polytope.)

Next, a vector $\mathbf{y} \in \mathcal{T}$ is termed an ***extreme homogeneous solution*** if and only if it is a basic feasible solution to the ***normalized homogeneous system*** $\mathbf{Ay} = \mathbf{0}$, $\mathbf{1'y} = 1$, $\mathbf{y} \geqq \mathbf{0}$. Let $\bar{\mathcal{T}} = \{\mathbf{y} | \mathbf{Ay} = \mathbf{0}, \mathbf{1'y} = 1, \mathbf{y} \geqq \mathbf{0}\}$ represent the set of normalized homogeneous solutions. Clearly $\bar{\mathcal{T}} \subseteq \mathcal{T}$. Here the ***normalizing constraint*** $\mathbf{1'y} = 1$ has been added to $\mathcal{T}$ so that $\mathbf{y} = \mathbf{0}$ cannot be a basic feasible or extreme homogeneous solution within $\bar{\mathcal{T}}$. Under this restriction we now offer

7.2.3. THEOREM. Let $\mathcal{I} = \{\mathbf{x} | \mathbf{Ax} = \mathbf{b}, \mathbf{x} \geqq \mathbf{0}, \mathbf{x} \epsilon R^n\} \neq \phi$ with $\mathcal{T}$, $\bar{\mathcal{T}}$ defined as above and $\{\mathbf{y}_1, \ldots, \mathbf{y}_\ell\}$ the set of extreme homogeneous (basic feasible) solutions of $\bar{\mathcal{T}} \subseteq \mathcal{T}$.

(1) Either $\hat{\mathbf{y}} = \mathbf{0} \epsilon \mathcal{T}$ is a unique homogeneous solution

or

(2) every $\mathbf{y} \epsilon \mathcal{T}$ is a member of $coni\{\mathbf{y}_1, \ldots, \mathbf{y}_\ell\}$.

To briefly rationalize this result suppose $\mathbf{y} = \mathbf{0} \epsilon \mathcal{T}$ is not a unique homogeneous solution associated with $\mathcal{I}$. Hence there must exist $\mathbf{y}(\neq \mathbf{0}) \epsilon \mathcal{T}$ with the following property: if $\bar{\mathbf{y}} \epsilon \bar{\mathcal{T}}$, then any $\mathbf{y} \epsilon \mathcal{T}$ (including $\mathbf{0}$) can be written as $\lambda \bar{\mathbf{y}}$, $0 < \lambda \epsilon R$, by virtue of the normalization constraint $\mathbf{1'y} = \mathbf{1}'(\lambda \bar{\mathbf{y}}) = \lambda$. Let us now apply the preceding resolution theorem to $\bar{\mathcal{T}}$. Since the homogeneous system associated with $\bar{\mathcal{T}}$ is $\mathbf{Ay} = \mathbf{0}$, $\mathbf{1'y} = 0$, $\mathbf{y} \geqq \mathbf{0}$, clearly $\mathbf{y} = \mathbf{0}$ is the only homogeneous solution satisfying these requirements, *i.e.*, $\mathbf{y} = \mathbf{0}$ is the unique homogeneous solution associated with $\bar{\mathcal{T}}$. Then according to the resolution theorem, every $\mathbf{y} \epsilon \bar{\mathcal{T}}$ is an element of the convex hull of basic feasible or extreme homogeneous solutions of $\bar{\mathcal{T}}$. Hence by virtue of the above proportionality property, every $\mathbf{y} \epsilon \mathcal{T}$ can be expressed as a conical combination of extreme homogeneous solutions of $\bar{\mathcal{T}}$.

We may actually combine the results presented in theorems 7.2.2, 7.2.3 into a more general theorem which directly allows for the possibility that $\mathcal{I}$ is bounded. To this end we have, from theorems 7.2.2 and 7.2.3,

7.2.4. RESOLUTION THEOREM.

Let $\mathcal{S} = \{\mathbf{x} | \mathbf{A}\mathbf{x} = \mathbf{b}, \mathbf{x} \geqq \mathbf{0}, \mathbf{x} \in \mathbf{R}^n\} \neq \phi$ with $\mathcal{T}$, $\bar{\mathcal{T}}$ defined as above. Additionally, let $\{\mathbf{x}_1, \ldots, \mathbf{x}_k\}$ be the set of basic feasible solutions of $\mathcal{S}$ and $\{\mathbf{y}_1, \ldots, \mathbf{y}_\ell\}$ the set of extreme homogeneous solutions of $\bar{\mathcal{T}} \subseteq \mathcal{T}$.

(1) If $\hat{\mathbf{y}} = \mathbf{0} \in \mathcal{T}$ is a unique homogeneous solution or, equivalently, $\mathcal{T} = \{\mathbf{0}\}$, then $\mathcal{S} = co\{\mathbf{x}_1, \ldots, \mathbf{x}_k\}$.

(2) If $\mathcal{T} \neq \{\mathbf{0}\}$, then $\mathcal{S} = co\mathbf{x}_1, \ldots \mathbf{x}_k\} + coni\{\mathbf{y}_1, \ldots, \mathbf{y}_\ell\}$.

If case (1) holds, then $\mathcal{S}$ is bounded and every $\mathbf{x} \in \mathcal{S}$ is expressible as

$$\mathbf{x} = \sum_{i=1}^{k} \lambda_i \mathbf{x}_i, \ \sum_{i=1}^{k} \lambda_i = 1, \ 0 \leq \lambda_i \in \mathbf{R};$$

if case (2) obtains, $\mathcal{S}$ is unbounded and thus every $\mathbf{x} \in \mathcal{S}$ can be written as

$$\mathbf{x} = \sum_{i=1}^{k} \lambda_i \mathbf{x}_i + \sum_{j=1}^{\ell} \mu_j \mathbf{y}_j, \ \sum_{i=1}^{k} \lambda_i = 1, \ 0 \leq \lambda_i \in \mathbf{R}, \ 0 \leq \mu_j \in \mathbf{R}.$$

This particular resolution theorem will be encountered again (but in a slightly different form) in Chapter 8 when we address the geometric interpretation of basic feasible solutions as well as extreme homogeneous solutions.

In section 2 of Chapter 1 we examined the conditions under which a simultaneous linear equation system of the form $\mathbf{A}\mathbf{x} = \mathbf{C}$ is inconsistent. The analytical device employed in carrying out the derivation was Farkas' theorem of the alternative. Let us again utilize this theorem to structure a set of conditions which would indicate when a mixed simultaneous system of linear inequalities and equalities $\mathbf{A}_1\mathbf{x} \leqq \mathbf{b}_1$, $\mathbf{A}_2\mathbf{x} = \mathbf{b}_2$ has no solution. Here $\mathbf{A}_1$ is $(m \times n)$, $\mathbf{A}_2$ is $(p \times n)$, $\mathbf{b}_1 \in \mathbf{R}^m$, and $\mathbf{b}_2 \in \mathbf{R}^p$.

If the statement "$\mathbf{A}_1\mathbf{x} \leqq \mathbf{b}_1$, $\mathbf{A}_2\mathbf{x} = \mathbf{b}_2$ is consistent" serves as Farkas' (I) (see, for instance, theorem 1.2.5), then what should Farkas' (II) look like? To answer this question let us first convert $\mathbf{A}_1\mathbf{x} \leqq \mathbf{b}_1$, to a set of equalities by introducing a vector $\mathbf{w} \in R^m$ of so-called nonnegative slack variables, *i.e.*, $\mathbf{A}_1\mathbf{x} + \mathbf{w} = \mathbf{b}_1$, $\mathbf{w} \geqq \mathbf{0}$. Then, with $\mathbf{x} = \mathbf{u} - \mathbf{v}$ and $\mathbf{u}, \mathbf{v} \geqq \mathbf{0}$, Farkas' (I) can be written

$$\left.\begin{aligned} \mathbf{A}_1\mathbf{u} - \mathbf{A}_1\mathbf{v} + \mathbf{I}_m\mathbf{w} &= \mathbf{b}_1 \\ \mathbf{A}_2\mathbf{u} - \mathbf{A}_2\mathbf{v} \qquad &= \mathbf{b}_2 \end{aligned}\right\} \text{has a solution } \mathbf{u}, \mathbf{v}, \mathbf{w} \geqq \mathbf{0} \text{ or}$$

$$\text{(I}'\text{)} \quad \begin{bmatrix} \mathbf{A}_1 & -\mathbf{A}_1 & \mathbf{I}_m \\ \mathbf{A}_2 & -\mathbf{A}_2 & \mathbf{0} \end{bmatrix} \begin{bmatrix} \mathbf{u} \\ \mathbf{v} \\ \mathbf{w} \end{bmatrix} = \begin{bmatrix} \mathbf{b}_1 \\ \mathbf{b}_2 \end{bmatrix} \text{ has a nonnegative solution.}$$

From Farkas' (II), the alternative to (I′) is

$$\begin{bmatrix} \mathbf{A}_1' & \mathbf{A}_2' \\ -\mathbf{A}_1' & -\mathbf{A}_2' \\ \mathbf{I}_m & \mathbf{0} \end{bmatrix} \begin{bmatrix} \mathbf{y}_1 \\ \mathbf{y}_2 \end{bmatrix} \geqq \mathbf{0}, \ (\mathbf{b}_1', \mathbf{b}_2') \begin{bmatrix} \mathbf{y}_1 \\ \mathbf{y}_2 \end{bmatrix} < 0 \quad \text{or}$$

$$\text{(II}'\text{)} \quad \left.\begin{aligned} \mathbf{A}_1'\mathbf{y}_1 + \mathbf{A}_2'\mathbf{y}_2 &= \mathbf{0} \\ \mathbf{y}_1 &\geqq \mathbf{0} \\ \mathbf{b}_1'\mathbf{y}_1 + \mathbf{b}_2'\mathbf{y}_2 &< 0 \end{aligned}\right\}$$

has a solution $\mathbf{y}' = (\mathbf{y}_1', \mathbf{y}_2') \in R^{m+p}$,

where the vector $\mathbf{y}_1(\mathbf{y}_2)$ is associated with the system of inequalities (equalities). Hence either $\mathbf{A}_1\mathbf{x} \leqq \mathbf{b}_1$, $\mathbf{A}_2\mathbf{x} = \mathbf{b}_2$ is consistent, or there exists a vector $\mathbf{y}' = (\mathbf{y}_1', \mathbf{y}_2')$, with $\mathbf{y}_1 \geq \mathbf{0}$, which is orthogonal to the columns of $\mathbf{A}_1, \mathbf{A}_2$ and which makes an obtuse angle ($\geq \pi/2$) with $\mathbf{b}_1, \mathbf{b}_2$. These observations are summarized in

7.2.5. THEOREM. The mixed simultaneous linear system $\mathbf{A}_1\mathbf{x} \leqq \mathbf{b}_1$, $\mathbf{A}_2\mathbf{x} = \mathbf{b}_2$, where $\mathbf{A}_1$ is $(m \times n)$, $\mathbf{A}_2$ is $(p \times n)$, $\mathbf{b}_1 \in R^m$, and $\mathbf{b}_2 \in R^p$, is inconsistent if and only if the linear system

$$\begin{aligned} \mathbf{A}_1'\mathbf{y}_1 + \mathbf{A}_2'\mathbf{y}_2 &= \mathbf{0} \\ \mathbf{y}_1 &\geqq \mathbf{0} \\ \mathbf{b}_1'\mathbf{y}_1 + \mathbf{b}_2'\mathbf{y}_2 &< 0 \end{aligned}$$

has a solution $\mathbf{y}' = (\mathbf{y}_1', \mathbf{y}_2') \in R^{m+p}$.

We may view the inconsistency of system $\mathbf{A}_1\mathbf{x} \leqq \mathbf{b}_1$, $\mathbf{A}_2\mathbf{x} = \mathbf{b}_2$ in a somewhat different light by working with its ***linear combination***. That is, forming

$$\begin{aligned} \mathbf{y}_1'\mathbf{A}_1\mathbf{x} &\leq \mathbf{y}_1'\mathbf{b}_1,\ \mathbf{y}_1 \geqq \mathbf{0}, \\ \mathbf{y}_2'\mathbf{A}_2\mathbf{x} &= \mathbf{y}_2'\mathbf{b}_2 \end{aligned}$$

and summing yields

$$(\mathbf{y}_1'\mathbf{A}_1 + \mathbf{y}_2'\mathbf{A}_2)\mathbf{x} \leq \mathbf{y}_1'\mathbf{b}_1 + \mathbf{y}_2'\mathbf{b}_2 < 0,\ \mathbf{y}_1 \geqq \mathbf{0}. \tag{7.13}$$

Clearly every solution $\mathbf{x}$ of the system $\mathbf{A}_1\mathbf{x} \leq \mathbf{b}_1$, $\mathbf{A}_2\mathbf{x} = \mathbf{b}_2$ must also satisfy their linear combination (7.13). Let $\mathbf{y}_1'\mathbf{b}_1 + \mathbf{y}_2'\mathbf{b}_2 = \alpha < 0$. Since $\mathbf{A}_1'\mathbf{y}_1 + \mathbf{A}_2'\mathbf{y}_2 = \mathbf{0}$, we see immediately from (7.13) that the system $\mathbf{A}_1\mathbf{x} \leqq \mathbf{b}_1$, $\mathbf{A}_2\mathbf{x} = \mathbf{b}_2$ is inconsistent if and only if its linear combination reads

$$\mathbf{0}'\mathbf{x} = 0 \leq \alpha,\ \ \alpha < 0. \tag{7.13.1}$$

The importance of theorem 7.2.5 is that it can be applied to the case where a check on the inconsistency of a system such as $\mathbf{A}\mathbf{x} = \mathbf{b}$, $\mathbf{x} \geqq \mathbf{0}$ is desired. All we need to do is rewrite this system as

$$\begin{aligned} -\mathbf{x} &\leqq \mathbf{0}, \\ \mathbf{A}\mathbf{x} &= \mathbf{b} \end{aligned}$$

and then directly apply the theorem. That is, $\mathbf{A}\mathbf{x} = \mathbf{b}$, $\mathbf{x} \geqq \mathbf{0}$ is inconsistent if and only if

$$\begin{aligned} -\mathbf{y}_1 + \mathbf{A}\mathbf{y}_2 &= \mathbf{0} \\ \mathbf{y}_1 \quad &\geqq \mathbf{0} \\ \mathbf{b}'\mathbf{y}_2 &< 0 \end{aligned}$$

or

$$\begin{aligned} \mathbf{A}\mathbf{y}_2 &\geqq \mathbf{0}, \\ \mathbf{b}'\mathbf{y}_2 &< 0. \end{aligned}$$

7.3. Complementary Slackness in Pairs of Dual Systems

Consider the general dual homogeneous systems (I), (II) presented in section 5.1 of Chapter 5. In systems such as these, if, say, the inequality $c_{11}x_1 + c_{12}x_2 + d_{11}y_1 + d_{12}y_2 \geqq 0$ in (I) is satisfied as a strict inequality (>0) at a particular solution of (I) and (II), then this inequality is said to be a ***slack inequality***. Moreover, given that this inequality is slack at the said solution, its complementary inequality $v_1 \geqq 0$ in (II) must hold as a strict equality (=0) and conversely. This assertion is verified by the following

7.3.1. COMPLEMENTARY SLACKNESS THEOREM [Tucker, 1956]. Given $(m \times n)$ matrices **A**, **B**, **C**, and **D** and solutions $\mathbf{x}, \mathbf{y} \in R^n$ and $\mathbf{u}, \mathbf{v} \in R^m$ in the general dual homogeneous systems

$$\text{(I)} \begin{cases} \mathbf{A}\mathbf{x} + \mathbf{B}\mathbf{y} = \mathbf{0} \\ \mathbf{C}\mathbf{x} + \mathbf{D}\mathbf{y} \leqq \mathbf{0} \\ \mathbf{x} \geqq \mathbf{0} \\ \mathbf{y} \text{ unrestricted} \end{cases} \qquad \text{(II)} \begin{cases} \mathbf{u} \text{ unrestricted} \\ \mathbf{v} \geqq \mathbf{0} \\ \mathbf{A}'\mathbf{u} + \mathbf{C}'\mathbf{v} \geqq \mathbf{0} \\ \mathbf{B}'\mathbf{u} + \mathbf{D}'\mathbf{v} = \mathbf{0} \end{cases}$$

each of the $m+n$ pairs of complementary inequalities

$$\begin{aligned} v_i \geq 0 &\longleftrightarrow \sum_j c_{ij}x_j + \sum_k c_{ik}y_k \leq 0, i=1,\ldots,m; \\ x_j \geq 0 &\longleftrightarrow \sum_h a_{hj}u_h + \sum_i c_{ik}v_i \geq 0, j=1,\ldots,n; \end{aligned} \tag{7.14}$$

contains exactly one inequality that is slack.

PROOF. Let $\mathbf{x}$, $\mathbf{y}$ and $\mathbf{u}$, $\mathbf{v}$ be any solutions of the dual systems (I), (II). By multiplying complementary items in these systems and summing yields

$$\begin{aligned}&(1)\ \mathbf{u}'(\mathbf{Ax}+\mathbf{By}) = 0\\&(2)\ \mathbf{v}'(\mathbf{Cx}+\mathbf{Dy}) \leq 0\\&(3)\ \mathbf{x}'(\mathbf{A}'\mathbf{u}+\mathbf{C}'\mathbf{v}) \geq 0\\&(4)\ \mathbf{y}'(\mathbf{B}'\mathbf{u}+\mathbf{D}'\mathbf{v}) = 0.\end{aligned} \tag{7.15}$$

Adding (7.15.1), (7.15.3) and (7.15.2), (7.15.4) we have $\mathbf{u}'\mathbf{By}+\mathbf{x}'\mathbf{C}'\mathbf{v} \geq 0$ and $\mathbf{v}'\mathbf{Cx}+\mathbf{y}'\mathbf{B}'\mathbf{u} \leq 0$ respectively so that $\mathbf{u}'\mathbf{By}+\mathbf{v}'\mathbf{Cx} = 0$. Combining this with (7.15.1), (7.15.4) we get

$$\mathbf{u}'\mathbf{Ax} = -\mathbf{u}'\mathbf{By} = \mathbf{v}'\mathbf{Cx} = \mathbf{v}'\mathbf{Dy}.$$

When use is made of these equations in (7.15.2), (7.15.3) we obtain the ***weak complementary slackness conditions***

$$\begin{aligned}\mathbf{v}'(\mathbf{Cx}+\mathbf{Dy}) &= \sum_i v_i\Big(\sum_j c_{ij}x_j + \sum_k d_{ik}y_k\Big) = 0.\\ \mathbf{x}'(\mathbf{A}'\mathbf{u}+\mathbf{C}'\mathbf{v}) &= \sum_j x_j\Big(\sum_h a_{hj}u_h + \sum_i c_{ij}v_k\Big) = 0.\end{aligned} \tag{7.16}$$

These two equations show that in each pair of dual inequalities in (7.14) ***at least one*** sign of equality (=) must hold for all solutions. Thus each pair of dual inequalities contains ***at most one*** inequality that is slack.

By theorem 5.7 (Tucker's fourth existence theorem), there exist solutions $\mathbf{x}$, $\mathbf{y}$ and $\mathbf{u}$, $\mathbf{v}$ of the dual systems such that the ***strong complementary slackness conditions*** hold. In this instance each pair of complementary

$$v_i + \left(-\sum_j c_{ij}x_j + \sum_k d_{ik}y_k \right) > 0, i=1,\dots,m,$$
$$x_j + \left(\sum_h a_{hj}u_h + \sum_i c_{ij}v_k \right) > 0, j=1,\dots,n \tag{7.17}$$

inequalities contains ***at least one*** inequality that is slack. Combining (7.16), (7.17) we see that each pair of complementary inequalities in the dual systems contains ***exactly one*** inequality that is slack. Q.E.D.

By virtue of this theorem the essence of the notion of complementary slackness may be summarized as: the set of slack inequalities in one of a pair of dual systems is exactly complementary to the set of slack inequalities in the other system; opposite each slack inequality there is a non-slack inequality and conversely.

In a similar vein we have

7.3.2. THEOREM [Tucker, 1956]. Let $\mathbf{K}$ be an $(m \times n)$ skew-symmetric matrix ($\mathbf{K} = -\mathbf{K}'$). Then in the self-dual system

$$\mathbf{Kw} \geqq \mathbf{0},\ \mathbf{w} \geqq 0$$

each of the n pairs of complementary inequalities

$$\sum_j k_{ij}w_j \geq 0 \longleftrightarrow w_i \geq 0,\ i=1,\dots,n, \tag{7.18}$$

contains exactly one inequality that is slack.

PROOF. For any solution $\mathbf{w} \geqq \mathbf{0}$, since $\mathbf{w}'\mathbf{Kw} = \mathbf{w}'\mathbf{K}'\mathbf{w} = \mathbf{w}'\mathbf{Kw}$, it follows that

$$\mathbf{w}'\mathbf{Kw} = \sum_i \sum_j w_i k_{ij} w_j = 0.$$

Hence in each pair of complementary inequalities (7.18) ***at least one*** sign of equality (=) must hold for all solutions $\mathbf{w} \geqq 0$. Thus each pair of complementary inequalities contains ***at most one*** inequality that is slack.

By theorem 5.8 (Tucker's fifth existence theorem), there exists a solution **w** such that

$$\sum_j k_{ij} w_j + w_i > 0, i=1,\ldots,n, \tag{7.19}$$

i.e., each pair of complementary inequalities contains ***at least one*** inequality that is slack. Taking (7.18), (7.19) together, each pair of complementary inequalities contains ***exactly one*** inequality that is slack. Q.E.D.

7.4 Exercises

1. Find a nondegenerate basic feasible solution to

$$\begin{aligned} x_1 + 2x_2 + x_3 - x_4 &= 7 \\ 3x_1 + x_2 \qquad\quad + x_4 &= 9 \\ x_1, x_2 \geq 0. \end{aligned}$$

2. Find the collection of vertices of $\mathcal{S} = \{\mathbf{x} \mid \mathbf{Ax} = \mathbf{b}, \mathbf{x} \geqq \mathbf{0}\}$, where

$$\mathbf{A} = \begin{bmatrix} 1 & 1 & 2 & 0 \\ 2 & 0 & -1 & 1 \end{bmatrix}, \quad \mathbf{b} = \begin{bmatrix} 2 \\ 4 \end{bmatrix}.$$

3. Let $\mathcal{T} = \{\mathbf{y} \mid \mathbf{Ay} = \mathbf{0}, \mathbf{y} \geqq \mathbf{0}\}$. Show that if $\mathbf{x} \in \mathcal{S}$ in problem two, then $\mathbf{x} = \hat{\mathbf{x}} + \hat{\mathbf{y}}$, where $\hat{\mathbf{x}}$, $\hat{\mathbf{y}}$ are as defined in theorem 7.2.2.

4. Find an extreme homogeneous solution to $\mathbf{Ay} = \mathbf{0}$ using $\mathcal{T}$ from the preceding problem.

5. Verify that theorem 7.2.4 holds using $\mathcal{S}$ from problem two.

6. Demonstrate using theorem 7.2.5 that the system

$$\begin{aligned} 3x_1 + x_2 + x_3 \qquad &= -5 \\ \tfrac{1}{2}x_1 + 4x_2 \qquad - x_4 &= 2 \end{aligned}$$

is inconsistent.

CHAPTER 8
EXTREME POINTS AND DIRECTIONS FOR CONVEX SETS

8.1 Extreme Points and Directions for General Convex Sets.

Let $\mathcal{S}$ be a convex set in $\mathbf{R}^n$. Two vectors (points) $\mathbf{x}_1, \mathbf{x}_2 \in \mathcal{S}$ are said to be ***distinct*** if $\mathbf{x}_1 \neq \alpha \mathbf{x}_2$ for any $0 < \alpha \in \mathbf{R}$ (*i.e*, one vector cannot be written as a positive multiple of the other). In this regard, a vector $\mathbf{x} \in \mathcal{S}$ is an ***extreme point*** of $\mathcal{S}$ if it is not the mean of two other distinct vectors in $\mathcal{S}$, *e.g.*, if extreme $\mathbf{x}$ is the mean of $\mathbf{x}_1$, $\mathbf{x}_2 \in \mathcal{S}$, then $\mathbf{x} = \mathbf{x}_1 = \mathbf{x}_2$. Equivalently, a point $\mathbf{x} \in \mathcal{S}$ is extreme if $\mathbf{x}$ cannot be expressed as a positive convex combination of two distinct points in $\mathcal{S}$. Thus $\mathbf{x}$ is an extreme point of $\mathcal{S}$ if and only if $\mathbf{x} = \lambda \mathbf{x}_1 + (1-\lambda)\mathbf{x}_2$, $0 < \lambda < 1$, and $\mathbf{x}_1$, $\mathbf{x}_2 \in \mathcal{S}$ implies $\mathbf{x} = \mathbf{x}_1 = \mathbf{x}_2$. Hence there is no way to express $\mathbf{x}$ as a positive convex combination of $\mathbf{x}_1$, $\mathbf{x}_2$ except by taking $\mathbf{x} = \mathbf{x}_1 = \mathbf{x}_2$. Clearly an extreme point of a convex set $\mathcal{S}$ cannot lie between and two other points of the set.

We note briefly that other characterizations of an extreme point are possible, *e.g.*:

(a) For a convex set $\mathcal{S}$ in $\mathbf{R}^n$, $\mathbf{x} \in \mathcal{S}$ is an extreme point if and only if $\mathbf{x} = \lambda \mathbf{x}_1 + (1-\lambda)\mathbf{x}_2$, $0 \leqq \lambda \leqq 1$, and $\mathbf{x}_1, \mathbf{x}_2 \in \mathcal{S}$ with $\mathbf{x}_1 \neq \mathbf{x}_2$ imply that either $\lambda = 0$ or $\lambda = 1$.

(b) For a convex set $\mathcal{S}$ in $\mathbf{R}^n$, $\mathbf{x} \in \mathcal{S}$ is an extreme point if and only if $\mathcal{S} \backslash \{\mathbf{x}\}$ is convex.

Under what conditions does a convex subset of $\mathbf{R}^n$ possess an extreme point? To answer this question we state the following

8.1.1. EXISTENCE THEOREM. A nonempty compact convex set $\mathcal{S}$ in $\mathbf{R}^n$ has at least one extreme point.

PROOF. Since $\mathcal{S}$ is compact and the function $\|\mathbf{x}\|$ is continuous on $\mathcal{S}$, Weierstrass's theorem (see footnote 2 of chapter 3) informs us that there exists an element $\mathbf{x}_0 \epsilon \mathcal{S}$ such that $\|\mathbf{x}_0\| = max\{\|\mathbf{x}\| \,|\, \mathbf{x} \epsilon \mathcal{S}\}$. With $\mathcal{S}$ convex, we may form $\mathbf{x}_0 = \lambda \mathbf{x}_1 + (1-\lambda)\mathbf{x}_2 = \mathbf{x}_2 + \lambda(\mathbf{x}_1 - \mathbf{x}_2)$, $0 < \lambda < 1$, where $\mathbf{x}_1, \mathbf{x}_2 \epsilon \mathcal{S}$. Then

$$\|\mathbf{x}\|^2 \leq \|\mathbf{x}_0\|^2 = \mathbf{x}_0'\mathbf{x}_0 = \|\mathbf{x}_2\|^2 + 2\lambda\mathbf{x}_2'(\mathbf{x}_1 - \mathbf{x}_2) + \lambda^2 \|\mathbf{x}_1 - \mathbf{x}_2\|^2. \tag{8.1}$$

For $\mathbf{x} = \mathbf{x}_2$, the preceding inequality becomes

$$0 \leq 2\mathbf{x}_2' \,(\mathbf{x}_1 - \mathbf{x}_2) + \lambda \|\mathbf{x}_1 - \mathbf{x}_2\|^2. \tag{8.1.1}$$

Since the inner product and norm operators are symmetrical, (8.1.1) may be rewritten as

$$0 \leq 2\mathbf{x}_1' \,(\mathbf{x}_2 - \mathbf{x}_1) + \lambda \|\mathbf{x}_2 - \mathbf{x}_1\|^2. \tag{8.1.2}$$

Adding (8.1.1), (8.1.2) yields

$$\|\mathbf{x}_1 - \mathbf{x}_2\|^2 \leq \lambda \|\mathbf{x}_1 - \mathbf{x}_2\|^2.$$

And since $0 < \lambda < 1$, we must have $\|\mathbf{x}_1 - \mathbf{x}_2\| = 0$ or $\mathbf{x}_1 = \mathbf{x}_2$. Q.E.D.

We next state

8.1.2. THEOREM. If $\mathcal{S}$ is a nonempty compact convex set in $\mathbf{R}^n$, then any tangent hyperplane to $\mathcal{S}$ contains at least one extreme point of $\mathcal{S}$.

PROOF. For $\mathcal{H} = \{\mathbf{x} | \mathbf{C}'\mathbf{x} = \alpha,\ \mathbf{x} \epsilon \mathbf{R}^n\}$ a hyperplane tangent to $\mathcal{S}$ at $\mathbf{x}_0$, the set $\mathcal{T} = \mathcal{H} \cap \mathcal{S}$ is a nonempty compact convex set and thus, by the preceding theorem, has at least one extreme point. If $\mathbf{x}_0 =$

$\lambda\mathbf{x}_1 + (1-\lambda)\mathbf{x}_2$, $0 < \lambda < 1$, with $\mathbf{x}_1$, $\mathbf{x}_2 \in \mathcal{S}$ is an extreme point of $\mathcal{T}$, then, since $\mathcal{S}$ is a subset of, say $[\mathcal{H}^+]$, it follows that $\mathbf{x}_1, \mathbf{x}_2 \in [\mathcal{H}^+]$ as well. Moreover, $C'\mathbf{x}_0 = \lambda C'\mathbf{x}_1 + (1-\lambda)C'\mathbf{x}_2 = \alpha$ only if $C'\mathbf{x}_1 = C'\mathbf{x}_2 = \alpha$. But this means that $\mathbf{x}_1, \mathbf{x}_2$ are on $\mathcal{H}$ and thus $\mathbf{x}_1, \mathbf{x}_2 \in \mathcal{T}$. With $\mathbf{x}_0$ an extreme point of $\mathcal{T}$ we must have $\mathbf{x} = \mathbf{x}_1 = \mathbf{x}_2$ so that $\mathbf{x}_0$ is also an extreme point of $\mathcal{S}$. Q.E.D.

The importance of extreme points of a compact convex set $\mathcal{S}$ is that they form a minimal subset $\mathcal{E}$ whose convex hull equals $\mathcal{S}$($\mathcal{E}$ is "minimal" in that if compact and convex $\mathcal{S} = co(\mathcal{E})$, then $\mathcal{S} \subset \mathcal{E}$). That $\mathcal{S}$ can be represented in this fashion by its set of extreme points $\mathcal{E}$ is asserted by the following fundamental

8.1.3. REPRESENTATION THEOREM [Krein-Milman, 1940]. A nonempty compact convex set $\mathcal{S}$ in $\mathbf{R}^n$ is the convex hull of its set $\mathcal{E}$ of extreme points, *i.e.*, $co(\mathcal{E}) = \mathcal{S}$.

PROOF. Since $\mathcal{E} \subseteq \mathcal{S}$, it follows that $co(\mathcal{E}) \subseteq \mathcal{S}$. To prove the reverse set inclusion, let us assume to the contrary that $\mathcal{S} \not\subset co(\mathcal{E})$. Hence we may choose a vector $\mathbf{x}_0 \in \mathcal{S}$ such that $\mathbf{x}_0 \notin co(\mathcal{E})$. By the strong separation theorem 3.2.2, there exists a hyperplane $\mathcal{H} = \{\mathbf{x} | C'\mathbf{x} = \alpha, \mathbf{x} \in \mathbf{R}^n\}$ which strictly separates $\mathbf{x}_0$ and $co(\mathcal{E})$, *i.e*, for any $\mathbf{x} \in co(\mathcal{E})$, $C'\mathbf{x} \leq \alpha < C'\mathbf{x}_0$. With $\mathcal{S}$ compact and with $C'\mathbf{x}$ continuous on $\mathcal{S}$, there exists an $\bar{\mathbf{x}} \in \mathcal{S}$ such that $c = C'\bar{\mathbf{x}} = max\{C'\mathbf{x} | \mathbf{x} \in \mathcal{S}\} \geq C'\mathbf{x}_0 > \alpha$ (we invoke Weierstrass's theorem).

Let us now introduce the hyperplane $\mathcal{H}_c = \{\mathbf{x} | C'\mathbf{x} = c,\ \mathbf{x} \in \mathbf{R}^n\}$. Since $\bar{\mathbf{x}} \in \mathcal{H}_c$ and $\mathbf{x} \in [\mathcal{H}_c^-]$ for all other $\mathbf{x} \in \mathcal{S}$, it follows from theorem 3.2.3 that $\mathcal{H}_c$ is a supporting hyperplane to $\mathcal{S}$ at $\bar{\mathbf{X}}$. Moreover, by theorem 8.1.2, $\mathcal{H}_c$ admits at least one extreme point $\hat{\mathbf{x}}$ of $\mathcal{S}$. Hence $\hat{\mathbf{x}} \in \mathcal{E}$ and consequently $\hat{\mathbf{x}} \in co(\mathcal{E})$. Thus $C'\hat{\mathbf{x}} \leq \alpha < c$. But this contradicts the fact that $\hat{\mathbf{x}} \in \mathcal{H}_c$ (since $\hat{\mathbf{x}} \in \mathcal{H}_c$ implies $C'\hat{\mathbf{x}} = c$). Hence we must have $\mathcal{S} \subseteq co(\mathcal{E})$ and thus $co(\mathcal{E}) = \mathcal{S}$ as required. Q.E.D.

The set $\mathcal{E}$ of extreme points of a compact convex set $\mathcal{S}$ in $\mathbf{R}^n$ need not be closed. If $\mathcal{S}$ is closed, there is no guarantee that $co(\mathcal{S})$ is closed. To avoid this difficulty we may choose to replace $co(\mathcal{S})$ in the preceding theorem by $\overline{co(\mathcal{S})}$, the closed convex hull of $\mathcal{S}$. (Note that if $\mathcal{S}$ is closed and bounded, then $co(\mathcal{S})$ is closed and bounded.)

The set $\mathcal{S}$ of extreme points of a convex set $\mathcal{I}$ in $\mathbf{R}^n$ may not be a finite set. However, if $\mathcal{I}$ is deemed a finite subset of $\mathbf{R}^n$, then the convex hull $co(\mathcal{I})$ of $\mathcal{I}$ has at most a finite number of extreme points. To see this we state

8.1.4. THEOREM. If $\mathcal{I}$ is a finite subset of $\mathbf{R}^n$ and $\mathcal{S}$ is the collection of all extreme points of $co(\mathcal{I})$, then $\mathcal{S} \subseteq \mathcal{I}$.

PROOF. Select an $\mathbf{x} \notin \mathcal{I}$. For $\mathbf{x} \in co(\mathcal{I})$, there exists a finite set of points $\mathbf{x}_i \in \mathcal{I}$, $i=1,\ldots,r > 1$, such that

$$\mathbf{x} = \sum_{i=1}^{r} \lambda_i \mathbf{x}_i,\ \lambda_i \geq 0, \quad \sum_{i=1}^{r} \lambda_i = 1.$$

Let us assume that $\lambda_r \neq 0$, $\Sigma_{i=1}^{r-1} \lambda_i = 1 - \lambda_r \neq 0$. For $\lambda = \lambda_r$, $\mathbf{x}_1 = \Sigma_{i=1}^{r-1} [\lambda_i/(1-\lambda)]\mathbf{x}_i$, and $\mathbf{x}_2 = \mathbf{x}_r$ we obtain

$$\mathbf{x} = (1-\lambda) \sum_{i=1}^{r-1} \frac{\lambda_i}{1-\lambda} \mathbf{x}_i + \lambda_r \mathbf{x}_r = (1-\lambda)\mathbf{x}_1 + \lambda \mathbf{x}_2, \quad 0 < \lambda < 1, \quad \mathbf{x}_1 \neq \mathbf{x}_2,$$

where now both $\mathbf{x}_1, \mathbf{x}_2 \in co(\mathcal{I})$. Since $\mathbf{x}_1, \mathbf{x}_2$ are distinct points of $co(\mathcal{I})$, $\mathbf{x}$ cannot be an extreme point of the same. Hence $\mathcal{S} \subseteq \mathcal{I}$. Q.E.D.

Theorem 8.1.3 above informs us that any point within a compact (closed and bounded) convex set $\mathcal{I}$ in $\mathbf{R}^n$ can be represented as a convex combination of its extreme points. However, this may not be true if $\mathcal{I}$ is unbounded. In this regard, to handle the case of $\mathcal{I}$ unbounded, let us introduce the concept of an extreme direction.

First, for $\mathcal{I}$ a closed convex set in $\mathbf{R}^n$, a vector $\mathbf{d}(\neq \mathbf{0})$ is a ***recession***

direction of $\mathcal{S}$ if for each $\mathbf{x} \in \mathcal{S}$, $\mathbf{x} + \lambda\mathbf{d} \in \mathcal{S}$ for all $0 \leq \lambda \in \mathbf{R}$, *i.e.*, starting at $\mathbf{x}$ we can move or recede along $\mathbf{d}$ for any nonnegative step length λ and remain within $\mathcal{S}$. Clearly if $\mathcal{S}$ is bounded, it has no recession directions. Then a ***recession ray or half-line*** emanating from vertex $\mathbf{x}$ and pointing in the recession direction $\mathbf{d}$ is the subset $\{\mathbf{x} + \lambda\mathbf{d},\ \lambda \geq 0,\ \mathbf{d} \neq \mathbf{0}\}$ of $\mathcal{S}$. Moreover, the recession directions of half-lines do not depend upon any particular vector $\mathbf{x}$, *i.e.*, recession half-lines in $\mathcal{S}$ starting at different vectors are simple translates of those starting at $\mathbf{x}$.

Next, for $\mathcal{S}$ again a closed convex set in $\mathbf{R}^n$, an ***extreme direction*** of $\mathcal{S}$ is a recession direction which cannot be represented as a positive linear combination of two other distinct recession directions of $\mathcal{S}$. (Recession directions $\mathbf{d}_1$, $\mathbf{d}_2 \in \mathcal{S}$ are ***distinct*** if $\mathbf{d}_1 \neq \alpha\mathbf{d}_2$ for any $0 < \alpha \in \mathbf{R}$.) Thus $\mathbf{d}$ is an extreme direction of $\mathcal{S}$ if and only if $\mathbf{d} = \lambda_1\mathbf{d}_1 + \lambda_2\mathbf{d}_2$ for scalars $\lambda_1, \lambda_2 > 0$ and $\mathbf{d}_1, \mathbf{d}_2 \in \mathcal{S}$ implies $\mathbf{d}_1 = \alpha\mathbf{d}_2$, $0 < \alpha < \mathbf{R}$. So if $\mathbf{d}_1, \mathbf{d}_2$ are both extreme directions of $\mathcal{S}$, then any other distinct direction $\mathbf{d}$ can be represented as $\mathbf{d} = \lambda_1\mathbf{d}_1 + \lambda_2\mathbf{d}_2$ for scalars $\lambda_1, \lambda_2 > 0$. Any recession half-line is $\mathcal{S}$ whose direction is an extreme direction will be termed an ***extreme half-line.***

Let us now tie some of the preceding concepts together. For $\mathcal{S}$ a closed convex set in $\mathbf{R}^n$, ***$\mathcal{S}$ recedes in the direction d*** if $\mathcal{S}$ consists of all recession half-lines pointing in the $\mathbf{d}$-direction and having vertices in $\mathcal{S}$. In particular, the set of all directions of recession of $\mathcal{S}$ will be termed the ***recession cone*** of $\mathcal{S}$ and denoted $0^+\mathcal{S}$, i.e., $0^+\mathcal{S}, = \{\mathbf{d} | \mathbf{x} + \lambda\mathbf{d} \in \mathcal{S},\ \lambda \geq 0,\ \mathbf{x} \in \mathcal{S},\ \mathbf{d} \in \mathbf{R}^n\}$. Note that $\mathbf{0} \in 0^+\mathcal{S}$. Examples of convex sets and their associated recession cones in $\mathbf{R}^2$ are illustrated in Figure 8.1.

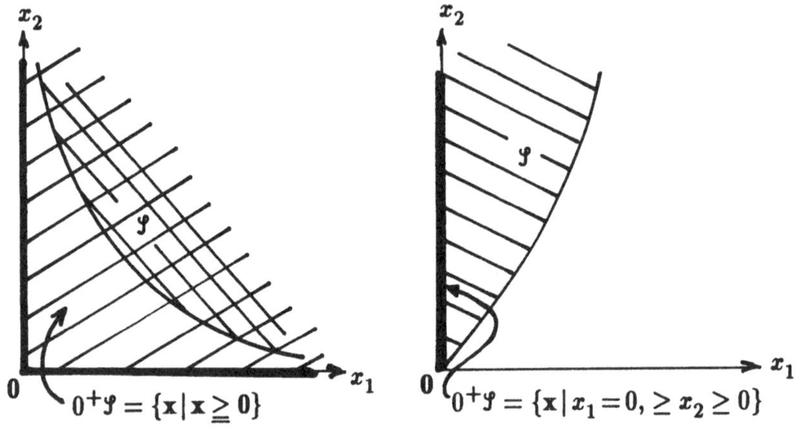

a. b.

Figure 8.1

Key properties of recession cones are the following:

(a) For $\mathcal{S} = \{\mathbf{x} | \mathbf{A}\mathbf{x} \geqq \mathbf{b}, \mathbf{x} \in \mathbf{R}^n\} \neq \phi$, $0^+\mathcal{S} = \{\mathbf{x} | \mathbf{A}\mathbf{x} \geqq \mathbf{0}, \mathbf{x} \in \mathbf{R}^n\}$.

(b) For $\mathcal{S}$ a nonempty convex set in $\mathbf{R}^n$, $0^+\mathcal{S}$ is a convex cone containing the origin $\mathbf{0}$; it is equivalent to the set of recession vectors $\mathbf{d}$ such that $\mathcal{S} + \mathbf{d} \subset \mathcal{S}$ (*i.e.*, $\mathbf{x} + \mathbf{d} \in \mathcal{S}$ for every $\mathbf{x} \in \mathcal{S}$). Moreover, $\mathbf{x} + m\mathbf{d} \in \mathcal{S}$ for every $\mathbf{x} \in \mathcal{S}$ and positive integer m.

(c) Let $\mathcal{C}$ be a nonempty closed convex cone in $\mathbf{R}^n$. Then $0^+\mathcal{C}$ is closed and consists of all possible limits of sequences of the form $\{\lambda_k \mathbf{x}_k\}$, where $\mathbf{x}_k \in \mathcal{S}$ and $\lambda_k \rightarrow 0+$. It is important to remember that $0^+\mathcal{C}$ may not be closed when $\mathcal{C}$ itself is not closed.

(d) Let $\mathcal{S}$ be a nonempty closed convex set in $\mathbf{R}^n$ with recession direction $\mathbf{d}(\neq \mathbf{0}) \in \mathcal{S}$. If there exists a vertex $\mathbf{x}$ such that the recession half-line $\{\mathbf{x} + \lambda\mathbf{d}, \lambda \geq 0, \mathbf{d} \neq \mathbf{0}\}$ is contained in $\mathcal{S}$, then the same assertion is true for every $\mathbf{x} \in \mathcal{S}$, *i.e.*, $\mathbf{d} \in 0^+\mathcal{S}$. Moreover, $\{\mathbf{x} + \lambda\mathbf{d}, \lambda \geq 0, \mathbf{d} \neq \mathbf{0}\} \subset ri(\mathcal{S})$ for each $\mathbf{x} \in ri(\mathcal{S})$ so that $\mathbf{x} \in 0^+(ri(\mathcal{S}))$.

(e) For any nonempty convex set $\mathcal{S}$ in $\mathbf{R}^n$, $0^+(ri(\mathcal{S})) = 0^+(\bar{\mathcal{S}})$. And for any $\mathbf{x} \in ri(\mathcal{S})$, recession direction $\mathbf{d} \in 0^+(\bar{\mathcal{S}})$ if and only if $\mathbf{x} + \lambda \mathbf{d} \in \mathcal{S}$ for every $\lambda > 0$.

(f) Any nonempty closed convex set $\mathcal{S}$ in $\mathbf{R}^n$ is bounded if and only if $0^+\mathcal{S} = \{\mathbf{0}\}$.

(g) For $\mathcal{S}$ a closed convex set in $\mathbf{R}^n$ which contains the origin, $0^+\mathcal{S} = \{\mathbf{d} | (1/\delta)\mathbf{d} \in \mathcal{S} \text{ for all } \delta > 0\}$.

(h) If $\{\mathcal{S}_k\}$ is an arbitrary sequence of closed convex sets in $\mathbf{R}^n$ whose intersection is not empty, then $0^+(\cap_k \mathcal{S}_k) = \cap_k 0^+(\mathcal{S}_k)$.

(i) For $\mathcal{S}$ a nonempty convex set in $\mathbf{R}^n$, the set $(-0^+\mathcal{S}) \cap 0^+\mathcal{S}$ is the ***lineality space*** of $\mathcal{S}$; it consists of the null vector and all recession directions $\mathbf{d}(\neq \mathbf{0})$ such that, for each vertex $\mathbf{x} \in \mathcal{S}$, $\mathbf{x} + \lambda \mathbf{d} \in \mathcal{S}$, $\lambda \geq 0$. The recession directions in the lineality space of $\mathcal{S}$ are called ***directions in which $\mathcal{S}$ is linear.*** In fact, the lineality space of $\mathcal{S}$ is the same as the set of recession directions $\mathbf{d}$ such that $\mathcal{S} + \mathbf{d} = \mathcal{S}$.

The lineality space of $\mathcal{S}$ is the largest subspace contained in the convex cone $0^+\mathcal{S}$. Its dimension is termed the ***lineality*** of $\mathcal{S}$(denoted $lin(\mathcal{S})$). The ***rank*** of $\mathcal{S} = dim(\mathcal{S}) - lin(\mathcal{S})$. Since the convex sets of rank zero are the affine sets, the rank of a closed convex set $\mathcal{S}$ equals its dimension if and only if $\mathcal{S}$ contains no recession directions.

(j) For $\mathcal{S} = \{\mathbf{x} | \mathbf{A}\mathbf{x} \geq \mathbf{b},\ \mathbf{x} \in \mathbf{R}^n\}$, the lineality space of $\mathcal{S}$ is the set $\{\mathbf{x} | \mathbf{A}\mathbf{x} = \mathbf{0},\ \mathbf{x} \in \mathbf{R}^n\}$.

(k) Let $\mathcal{S}_1, \mathcal{S}_2$ be nonempty closed convex sets in $\mathbf{R}^n$. If either $\mathcal{S}_1$ or $\mathcal{S}_2$ is bounded, then $\mathcal{S}_1 + \mathcal{S}_2$ is closed and $0^+(\mathcal{S}_1 + \mathcal{S}_2) = 0^+\mathcal{S}_1 + 0^+\mathcal{S}_2$.

(l) If $\mathcal{S}_i$, $i=1,\ldots,r$, are nonempty closed convex sets in $\mathbf{R}^n$ all having the same recession cone $0^+\mathcal{S}$, then the convex set

$$\mathcal{K} = co\left(\cup_{i=1}^{r} \mathcal{S}_i\right) \text{ is closed with } 0^+\mathcal{K} = 0^+\mathcal{S}.$$

At times we shall find it convenient to "generate" a cone from a nonempty subset $\mathcal{S}$ of $\mathbf{R}^n$ (Rockafellar, 1970). In this regard, let $\mathcal{S}$ be an arbitrary subset of $\mathbf{R}^n$ and let $\mathcal{C}$ be the set of all positive linear combinations of points in $\mathcal{S}$. Then $\mathcal{C}$ is the smallest convex cone which includes $\mathcal{S}$. If $\mathcal{S}$ is convex, then the smallest convex cone which includes $\mathcal{S}$ has the simple form $\mathcal{C} = \{\lambda\mathbf{x} | \lambda > 0, \mathbf{x} \in \mathcal{S}\}$. The convex cone obtained by adjoining the origin to $\mathcal{C}$ is known as the ***convex cone generated by*** $\mathcal{S}$ and will be denoted $cone(\mathcal{S})$. (Clearly the "convex cone generated by $\mathcal{S}$" coincides with the "smallest convex cone containing $\mathcal{S}$" only if the latter contains the origin.) If $\mathcal{S}$ is nonempty, $cone(\mathcal{S})$ consists of all conical (nonnegative linear) combinations of points from $\mathcal{S}$. In addition, if $ray(\mathcal{S})$ denotes the union of the origin and a set of recession half-lines $\{\lambda\mathbf{d} | \lambda \geq 0\}$ generated by the non-null vectors whose directions $\mathbf{d}(\neq\mathbf{0}) \in \mathcal{S}$, then $co(ray(\mathcal{S})) = cone(\mathcal{S})$.

8.2. Convex Hulls Revisited

In chapter two we considered the notion of the convex hull $co(\mathcal{S})$ of a set $\mathcal{S}$ in $\mathbf{R}^n$. There $co(\mathcal{S})$ was defined as the collection of all convex combinations of vectors from (arbitrary) $\mathcal{S}$ and was found to be the smallest convex set containing $\mathcal{S}$. Following Rockafellar (1970) we may now extend this definition somewhat by considering convex hulls of sets which consist of both points and recession directions. This broadened definition will allow us to address the generation of convex cones (and other unbounded sets) as well as the specification of ordinary convex hulls.

To this end let $\mathcal{S}_1$ be a set of points and $\mathcal{S}_2$ a set of recession directions respectively in $\mathbf{R}^n$. The ***convex hull*** $co(\mathcal{S})$ of $\mathcal{S} = \mathcal{S}_1 \cup \mathcal{S}_2$ is the smallest convex set $\mathcal{T}$ in $\mathbf{R}^n$ such that $\mathcal{S}_1 \subset \mathcal{T}$ and $\mathcal{T}$ recedes in all the directions of $\mathcal{S}_2$. If we define:

$ray(\mathcal{S}_2) \equiv$ the origin along with the recession rays generated by all vectors whose recession directions belong to $\mathcal{S}_2$;

$cone(\mathcal{S}_2) \equiv$ the convex cone generated by all the vectors whose recession directions belong to $\mathcal{S}_2$, *i.e.*, $co(ray(\mathcal{S}_2)) = cone(\mathcal{S}_2)$,

then

$$\mathcal{T} = co(\mathcal{S}_1 + ray(\mathcal{S}_2)) = co(\mathcal{S}_1) + co(ray(\mathcal{S}_2)) = co(\mathcal{S}_1) + cone(\mathcal{S}_2).$$

Hence a vector $\mathbf{x}$ is an element of $co(\mathcal{S})$ if and only if it is expressible as a convex combination of m points and recession directions of $\mathcal{S}$. So if $\mathbf{x}_1, \ldots, \mathbf{x}_k$ are vectors from $\mathcal{S}_1$ and $\mathbf{x}_{k+1}, \ldots, \mathbf{x}_m$ are arbitrary recession directions in $\mathcal{S}_2 (1 \leq k \leq m)$, then

$$\begin{aligned} co(\mathcal{S}) = \Big\{ \mathbf{x} \mid \mathbf{x} = \sum_{i=1}^{k} \lambda_i \mathbf{x} + \sum_{i=1}^{k+1} \mu_i \mathbf{x}_i, \mathbf{x}_i \in \mathcal{S}_1, 0 \leq \lambda_i \in \mathbf{R}, i = 1, \ldots, k, \\ \sum_{i=1}^{k} \lambda_i = 1; \mathbf{x}_i \in \mathcal{S}_2, 0 \leq \mu_i \in \mathbf{R}, i = k+1, \ldots, \text{with } m \\ \text{an arbitrary positive integer} \Big\} \\ = co(\mathcal{S}_1) + cone(\mathcal{S}_2). \end{aligned}$$

The notion of a recession direction enables us to expand upon two additional set-theoretic concepts that we encountered earlier, namely those pertaining to the generation of convex cones and affine hulls. Specifically, given a set $\mathcal{Y}$ in $\mathbf{R}^n$, the ***convex cone*** $\mathcal{C}$ generated by $\mathcal{Y}$ can be represented as $co(ray(\mathcal{S})) = cone(\mathcal{S})$; it amounts to the convex hull of the set $\mathcal{S}$ consisting of the origin and all vectors whose recession directions belong to $\mathcal{Y}$. Hence a convex combination $\mathbf{x}$ of m elements in $cone(\mathcal{S})$ is necessarily a convex combination of $\mathbf{0}$ and $m-1$ recession directions in $\mathcal{S}$; it is simply a conical combination of $m-1$ vectors of $\mathcal{Y}$.

Next, the ***affine hull*** $aff(\mathcal{S})$ of a mixed set $\mathcal{S}$ of points and recession directions in $\mathbf{R}^n$ is defined as $aff(co(\mathcal{S}))$; it represents the smallest affine set which contains all points in $\mathcal{S}$ and recedes in all the directions of $\mathcal{S}$. Clearly $aff(\mathcal{S}) = co(\mathcal{S}) = \phi$ if $\mathcal{S}$ contains exclusively recession directions. Moreover, a nonempty subset $\mathcal{S}$ of $\mathbf{R}^n$ is said to be ***affinely independent*** if $dim(aff(\mathcal{S})) = m-1$, where m is the total number of points and recession directions in $\mathcal{S}$.

Again armed with the concept of a recession direction we may now address an expanded version of representation theorem 8.1.2, *i.e.*,

8.1.5. REPRESENTATION THEOREM. A nonempty closed convex set $\mathcal{S}$ in $\mathbf{R}^n$ is the convex hull of its set $\mathcal{G}$ of extreme points and extreme directions, *i.e.*, $co(\mathcal{G}) = \mathcal{S}$.

In this regard, every extreme point of $\mathcal{S}$ is a point of $\mathcal{G}$ and every extreme direction of $\mathcal{S}$ is a recession direction in $\mathcal{G}$. Moreover, we also offer a much more comprehensive version of

8.2.1. CARATHÉODORY'S THEOREM [Rockafellar, 1970]. Let $\mathcal{S}$ be any set of points and recession directions in $\mathbf{R}^n$. Then $\mathbf{x} \in co(\mathcal{S})$ if and only if $\mathbf{x}$ can be expressed as a convex combination of $n+1$ of the points and recession directions of $\mathcal{S}$ (not necessarily distinct).

Three consequences of this theorem are:

8.1.7. COROLLARY. Let $\mathcal{S}_i$, $i=1,\ldots,r$, be an arbitrary collection of convex sets in $\mathbf{R}^n$ with $\mathcal{S} = co(\cup_{i=1}^r \mathcal{S}_i)$. Then every point of $\mathcal{S}$ can be expressed as a convex combination of $n+1$ or fewer affinely independent points, each belonging to a different $\mathcal{S}_i$.

8.1.8. COROLLARY. Let $\mathcal{S}_i, i=1,\ldots,r$, be an arbitrary collection of nonempty convex sets in $\mathbf{R}^n$ and let $\mathcal{C}$ be the convex cone generated by the union of the $\mathcal{S}_i$. Then every non-null vector of $\mathcal{C}$ can be expressed as a conical combination of n or fewer linearly independent vectors, each belonging to a different $\mathcal{S}_i$.

8.1.9. COROLLARY. Let $\mathcal{C}$ be a closed convex cone containing more than just the origin. In addition, let $\mathcal{T}$ be the set of vectors in $\mathcal{C}$ such that each extreme half-line of $\mathcal{C}$ is generated by some point $\mathbf{x} \in \mathcal{T}$. Then $\mathcal{C}$ is the convex cone generated by $\mathcal{T}$, *i.e.*, $\mathcal{C} = co(\mathcal{S})$, where $\mathcal{S}$ consists of the origin and the directions of the vectors in $\mathcal{T}$. (Here the origin is the unique extreme point of $\mathcal{C}$ and the directions of the vectors in $\mathcal{T}$ are the extreme directions of $\mathcal{C}$.)

8.3. Faces of Polyhedral Convex Sets: Extreme Points, Facets, and Edges

In this section we shall restrict our discussion to a specialized class of convex sets in $\mathbf{R}^n$, namely those which may be characterized as "polyhedral." Any such set is generated by intersecting a finite number of closed half-spaces (not all of which pass through the origin) and thus amounts to a closed convex set in $\mathbf{R}^n$. While any closed convex set in $\mathbf{R}^n$ is the intersection of "all" supporting closed half-spaces containing it (see theorem 3.2.4), a polyhedral convex set can be represented by intersecting only a "finite number" m of closed half-spaces. In this regard, a nonempty set $\mathcal{S}$ in $\mathbf{R}^n$ is a ***polyhedral convex set*** (or simply a ***convex polyhedron***) if $\mathcal{S} = \{\mathbf{x} | \mathbf{a}_i'\mathbf{x} \leq b_i, \mathbf{a}_i' \neq \mathbf{0}', b_i \in \mathbf{R}, i=1,\ldots,m, \mathbf{x} \in \mathbf{R}^n\}$. For $\mathbf{a}_i'$ the ith row of the $(m \times n)$ matrix A and $b_i, i=1,\ldots,m$, the ith component of the $(m \times 1)$ vector $\mathbf{b}$, we may alternatively write $\mathcal{S} = \{\mathbf{x} | \mathbf{A}\mathbf{x} \leqq \mathbf{b}, \mathbf{x} \in \mathbf{R}^n\}$. Here $\mathcal{S}$ serves as a ***representation*** for a convex polyhedron.

If a convex polyhedron in $\mathbf{R}^n$ is bounded, then it is termed a ***convex polytope*** (Figure 8.2). Equivalently, a convex polytope can be viewed as the convex hull generated or spanned by finitely many points in $\mathbf{R}^n$, *i.e.*, for the finite set $\{\mathbf{x}_1,\ldots,\mathbf{x}_m\}$, $co\{\mathbf{x}_1,\ldots,\mathbf{x}_m\}$ is a convex polytope consisting of all convex combinations of $\{\mathbf{x}_1,\ldots,\mathbf{x}_m\}$.

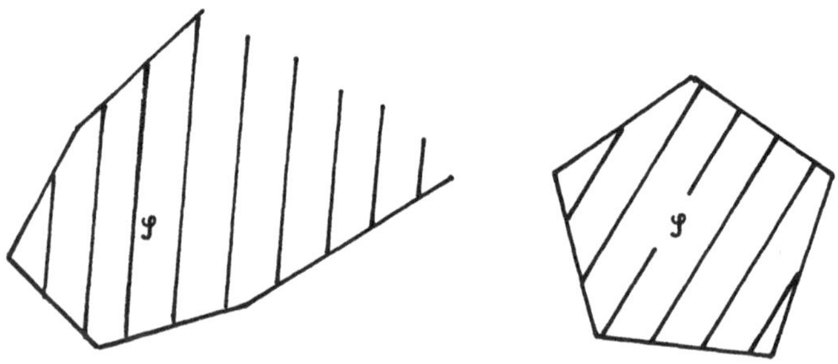

$\mathcal{S}$ is a convex polyhedron; not a convex polytope.

a.

$\mathcal{S}$ is a convex polyhedron which is also a convex polytope

b.

Figure 8.2

Generally speaking, a polyhedral convex set $\mathcal{S}$ in $\mathbf{R}^n$ can be represented by a finite number of inequalities and/or equalities (*e.g.*, the equality $a_{11}x_1 + a_{12}x_2 = b_1$ can be replaced by the two inequalities $a_{11}x_1 + a_{12}x_2 \leq b_1$, $-a_{11}x_1 - a_{12}x_2 \leq -b_1$). Hence we may also express a convex polyhedron as $\mathcal{S} = \{\mathbf{x}|\mathbf{A}\mathbf{x} = \mathbf{b}, \mathbf{x}\in\mathbf{R}^n\}$. Oftentimes the m inequalities indicated above are accompanied by a set of nonnegativity conditions $x_i \geq 0$, $i=1,\ldots,n$. In this instance, $\mathcal{S} = \{\mathbf{x}|\mathbf{A}\mathbf{x} \leqq \mathbf{b},\ \mathbf{x} \geqq \mathbf{0},\ \mathbf{x}\in\mathbf{R}^n\}$ or possibly $\mathcal{S} = \{\mathbf{x}|\mathbf{A}\mathbf{x} = \mathbf{b},\ \mathbf{x} \geqq \mathbf{0},\ \mathbf{x}\in\mathbf{R}^n\}$. Moreover any unrestricted variable (one which may be positive, negative, or zero) can be replaced by the difference between two nonnegative variables (*e.g.*, x_k unrestricted in sign is equivalent to $x_k = x_k' - x_k''$, where $x_k', x_k'' \geq 0$).

Extreme points were discussed in considerable detail in section 8.1 above. To briefly reiterate: a vector $\mathbf{x}$ is an extreme point of a (general) convex set $\mathcal{S}$ in $\mathbf{R}^n$ if $\mathbf{x}$ cannot be written as a positive convex combination of two distinct points in $\mathcal{S}$, *i.e.*, $\mathbf{x}$ is not an interior point of any closed line segment in $\mathcal{S}$. When $\mathcal{S}$ in $\mathbf{R}^n$ is a polyhedral convex set (for concreteness, let $\mathcal{S} = \{\mathbf{x}|\mathbf{A}\mathbf{x} \leqq \mathbf{b},\ \mathbf{x}\in\mathbf{R}^n\}$), this definition, while still valid, lends itself to further geometric interpretation. To set the stage for what follows let us review the notion of a supporting hyperplane for $\mathcal{S}$. Specifically, let $\mathcal{S}$ be a convex polyhedron in $\mathbf{R}^n$ and, for some vector $\mathbf{C}(\neq\mathbf{0})\in\mathbf{R}^n$, let $\mathcal{H} = \{\mathbf{x}|\mathbf{C}'\mathbf{x} = \alpha,\ \alpha\in\mathbf{R},\ \mathbf{x}\in\mathbf{R}^n\}$ be a hyperplane such that: (1) $\mathbf{C}'\mathbf{x} \geq \alpha$ (or $\leq \alpha$), $\mathbf{x}\in\mathcal{S}$ ($\mathcal{H}$ has $\mathcal{S}$ on one of its sides); and (2) there is a point $\bar{\mathbf{x}}$ such that $\bar{\mathbf{x}}\in\mathcal{S}$ and $\mathbf{C}'\bar{\mathbf{x}} = \alpha$ (*i.e.*, $\bar{\mathbf{x}}\in\mathcal{S}\cap\mathcal{H}$). Then $\mathcal{H}$ is termed a ***supporting hyperplane*** for $\mathcal{S}$ at the point $\bar{\mathbf{x}}$. (In Figure 8.3.a, $\mathcal{H}_1$ is a supporting hyperplane for $\mathcal{S}$ in $\mathbf{R}^2$ at $\mathbf{x}_1$ while $\mathcal{H}_2$ is also a supporting hyperplane for $\mathcal{S}$ at $\mathbf{x}_7 = \lambda\mathbf{x}_1 + (1-\lambda)\mathbf{x}_2,\ 0 \leq \lambda \leq 1$.)

The polyhedral convex set $\mathcal{S}$ specified above is made up of the m supporting closed (defining) half-spaces $\mathbf{a}_i'\mathbf{x} \leq b_i$, each with its associated supporting hyperplane $\mathbf{a}_i'\mathbf{x} = b_i, i=1,\ldots,m$. In this regard, a vector $\bar{\mathbf{x}}\in\mathcal{S}$ is termed an ***extreme point*** (or ***vertex*** or ***corner point***) of $\mathcal{S}$ if and only if $\bar{\mathbf{x}}$ lies on n linearly independent supporting hyperplanes of $\mathcal{S}$, *i.e.*, if at $\bar{\mathbf{x}}$ we have $\mathbf{a}_i'\bar{\mathbf{x}} = b_i, i=1,\ldots,k$, and $\mathbf{a}_i'\bar{\mathbf{x}} < b_i, i=k+1,\ldots,m$ (assume the first k supporting hyperplanes are satisfied at $\bar{\mathbf{x}}$ or the first k inequalities are "tight" there), then $\bar{\mathbf{x}}$ is a unique solution to the system $\mathbf{a}_i'\mathbf{x} = b_i,\ i=1,\ldots,k$, if and only if the set $\{\mathbf{a}_1,\ldots,\mathbf{a}_k\}$ contains n linearly independent vectors or the $(n\times k)$ matrix $[\mathbf{a}_1,\ldots,\mathbf{a}_k]$ has (column) rank n. If more than n supporting hyperplanes pass through $\bar{\mathbf{x}}$, then this extreme point is said to be ***degenerate.*** Hence a nondegenerate extreme point of $\mathcal{S}$ is one which satisfies exactly n supporting hyperplanes of the $\mathbf{A}\mathbf{x} \leqq \mathbf{b}$ system. (In Figure 8.3.a extreme points $\mathbf{x}_2$, $\mathbf{x}_3$, $\mathbf{x}_4$, and $\mathbf{x}_6$ are nondegenerate.)

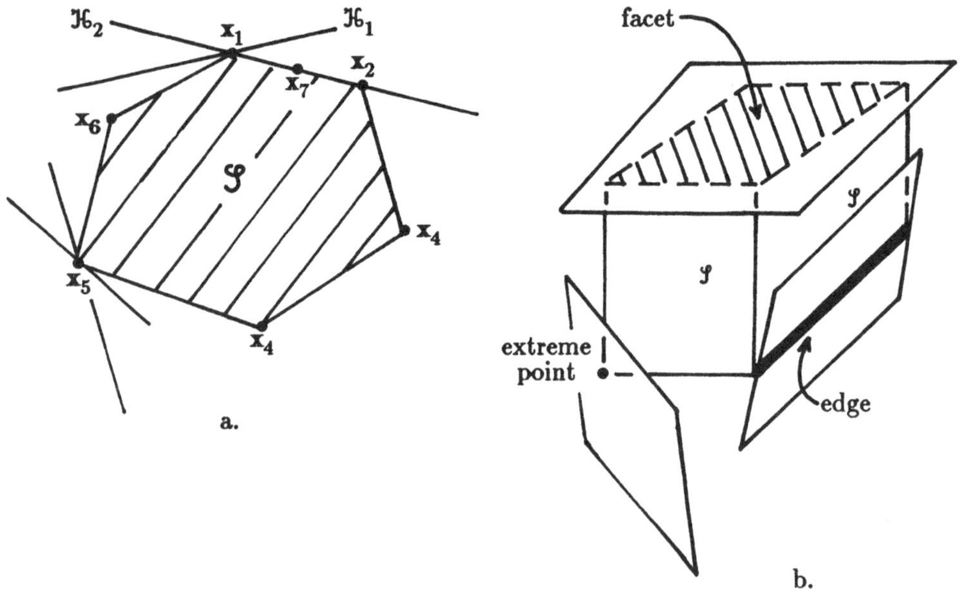

Figure 8.3

If polyhedral $\mathcal{S} = \{\mathbf{x}|\mathbf{Ax} = \mathbf{b},\ \mathbf{x} \geqq \mathbf{0},\ \mathbf{x} \epsilon \mathbf{R}^n\}$, with $\mathbf{A}$ of order $(m \times n)$ and $\rho(\mathbf{A}) = m < n$, then an extreme point $\bar{\mathbf{x}} \epsilon \mathcal{S}$ occurs at the intersection of the m linearly independent equations $\mathbf{Ax} = \mathbf{b}$ and $n-m$ additional hyperplane equalities taken from the nonnegativity conditions $\mathbf{x} \geqq \mathbf{0}$. And if $\mathcal{S} = \{\mathbf{x}|\mathbf{Ax} \leqq \mathbf{b},\ \mathbf{x} \geqq \mathbf{0},\ \mathbf{x} \epsilon \mathbf{R}^n\}$, where again $\mathbf{A}$ is $(m \times n)$ and $\rho(\mathbf{A}) = m < n$, let $\mathbf{D}' = [\mathbf{A}',\ \mathbf{I}_n]$ represent the coefficient matrix associated with the $m+n$ supporting hyperplane equalities in $\mathcal{S}$. If $\rho(\mathbf{D}) = m+n$ (*i.e.*, $\mathbf{D}$ is of full row rank), then these $m+n$ supports are linearly independent and thus any extreme point $\bar{\mathbf{x}} \epsilon \mathcal{S}$ must lie on some subset of n such supports.

That the preceding definition of a nondegenerate extreme point of $\mathcal{S}$ is equivalent to the one given earlier can be easily seen by noting that when $\bar{\mathbf{x}} \in \mathcal{S}$ lies on n linearly independent supporting hyperplanes of $\mathcal{S}$ and $\bar{\mathbf{x}} = \lambda \mathbf{x}_1 + (1-\lambda)\mathbf{x}_2$, $0 < \lambda < 1$ and $\mathbf{x}_1, \mathbf{x}_2 \in \mathcal{S}$, then clearly $\mathbf{x}_1, \mathbf{x}_2$ lie on the same set of hyperplanes. Since the simultaneous solution of these n supports is unique, it follows that $\bar{\mathbf{x}} = \mathbf{x}_1 = \mathbf{x}_2$ so that $\bar{\mathbf{x}}$ must be an extreme point.

A convex polyhedron $\mathcal{S} = \{\mathbf{x} | \mathbf{A}\mathbf{x} \leqq \mathbf{b}\}$ in $\mathbf{R}^n$ is said to be ***pointed*** if it has extreme points or vertices. A necessary and sufficient condition for a convex polyhedron to be pointed is that it contains no line, *i.e.*, the associated homogeneous system of equations $\mathbf{AU} = \mathbf{0}$ admits only the trivial solution $\mathbf{u} = \mathbf{0}$.

Additional considerations regarding convex polyhedra are the following:

(a) An inequality $\mathbf{a}_t'\mathbf{x} \leq b_t$ from system $\mathbf{A}\mathbf{x} \leqq \mathbf{b}$ is often termed an ***implicit equality*** if $\mathbf{a}_t'\mathbf{x} = b_t$ for all $\mathbf{x} \in \mathcal{S} = \{\mathbf{x} | \mathbf{A}\mathbf{x} \leqq \mathbf{b}, \ \mathbf{x} \in \mathbf{R}^n\}$. For convenience, let $\mathcal{I} = \{i | i=1,\ldots,m\}$. Then the index set of subscripts corresponding to implicit equalities will be denoted as $\mathcal{I}^{=} = \{i | \mathbf{a}_i'\mathbf{x} = b_i \text{ for all } \mathbf{x} \in \mathcal{S}, i \in \mathcal{I}\}$ while $\mathcal{I}^{\leq} = \mathcal{I} \backslash \mathcal{I}^{=} = \{i | \mathbf{a}_i'\mathbf{x} \leq b_i$ and $\mathbf{a}_i' < b_i$ for some $\mathbf{x} \in \mathcal{S}, i \in \mathcal{I}\}$. In addition, let $[\mathbf{A}^{=}, \mathbf{b}^{=}]$, $[\mathbf{A}^{\leq}, \mathbf{b}^{\leq}]$ represent the corresponding rows of $[\mathbf{A}, \mathbf{b}]$ respectively.

(b) An inequality $\mathbf{a}_t'\mathbf{x} \leq b_t$ from system $\mathbf{A}\mathbf{x} \leqq \mathbf{b}$ is deemed ***redundant*** (or ***superfluous***) if it can be omitted from the system without changing the structure of the polyhedral convex set $\mathcal{S} = \{\mathbf{x} | \mathbf{A}\mathbf{x} \leqq \mathbf{b}, \ \mathbf{x} \in \mathbf{R}^n\}$. Hence any redundant inequality is "implied" by the other inequalities of the $\mathbf{A}\mathbf{x} \leqq \mathbf{b}$ system.

(c) Let $\mathcal{S}$ be a nonempty polyhedral convex set in $\mathbf{R}^n$. The ***dimension of*** $\mathcal{S}$, *dim*$(\mathcal{S})$, is the dimension of the smallest affine space containing $\mathcal{S}$. So given a point $\bar{\mathbf{x}}(\neq \mathbf{0}) \in \mathcal{S}$ and a set of points $\{\mathbf{x}_1,\ldots,\mathbf{x}_r\}$ in $\mathcal{S}$, $dim(\mathcal{S})$ is the maximum number r such that the translated set $\{\mathbf{x}_1-\bar{\mathbf{x}},\ldots,\mathbf{x}_r-\bar{\mathbf{x}}\}$ is linearly independent. Looked at in another fashion, $dim(\mathcal{S})=r$ if the maximum number of affinely independent points in $\mathcal{S}$ is $r+1$. A convex polyhedron $\mathcal{S}$ in $\mathbf{R}^n$ is termed ***full-dimensional*** if $dim(\mathcal{S})=n$, and this holds if and only if $\mathcal{S}$ has an ***interior point***, *i.e.*, a point $\bar{\mathbf{x}}$ such that $\mathbf{a}_i'\bar{\mathbf{x}}<b_i$ for all i. For instance, if $\mathcal{S}=\{\mathbf{x}|\mathbf{A}\mathbf{x}=\mathbf{b},\mathbf{x}\geqq\mathbf{0},\mathbf{x}\in\mathbf{R}^n\}$, where $\mathbf{A}$ is of order $(m\times n)$ and $\rho(\mathbf{A})=m$, then $dim(\mathcal{S})\leq n-m$. If every solution to $\mathbf{A}\mathbf{x}=\mathbf{b}$ has at least m nonzero components, then $dim(\mathcal{S})=n-m$ exactly. Moreover, for some subsystem of implicit equalities $\bar{\mathbf{A}}\mathbf{x}\leqq\bar{\mathbf{b}}$ of $\mathbf{A}\mathbf{x}\leqq\mathbf{b}$ (*i.e.*, $\bar{\mathbf{A}}\mathbf{x}=\bar{\mathbf{b}}$ for all $\mathbf{x}\in\mathcal{S}$), let $\rho(\bar{\mathbf{A}})=\rho[\bar{\mathbf{A}},\bar{\mathbf{b}}]=k\leq m$. Then $dim(\mathcal{S})=n-k$, $0\leq k\leq n$.

(d) Let $\mathcal{S}=\{\mathbf{x}|\mathbf{A}\mathbf{x}\leqq\mathbf{b},\mathbf{x}\in\mathbf{R}^n\}$ (with $\mathbf{A}\mathbf{x}\leqq\mathbf{b}$ nonredundant) be a nonempty convex polyhedron in $\mathbf{R}^n$ and let $[\mathcal{H}_i^-]=\{\mathbf{x}|\mathbf{a}_i'\mathbf{x}\leq b_i,\mathbf{x}\in\mathbf{R}^n\}$ be a supporting closed half-space with associated supporting hyperplane $\mathcal{H}_i=\{\mathbf{x}|\mathbf{a}_i'\mathbf{x}=\mathbf{b}_i,\mathbf{x}\in\mathbf{R}^n\}$. A ***face*** $\mathcal{F}$ of $\mathcal{S}$ is a subset of points in $\mathcal{S}$ determined by some nonempty collection of defining supports of $\mathcal{S}$. Hence $\mathcal{F}$ is either $\mathcal{S}$ itself, the null set, or the intersection of $\mathcal{S}$ with $\mathcal{H}_i$. In this latter instance $\mathcal{F}=\{\mathbf{x}|\mathbf{a}_i'\mathbf{x}=b_i,\mathbf{x}\in\mathcal{S}\}$. Here the inequality $\mathbf{a}_i'\mathbf{x}\leq b_i$ is said to "define" or "represent" $\mathcal{F}$. In addition, $\mathcal{F}$ is termed ***proper*** if $\mathcal{F}\neq\phi$ and $\mathcal{F}\neq\mathcal{S}$. If $\mathcal{F}$ is a proper face of $\mathcal{S}$, $dim(\mathcal{F})<dim(\mathcal{S})$. In fact, given a proper face $\mathcal{F}$ for $\mathcal{S}$, if r represents the maximum number of linearly independent support equalities holding at "all" feasible points of $\mathcal{F}$, then $dim(\mathcal{F})=n-r$. All of the following are types of faces:

(i) ***facet*** -- a face whose dimension is one less than $dim(\mathcal{S})$ (since it is the largest face in $\mathcal{S}$ distinct from $\mathcal{S}$).

(ii) ***extreme point*** -- a face of dimension zero (since it occurs at the intersection of n linearly independent supporting hyper-planes of $\mathcal{S}$).

(iii) ***edge*** -- a face of dimension one (since it is formed by intersecting $n-1$ linearly independent supporting hyperplanes of $\mathcal{S}$).

As Figure 8.3.b reveals, a hyperplane defining a facet corresponds to a supporting closed half-space of $\mathcal{S}$ while an edge is always a line segment joining two extreme points. (Hence an extreme point is located at the intersection of two a more distinct edges of $\mathcal{S}$.) Furthermore, $\mathcal{S}$ has only finitely many faces, each face is a nonempty convex polyhedron, and, in this light, if $\mathcal{F}$ is a face of $\mathcal{S}$ and $\bar{\mathcal{F}} \subseteq \mathcal{F}$, then $\bar{\mathcal{F}}$ is a face of $\mathcal{S}$ if and only if $\bar{\mathcal{F}}$ is a face of $\mathcal{F}$.

(e) Using the definition of a face just given in (d), we can alternatively characterize a proper face $\mathcal{F}$ of a convex polyhedron $\mathcal{S}$ in R^n as:

(i) $\mathcal{F} = \{\mathbf{x} | \mathbf{a}_i'\mathbf{x} = b_i, \mathbf{x} \in \mathcal{S}\}$ is a nonempty proper face of $\mathcal{S}$ if and only if there exists a vector $\mathbf{a}_i \in R^n$ for which $\mathcal{F}$ is the set of points attaining $max\{\mathbf{a}_i'\mathbf{x} | \mathbf{x} \in \mathcal{S}\} = b_i$. Clearly $\mathbf{a}_i'\mathbf{x} = b_i$ supports $\mathcal{S}$.

(ii) $\mathcal{F}(\neq \phi)$ is a face of $\mathcal{S}$ if and only if $\mathcal{F} = \{\mathbf{x} | \bar{A}\mathbf{x} = \bar{\mathbf{b}}, \mathbf{x} \in \mathcal{S}\}$ for same subsystem $\bar{A}\mathbf{x} \leqq \bar{\mathbf{b}}$ of $A\mathbf{x} \leqq \mathbf{b}$. If $\rho(A) = n-k$, $\mathcal{S}$ has a face of dimension k and has no proper face of lower dimension.

(f) Let $\mathcal{S} = \{\mathbf{x} | A\mathbf{x} \leqq \mathbf{b}, \mathbf{x} \in R^n\}$ be a nonempty convex polyhedron with A of order $(m \times n)$. If $dim(\mathcal{S}) = n-m$ and $A\mathbf{x} \leqq \mathbf{b}$ is nonredundant, there is a one-to-one correspondence between the inequalities $\mathbf{a}_i'\mathbf{x} \leq b_i$, $i \in \mathcal{I}^{\leq}$, and the facets of $\mathcal{S}$. That is, since each facet $\mathcal{F}$ is associated with "one" of the m inequalities $\mathbf{a}_i'\mathbf{x} \leq b_i$ (in the sense that the points "on" the facet satisfy $\mathbf{a}_i'\mathbf{x} = b_i$), $i \in \mathcal{I}^{\leq}$, facet $\mathcal{F}$ has the representation $\mathcal{F} = \{\mathbf{x}|\mathbf{a}_i'\mathbf{x} = b_i, \mathbf{x} \in \mathcal{S}, i \in \mathcal{I}^{\leq}\}$, where the inequality $\mathbf{a}_i'\mathbf{x} \leq b_i$ "defines" $\mathcal{F}$.

(g) Let $\bar{\mathbf{x}}$ be a nondegenerate extreme point of the polyhedral convex set $\mathcal{S} = \{\mathbf{x}|\mathbf{A}\mathbf{x} \leqq \mathbf{b},\ \mathbf{x} \in \mathbf{R}^n\}$ and let us assume that the first n supporting hyperplanes associated with the supporting closed half-spaces in $\mathbf{A}\mathbf{x} \leqq \mathbf{b}$ are satisfied at $\bar{\mathbf{x}}$. Then an edge emanating from $\bar{\mathbf{x}}$ is a closed line segment of the form $\bar{\mathbf{x}} - \theta\mathbf{s}_k$, $0 \leq \theta \leq \hat{\theta}_k$, where $\hat{\theta}_k$ is the largest scalar such that $\bar{\mathbf{x}} - \theta\mathbf{s}_k \in \mathcal{S}$ and $\mathbf{s}_k \in \mathbf{R}^n$ is chosen so that

$$\mathbf{a}_i'\mathbf{s}_k = 0, i{=}1,\ldots,n,\ i \neq k;$$
$$\mathbf{a}_k'\ \mathbf{s}_k = 1.$$

Here $\mathbf{s}_k$ is orthogonal to each normal $\mathbf{a}_i$ of $\mathcal{H}_i$, $i{=}1,\ldots,n$, $i \neq k$. Under this scheme there are n different edges originating at $\bar{\mathbf{x}}$.

(h) Two extreme points of a convex polyhedron $\mathcal{S}$ in $\mathbf{R}^n$ are said to be ***adjacent*** if they are connected by an edge.

(i) For a convex polyhedron $\mathcal{S} = \{\mathbf{x}\,|\,\mathbf{A}\mathbf{x} \leqq \mathbf{b}, \mathbf{x} \in \mathbf{R}^n\}$, the ***lineality space*** of $\mathcal{S}$, $lin(\mathcal{S})$, is the solution set of the homogeneous system of linear equations $\mathbf{A}\mathbf{x} = \mathbf{0}$ or the linear space $\{\mathbf{x}|\mathbf{A}\mathbf{x} = \mathbf{0}, \mathbf{x} \in \mathbf{R}^n\}$.

(j) Given a convex polyhedron $\mathcal{S} = \{\mathbf{x}|\mathbf{A}\mathbf{x} \leqq \mathbf{b}, \mathbf{x} \in \mathbf{R}^n\}$, where $\mathbf{A}$ is $(m \times n)$, the ***affine hull*** of $\mathcal{S}$, $aff(\mathcal{S})$, is the set of all vectors $\mathbf{x}$ satisfying a subsystem of implicit equalities in $\mathbf{A}\mathbf{x} \leqq \mathbf{b}$, *i.e.*, $aff(\mathcal{S}) = \{\mathbf{x}|\mathbf{a}_i'\mathbf{x} \leq b_i, i \in \mathcal{I}^{=}, \mathbf{x} \in \mathbf{R}^n\}$. Under this specification we may alternatively represent $\mathcal{S}$ as $\mathcal{S} = aff(\mathcal{S}) \cap \{\mathbf{x}|\mathbf{a}_i'\mathbf{x} \leq b_i, i \in \mathcal{I}^{\leq}, \mathbf{x} \in \mathbf{R}^n\}$ provided that the inequalities in $\mathbf{A}\mathbf{x} \leqq \mathbf{b}$ corresponding to $i \in \mathcal{I}^{\leq}$ are nonredundant.

It is also true that $dim(\mathcal{S}) = dim\big(aff(\mathcal{S})\big)$.

One additional point concerning the representation of a convex polyhedron $\mathcal{S} = \{\mathbf{x} \mid \mathbf{Ax} \leqq \mathbf{b}, \ \mathbf{x} \in \mathbf{R}^n\} \neq \phi$ is in order. Specifically, let us address the issue of which of the inequalities in the system $\mathbf{Ax} \leqq \mathbf{b}$ are necessary (and sufficient) for the "minimal" description of $\mathcal{S}$. We may note first that for each facet $\mathcal{F}$ of $\mathcal{S}$, at least one of the inequalities defining $\mathcal{F}$ is necessary for the representation of $\mathcal{S}$. Moreover, any inequality $\mathbf{a}_i'\mathbf{x} \leq b_i$, $i \in \mathcal{I}^{\leq}$, that defines a face for $\mathcal{S}$ of dimension less than $dim(\mathcal{S}) - 1$ is superfluous to the representation of $\mathcal{S}$; any such inequality does not render a support for $\mathcal{S}$. In this regard, if $dim(\mathcal{S}) = n$, then $\mathcal{S}$ has a unique ***minimal representation*** given by a finite system of linear inequalities $\mathbf{a}_i'\mathbf{x} \leq b_i$, $i \in \mathcal{I}$, in the sense that each inequality defines a facet $\mathcal{F}$ of $\mathcal{S} = \{\mathbf{x} | \mathbf{a}_i'\mathbf{x} \leq b_i, i \in \mathcal{I}\}$. If $dim(\mathcal{S}) = n-k$, $0 \leq k \leq n$, then $\mathbf{Ax} \leqq \mathbf{b}$ contains a "maximal" subsystem of k linearly independent equalities $\bar{\mathbf{A}}\mathbf{x} = \bar{\mathbf{b}}$.

Given a polyhedral convex set $\mathcal{S} = \{\mathbf{x} | \mathbf{Ax} \leqq \mathbf{b}, \mathbf{x} \in \mathbf{R}^n\}$, if $\mathbf{b}$ is replaced by the null vector, then the resulting set is termed a ***polyhedral convex cone***. In this regard, a polyhedral convex cone arises as the intersection of a finite number of closed half-spaces whose defining hyperplanes pass through the origin and appears as $\mathcal{C} = \{\mathbf{x} | \mathbf{Ax} \leqq \mathbf{0}, \mathbf{x} \in \mathbf{R}^n\}$. Here the ith row of $\mathbf{A}$, $\mathbf{a}_i'$, is the normal to the hyperplane $\mathcal{H}_i = \{\mathbf{x} | \mathbf{a}_i'\mathbf{x} = 0, \mathbf{x} \in \mathbf{R}^n\}$ generating the ith closed half-space $[\mathcal{H}_i^-] = \{\mathbf{x} \mid \mathbf{a}_i'\mathbf{x} \leq 0, \mathbf{x} \in \mathbf{R}^n\}$. More formally,

8.3.1. THEOREM. A convex polyhedron $\mathcal{S} = \{\mathbf{x} | \mathbf{Ax} \leqq \mathbf{b}, \mathbf{x} \in \mathbf{R}^n\}$ is a polyhedral convex cone if and only if $\mathbf{b} = \mathbf{0}$, *i.e.*, every polyhedral cone is the solution set of a homogeneous system of linear inequalities.

PROOF. (sufficiency) That $\mathcal{C} = \{\mathbf{x} | \mathbf{Ax} \leqq \mathbf{0}, \mathbf{x} \in \mathbf{R}^n\}$ is a polyhedral convex cone follows from Minkowski's theorem 4.5.1. (necessity) If $\mathcal{S}$ is a cone, then $\mathbf{0} \in \mathcal{S}$ and thus $\mathbf{b} \geqq \mathbf{0}$. But if $\mathbf{x} \in \mathcal{S}$, then, for $t > 0$, homogeneity implies that $t\mathbf{x} \in \mathcal{S}$ so that we must have $\mathbf{b} = \mathbf{0}$ or $\mathbf{Ax} \leqq \mathbf{0}$ for all $\mathbf{x} \in \mathcal{C}$. Q.E.D.

We close this section by noting that:

(a) If a polyhedral convex set $\mathscr{S}$ in $\mathbf{R}^n$ contains the null vector, then the convex cone generated by $\mathscr{S}$ is polyhedral.

(b) Let $\mathscr{S}$ $(\neq\phi)$ be a compact convex set in $\mathbf{R}^n$ with $\mathscr{D}$ any convex set such that $\mathscr{S}\subset\mathscr{D}^o$. Then there exists a polyhedral convex set $\mathscr{K}$ in $\mathbf{R}^n$ such that $\mathscr{K}$ closely approximates $\mathscr{S}$ in the sense that $\mathscr{K}\subset\mathscr{D}^o$ and $\mathscr{S}\subset\mathscr{K}^o$.

8.4. Extreme Point Representation for Polyhedral Convex Sets

Let us now consider the polyhedral convex set $\mathscr{S}=\{\mathbf{x}|\mathbf{Ax}=\mathbf{b}, \mathbf{x}\geqq\mathbf{0}, \mathbf{x}\epsilon\mathbf{R}^n\}$, where $\mathbf{A}$ is of order $(m\times n)$ and $\rho(\mathbf{A})=m$. We developed in chapter seven the concept of a basic feasible solution to a simultaneous linear system of the form $\mathbf{Ax}=\mathbf{b}$ given that $\rho(\mathbf{A})=\rho[\mathbf{A},\mathbf{b}]=m$. That is, we extracted m linearly independent columns form $\mathbf{A}$ and formed an mth order nonsingular basis matrix $\mathbf{B}$ (hence $\mathbf{B}^{-1}$ exists) and subsequently obtained a unique basic feasible solution

$$\mathbf{x}=\begin{bmatrix}\mathbf{X}_B\\ \mathbf{0}\end{bmatrix}=\begin{bmatrix}\mathbf{B}^{-1}\mathbf{b}\\ \mathbf{0}\end{bmatrix}\geqq\mathbf{0},$$

where the vector of basic variables $\mathbf{X}_B$ has as its components the variables in $\mathbf{x}$ corresponding to the m linearly independent columns of $\mathbf{A}$. As we shall now see, a basic feasible solution is a useful analytical device for locating extreme points of $\mathscr{S}$; it provides us with a "necessary and sufficient characterization" of an extreme point. Specifically, we state the

8.4.1. EXTREME POINT THEOREM. A vector $\mathbf{x}\epsilon\mathbf{R}^n$ is an extreme point of the convex polyhedron $\mathscr{S}=\{\mathbf{x}|\mathbf{Ax}=\mathbf{b},\mathbf{x}\geqq\mathbf{0}\}$ ($\mathbf{A}$ is $(m\times n)$ and $\rho(\mathbf{A})=m$) if and only if $\mathbf{x}$ is a basic feasible solution to $\mathbf{Ax}=\mathbf{b}$.

PROOF. (sufficiency) What we shall demonstrate first is that if $\bar{\mathbf{x}} \epsilon \mathbf{R}^n$ is a basic feasible solution to $\mathbf{Ax} = \mathbf{b}$, then $\bar{\mathbf{x}}$ is an extreme point of $\mathcal{S}$. Given that $\bar{\mathbf{x}}$ is a basic feasible solution, if it is also an extreme point of $\mathcal{S}$, then it cannot be expressed as a convex combination of any other two distinct vectors in $\mathcal{S}$. Hence there do not exist vectors $\mathbf{x}_1, \mathbf{x}_2 \epsilon \mathcal{S}, \mathbf{x}_1 \neq \mathbf{x}_2$, different from $\bar{\mathbf{x}}$ such that

$$\bar{\mathbf{x}} = \theta \mathbf{x}_2 + (1-\theta)\mathbf{x}_1, \quad 0 < \theta < 1. \tag{8.2}$$

If $\mathbf{x}_1' = (\mathbf{y}_1', \mathbf{0}')$, $\mathbf{x}_2' = (\mathbf{y}_2', \mathbf{0}')$, where $\mathbf{y}_1, \mathbf{y}_2$ are both of order $(m \times 1)$, then (8.2) becomes

$$\begin{bmatrix} \mathbf{x}_B \\ \mathbf{0} \end{bmatrix} = \theta \begin{bmatrix} \mathbf{y}_2 \\ \mathbf{0} \end{bmatrix} + (1-\theta) \begin{bmatrix} \mathbf{y}_1 \\ \mathbf{0} \end{bmatrix}, \quad 0 < \theta < 1 .$$

With $\mathbf{x}_B = \theta \mathbf{y}_2 + (1-\theta)\ \mathbf{y}_1 = \mathbf{B}^{-1}\mathbf{b}$ and $\mathbf{A}\mathbf{x}_1 = \mathbf{B}\mathbf{y}_1 = \mathbf{b}$, $\mathbf{A}\mathbf{x}_2 = \mathbf{B}\mathbf{y}_2 = \mathbf{b}$, it follows that

$$\theta(\mathbf{y}_2 - \mathbf{y}_1) + \mathbf{y}_1 = \theta\,(\mathbf{y}_2 - \mathbf{y}_1) + \mathbf{B}^{-1}\mathbf{b} = \mathbf{B}^{-1}\mathbf{b}$$

and thus $\theta(\mathbf{y}_2 - \mathbf{y}_1) = \mathbf{0}$ or $\mathbf{y}_1 = \mathbf{y}_2$. Hence $\mathbf{x}_1 = \mathbf{x}_2 = \bar{\mathbf{x}}$ (*i.e.*, the representation of $\mathbf{b}$ in terms of basis vectors is unique), so that there do not exist distinct vectors $\mathbf{x}_1, \mathbf{x}_2 \epsilon \mathcal{S}$ such that (8.2) holds. Hence $\bar{\mathbf{x}}$ is an extreme point of $\mathcal{S}$.

(necessity) Next, if $\bar{\mathbf{x}}$ is an extreme point of $\mathcal{S}$, then there are at most m positive components of $\bar{\mathbf{x}}$ which have associated with them a set of linearly independent vectors from $\mathbf{A}$ that provide a basic feasible solution to $\mathbf{Ax} = \mathbf{b}$. Without loss of generality, let the first $k \leq m$ components of $\bar{\mathbf{x}}' = (\mathbf{y}', \mathbf{0}')$ be nonzero, where $\mathbf{y}$ is of order $(k \times 1)$. Then the vectors associated with these components correspond to the first k columns of $\mathbf{A}$, namely $\bar{\mathbf{a}}_1, \ldots, \bar{\mathbf{a}}_k$, so that

$$\mathbf{A\bar{x}} = \mathbf{A}_1\mathbf{y} = \sum_{j=1}^{k} x_j\bar{\mathbf{a}}_j = \mathbf{b},$$

where $\mathbf{A}_1$ is of order $(m \times k)$. If the columns of $\mathbf{A}_1$ yield a basic feasible solution to $\mathbf{Ax} = \mathbf{b}$, then they must be linearly independent. Let us assume that the columns of $\mathbf{A}_1$ are linearly dependent, *i.e.*, there exists at least one coefficient $\lambda_j \neq 0$ such that

$$\mathbf{A}_1\boldsymbol{\lambda} = \sum_{j=1}^{k} \lambda_j\bar{\mathbf{a}}_j = \mathbf{0}.$$

For some scalar $\alpha > 0$, $\alpha\mathbf{A}_1\boldsymbol{\lambda} = \mathbf{0}$ so that

$$\begin{aligned} \mathbf{A}_1\mathbf{y} + \alpha\mathbf{A}_1\boldsymbol{\lambda} &= \mathbf{A}_1(\mathbf{y}+\alpha\boldsymbol{\lambda}) = \mathbf{A}_1\mathbf{y}_1 = \mathbf{b}, \\ \mathbf{A}_1\mathbf{y} - \alpha\mathbf{A}_1\boldsymbol{\lambda} &= \mathbf{A}_1(\mathbf{y}-\alpha\boldsymbol{\lambda}) = \mathbf{A}_1\mathbf{y}_2 = \mathbf{b}. \end{aligned}$$

With α sufficiently small, both $\mathbf{y}_1, \mathbf{y}_2 \geqq \mathbf{0}$ and thus the vectors $\mathbf{x}_1' = (\mathbf{y}_1', \mathbf{0}')$, $\mathbf{x}_2' = (\mathbf{y}_2', \mathbf{0}')$ constitute feasible solutions to $\mathbf{Ax} = \mathbf{b}$. But $\bar{\mathbf{x}} = \frac{1}{2}\mathbf{x}_1 + \frac{1}{2}\mathbf{x}_2$ so that if the vectors $\mathbf{a}_1, \ldots, \mathbf{a}_k$ are linearly dependent, then $\bar{\mathbf{x}}$ is not an extreme point of $\mathcal{S}$ since it can be written as a convex combination of two distinct vectors $\mathbf{x}_1, \mathbf{x}_2 \in \mathcal{S}$. Since this contradicts the assumption that $\bar{\mathbf{x}}$ is an extreme point, the columns of $\mathbf{A}_1$ must be linearly independent so as to provide a basis for $\mathbf{R}^k$ and thus a basic feasible solution to $\mathbf{Ax} = \mathbf{b}$.

If $k = m$, then

$$\sum_{j=1}^{m} x_j\bar{\mathbf{a}}_j = \mathbf{Bx}_B = \mathbf{b} \text{ and } \mathbf{x}_B = \mathbf{B}^{-1}\mathbf{b} \tag{8.3}$$

so that a unique basic feasible solution obtains. If $k < m$, we may add to $\mathbf{A}_1$ any $m-k$ additional columns selected from those remaining in $\mathbf{A}$ to form

$$\sum_{j=1}^{k} \mathbf{x}_j \bar{\mathbf{a}}_j + \sum_{j=k+1}^{m} 0\,\bar{\mathbf{a}}_j = \mathbf{b} \tag{8.3.1}$$

and thus a basis for $\mathbf{R}^m$. But since these vectors are admitted at a zero level, it is evident that the resulting basic feasible solution to $\mathbf{Ax} = \mathbf{b}$ is not unique, *i.e.*, more than one basic feasible solution may correspond to the same extreme point. So for $x_j = 0, j = k+1,\ldots,m$, it is obvious that such a basic feasible solution is degenerate. Since a basis for $\mathbf{R}^m$ cannot contain more than m linearly independent vectors, there are at most m linearly independent columns in $\mathbf{A}$ and thus at most m positive components in $\bar{\mathbf{x}}$. Q.E.D.

In sum, every extreme point of $\mathcal{S}$ corresponds to a basic feasible solution to $\mathbf{Ax} = \mathbf{b}$ (although it may be degenerate) and every basic feasible solution to the same represents an extreme point of $\mathcal{S}$.

It should be intuitively obvious that

8.4.2. COROLLARY. A convex polyhedron $\mathcal{S} = \{\mathbf{x} | \mathbf{Ax} = \mathbf{b}, \mathbf{x} \geq \mathbf{0}, \mathbf{x} \epsilon \mathbf{R}^n\}$ has a finite number of extreme points.

PROOF. The maximum number of possible ways to choose m linearly independent columns from an $(m \times n)$ matrix $\mathbf{A}$ to form a basis matrix $\mathbf{B}$ is given by the combinatorial expression $n!/m!(n-m)!$. Clearly the number of extreme points in $\mathcal{S}$ cannot exceed this value. Q.E.D.

In section 7.2 of the preceding chapter we found that by selecting a nonbasis vector $\mathbf{r}_j$ to enter the basis $\mathcal{B} = \{\mathbf{b}_1,\ldots,\mathbf{b}_m\}$ and then applying the exit criterion (7.10.1) to determine the vector $\mathbf{b}_r$ to be removed from $\mathcal{B}$, we could easily move from one basic feasible solution to another. The new basic feasible solution was denoted

$$\hat{\mathbf{x}} = \begin{bmatrix} \mathbf{x}_{\hat{B}} \\ \mathbf{0} \end{bmatrix} = \begin{bmatrix} \hat{\mathbf{B}}^{-1}\mathbf{b} \\ \mathbf{0} \end{bmatrix},$$

where $\hat{\mathbf{B}} = [\hat{\mathbf{b}}_1, \ldots, \hat{\mathbf{b}}_m]$ with

$$\hat{\mathbf{b}}_i = \mathbf{b}_i,\ i \neq r;\ \ \hat{\mathbf{b}}_r = \mathbf{r}_j,$$

and

$$x_{\hat{B}i} = x_{Bi} - \hat{\theta}\lambda_{ij},\ i \neq r;\ x_{\hat{B}r} = \hat{\theta} = x_{Br} \Big/ \lambda_{rj}.$$

How can we be sure that this new basic feasible solution is actually an extreme point of $\mathcal{S} = \{\mathbf{x} | \mathbf{A}\mathbf{x} = \mathbf{b}, \mathbf{x} \geqq \mathbf{0}, \mathbf{x} \in \mathbf{R}^n\}$? The answer depends upon whether the vectors within $\hat{\mathbf{B}}$ are linearly independent. If they are, the answer must be yes. Let us assume to the contrary that they are linearly dependent. Then, by definition, there exists an $(m \times 1)$ vector $\mathbf{t} \neq \mathbf{0}$ such that $\hat{\mathbf{B}}\mathbf{t} = \Sigma_{i=1}^m t_i \hat{\mathbf{b}}_i = \mathbf{0}$ or

$$\sum_{\substack{i=1 \\ i \neq r}}^{m} t_i \mathbf{b}_i + t_r \mathbf{r}_j = \mathbf{0},$$

where $\hat{\mathbf{b}}_i = \mathbf{b}_i,\ i \neq r;\ \hat{\mathbf{b}}_r = \mathbf{r}_j$. Since the columns of basis matrix $\mathbf{B}$ corresponding to the original basis $\mathcal{B}$ are linearly independent, any subset $\mathbf{b}_i, i=1, \ldots, m,\ i \neq r$, of $\mathcal{B}$ must also be linearly independent so that in the previous expression $t_r \neq 0$. Hence

$$\mathbf{r}_j = \sum_{\substack{i=1 \\ i \neq r}}^{m} (-t_i/t_r)\mathbf{b}_i.$$

Subtracting this equation from (7.8) yields

$$\begin{aligned} \mathbf{0} &= \sum_{i=1}^{m} \lambda_{ij}\mathbf{b}_i - \sum_{\substack{i=1 \\ i \neq r}}^{m} (-t_i/t_r)\, \mathbf{b}_i \\ &= \sum_{\substack{i=1 \\ i \neq r}}^{m} \lambda_{ij}\mathbf{b}_i + \lambda_{rj}\mathbf{b}_r + \sum_{\substack{i=1 \\ i \neq r}}^{m} (t_i/t_r)\mathbf{b}_i \\ &= \lambda_{rj}\mathbf{b}_r + \sum_{\substack{i=1 \\ i \neq r}}^{m} (\lambda_{ij} + t_i/t_r)\mathbf{b}_i. \end{aligned}$$

Since $\mathfrak{B}$ is linearly independent, it must be the case that $\lambda_{rj} = 0$. But this contradicts the assumption made in (7.10.1) above that $\lambda_{rj} > 0$ for otherwise, by theorem 7.2.1, $\mathbf{r}_j$ could not replace $\mathbf{b}_r$ in $\mathfrak{B}$. Hence the columns of $\hat{\mathbf{B}}$ must be linearly independent and thus correspond to an extreme point of $\mathcal{S}$.

For polyhedral convex sets in $\mathbf{R}^n$, the counterpart to existence theorem 8.1.1 is

8.4.3. EXISTENCE THEOREM. The convex polyhedron $\mathcal{S} = \{\mathbf{x}|\mathbf{A}\mathbf{x} = \mathbf{b}, \mathbf{x} \geqq \mathbf{0}, \mathbf{x} \in \mathbf{R}^n\} \neq \phi$ ($\mathbf{A}$ is $(m \times n)$ and $\rho(\mathbf{A}) = m$) has at least one extreme point.

PROOF. According to theorem 7.1, if $\mathbf{A}\mathbf{x} = \mathbf{b}, \mathbf{x} \geqq \mathbf{0}$ has a feasible solution, then it also has a basic feasible solution. And since we know that there exists a one-to-one correspondence between basic feasible and extreme point solutions (by the preceding theorem), the proof is complete. Q.E.D.

We mentioned earlier that the convex hull of a finite number of points is termed the convex polytope spanned by these points. Clearly a convex polytope generated by m points cannot have more than m extreme points (since it is defined as the set of all convex combinations of these m points). Under what circumstances can a point within a convex polytope be represented as a convex combination of its extreme points? As we shall now demonstrate, any nonempty compact convex set $\mathcal{S}$ in $\mathbf{R}^n$ with a finite number of extreme points is the convex hull (set of all convex combinations) of the extreme points. Specifically, we have the following special case of the Krein-Milman (representation) theorem 8.1.3

8.4.4. REPRESENTATION THEOREM. If $\mathcal{S}$ is a convex polytope in $\mathbf{R}^n$ and $\mathcal{E}$ is its finite set of extreme points, then $co(\mathcal{E}) = \mathcal{S}$.

PROOF. Let $\mathcal{S} = co\{\mathbf{x}_1, \ldots, \mathbf{x}_m\}$. Clearly any extreme point of $\mathcal{S}$ is one of the $\mathbf{x}_i, i=1,\ldots,m$. From $\{\mathbf{x}_1,\ldots,\mathbf{x}_m\}$ choose a minimal subset $\{\mathbf{x}_1,\ldots,\mathbf{x}_r\}$, $r \leq m$, whose convex hull is $\mathcal{S}$. Our objective is to verify that each point in this minimal subset is extreme. From the linear combinations

$$\mathbf{x} = \sum_{i=1}^{r} \lambda_i \mathbf{x}_i,\ \bar{\mathbf{x}} = \sum_{i=1}^{r} \bar{\lambda} \mathbf{x}_i \qquad (\mathbf{x},\bar{\mathbf{x}} \in \mathcal{S})$$

we may express $\mathbf{x}_r$ as a convex combination of $\mathbf{x}, \bar{\mathbf{x}}$ or

$$\mathbf{x}_r = \mu\mathbf{x} + (1-\mu)\bar{\mathbf{x}} = \sum_{i=1}^{r}\left[\mu\lambda_i + (1-\mu)\bar{\lambda}_i\right]\mathbf{x}_i = \sum_{i=1}^{r}\theta_i\mathbf{x}_i,\ \ 0 < \mu < 1.$$

If $\theta_r < 1$, then $\mathbf{x}_r = (1-\theta_r)^{-1}\Sigma_{i=1}^{r-1}\theta_i\mathbf{x}_i$, contrary to the assumption that $\{\mathbf{x}_1,\ldots,\mathbf{x}_r\}$ is a minimal subset of $\{\mathbf{x}_1,\ldots,\mathbf{x}_m\}$. If $\theta_r = 1$, then $\theta_i = \mu\lambda_i + (1-\mu)\bar{\lambda}_i = 0$, $i \neq r$. But this implies that $\lambda_i = \bar{\lambda}_i = 0$ for $i \neq r$ so that $\mathbf{x} = \bar{\mathbf{x}} = \mathbf{x}_r$ as required. Q.E.D.

In sum, every convex polytope is the convex hull of its extreme points. (Conversely, if $\{\mathbf{x}_1,\ldots,\mathbf{x}_m\}$ is a finite set of points, then the convex hull of $\{\mathbf{x}_1,\ldots,\mathbf{x}_m\}$ is a convex polytope $\mathcal{S}$. The set of extreme points $\mathcal{E}$ is a subset of $\{\mathbf{x}_1,\ldots,\mathbf{x}_m\}$.)

8.5. Directions for Polyhedral Convex Sets

We noted above that, for $\mathcal{S}$ a closed convex set in $\mathbf{R}^n$, a vector $\mathbf{d}(\neq\mathbf{0})$ is a recession direction for $\mathcal{S}$ if for each $\mathbf{x}\in\mathcal{S}, \mathbf{x}+\lambda\mathbf{d}\in\mathcal{S}$ for all $0 \leq \lambda \in \mathbf{R}$. If $\mathcal{S}$ is now the polyhedral convex set $\{\mathbf{x}|\mathbf{A}\mathbf{x} \leqq \mathbf{b},\ \mathbf{x} \geqq \mathbf{0}, \mathbf{x}\in\mathbf{R}^n\} \neq \phi$, where $\mathbf{A}$ is of order $(m \times n)$, then a vector $\mathbf{d}(\neq\mathbf{0})$ is a ***recession direction*** of $\mathcal{S}$ if and only if

$$\mathbf{A}(\mathbf{x}+\lambda\mathbf{d}) \leqq \mathbf{b}$$
$$\mathbf{x}+\lambda\mathbf{d} \geqq \mathbf{0}$$

for each $0 \leq \lambda \in \mathbf{R}$ and $\mathbf{x} \in \mathcal{S}$. Since $\mathbf{x} \in \mathcal{S}$, *i.e.*, $\mathbf{Ax} \leqq \mathbf{b}$ and $\mathbf{x} \geqq \mathbf{0}$, the first inequality holds for λ arbitrarily large if and only if $\mathbf{Ad} \leqq \mathbf{0}$. Similarly, $\mathbf{x} + \lambda\mathbf{d} \geqq \mathbf{0}$ for arbitrarily large λ if and only if $\mathbf{d} \geqq \mathbf{0}$. In sum, $\mathbf{d}(\neq \mathbf{0})$ is a recession direction for polyhedral $\mathcal{S}$ if and only if $\mathbf{Ad} \leqq \mathbf{0}$ and $\mathbf{d} \geqq \mathbf{0}$.

For polyhedral $\mathcal{S} = \{\mathbf{x} | \mathbf{Ax} = \mathbf{b}, \mathbf{x} \geqq \mathbf{0}, \mathbf{x} \in \mathbf{R}^n\} \neq \phi$, $\mathbf{d}(\neq \mathbf{0})$ is a recession direction for $\mathcal{S}$ if and only if $\mathbf{Ad} = \mathbf{0}$ and $\mathbf{d} \geqq \mathbf{0}$ (this result obtains by applying the preceding argument to the two sets of linear inequalities $\mathbf{Ax} \leqq \mathbf{b}$, $-\mathbf{Ax} \leqq -\mathbf{b}$ implied by $\mathbf{Ax} = \mathbf{b}$).

Given $\mathcal{S} = \{\mathbf{x} | \mathbf{Ax} \leqq \mathbf{b}, \mathbf{x} \in \mathbf{R}^n\}$, the ***recession cone*** of $\mathcal{S}$ is the solution set for the homogeneous system of linear inequalities $\mathbf{Ad} \leqq \mathbf{0}$, $\mathbf{d} \geqq \mathbf{0}$, $\mathbf{d} \neq \mathbf{0}$; it is the polyhedral convex cone $0^+\mathcal{S} = \{\mathbf{d} | \mathbf{x} + \lambda\mathbf{d} \in \mathcal{S}, \lambda \geq 0, \mathbf{x} \in \mathcal{S}, \mathbf{d} \in \mathbf{R}^n\} = \{\mathbf{d} | \mathbf{Ad} \leqq \mathbf{0}, \mathbf{d} \geqq \mathbf{0}, \mathbf{d} \in \mathbf{R}^n\}$ whose nonnull elements are the recession directions of $\mathcal{S}$.

We may view the recession cone $0^+\mathcal{S}$ in an alternative light by noting that if $\mathcal{T}$ is a finite set of points and recession directions in $\mathbf{R}^n$ and $\mathcal{S}$ is represented as $co(\mathcal{T})$, then $0^+\mathcal{S} = co(\mathcal{T}_0)$, where $\mathcal{T}_0$ consists of the origin and the recession directions in $\mathcal{T}$.

It is also true for a polyhedral convex set $\mathcal{S}$ that:

(a) $\mathcal{S}$ is bounded if and only if $0^+\mathcal{S} = \{\mathbf{0}\}$.

(b) $\mathcal{S} + 0^+\mathcal{S} = \mathcal{S}$.

(c) The ***lineality space*** of $\mathcal{S}$ is the linear space
$lin(\mathcal{S}) = 0^+\mathcal{S} \cap (-0^+\mathcal{S}) = \{\mathbf{x} | \mathbf{Ax} = \mathbf{0}, \mathbf{x} \in \mathbf{R}^n\}$.

We noted earlier that an extreme direction for a closed convex set $\mathcal{S}$ in $\mathbf{R}^n$ is a recession direction which cannot be represented as a positive linear combination of two other distinct recession directions of $\mathcal{S}$. When $\mathcal{S}$ is a polyhedral convex set of the form $\{\mathbf{x} | \mathbf{Ax} = \mathbf{b}, \mathbf{x} \geqq \mathbf{0}, \mathbf{x} \in \mathbf{R}^n\}$, with $\mathbf{A}$ of order $(m \times n)$ and $\rho(\mathbf{A}) = m$, how shall we choose or characterize extreme directions? To answer this let us first establish the form of a specialized type of recession direction for $\mathcal{S}$ in $\mathbf{R}^n$. Following the presentation given in chapter seven, we may partition $\mathbf{A}$ as $[\mathbf{B}, \mathbf{R}]$ (with basis matrix $\mathbf{B}$ nonsingular) and $\mathbf{d}' = (\mathbf{d}_1', \mathbf{d}_2')$, where $\mathbf{d}_1$ is of order $(m \times 1)$ and $\mathbf{d}_2$ is of order $\big((n-m) \times 1\big)$. Then the requirement $\mathbf{Ad} = \mathbf{0}$ becomes

$$[\mathbf{B}, \mathbf{R}]\begin{bmatrix}\mathbf{d}_1\\ \mathbf{d}_2\end{bmatrix} = \mathbf{0} \quad \text{or} \quad \mathbf{d}_1 = -\mathbf{B}^{-1}\mathbf{R}\mathbf{d}_2.$$

Hence

$$\mathbf{d} = \begin{bmatrix}\mathbf{d}_1\\ \mathbf{d}_2\end{bmatrix} = \begin{bmatrix}-\mathbf{B}^{-1}\mathbf{R}\mathbf{d}_2\\ \mathbf{d}_2\end{bmatrix}.$$

If there exists a nonbasic column $\mathbf{r}_j$ of $\mathbf{R}$ such that $\mathbf{B}^{-1}\mathbf{r}_j \leqq \mathbf{0}$, $j=1,\ldots,n-m$, then set $\mathbf{d}_2 = \bar{d}_j\mathbf{e}_j, \bar{d}_j > 0$ ($\mathbf{e}_j$ is the jth unit column vector), so that

$$\bar{\mathbf{d}}_j = \bar{d}_j\begin{bmatrix}-\mathbf{B}^{-1}\mathbf{r}_j\\ \mathbf{e}_j\end{bmatrix} \geqq \mathbf{0} \tag{8.4}$$

is a recession direction for $\mathcal{S}$. To see this let $\bar{\mathbf{x}}$ be an extreme point of $\mathcal{S}$. Then $\bar{\mathbf{d}}_j(\neq\mathbf{0})$ is a recession direction for $\mathcal{S}$ if $\bar{\mathbf{x}}\in\mathcal{S}$ implies $\mathbf{x} = \bar{\mathbf{x}} + \lambda\bar{\mathbf{d}}_j \in \mathcal{S}$. Since $\mathbf{A}\bar{\mathbf{d}}_j = \mathbf{0}$, $\bar{\mathbf{d}}_j \geqq \mathbf{0}$, and $\mathbf{A}\bar{\mathbf{x}} = \mathbf{b}$, it follows that $\mathbf{Ax} = \mathbf{b}$ for $\mathbf{x} = \bar{\mathbf{x}} + \lambda\bar{\mathbf{d}}_j$ and for all λ. In addition, $\mathbf{x}$ is a nonnegative or feasible vector if $\lambda \geq 0$ and $\mathbf{B}^{-1}\mathbf{r}_j \leqq \mathbf{0}$. As we shall now demonstrate, recession directions structured as (8.4) serve as extreme direction of $\mathcal{S}$. Specifically,

8.5.1. THEOREM. Given the polyhedral convex set $\mathcal{S} = \{\mathbf{x} | \mathbf{Ax} = \mathbf{b},\ \mathbf{x} \geqq \mathbf{0}, \mathbf{x}\in\mathbf{R}^n\}$, let $\mathbf{A}$ be of order $(m \times n)$ with $\rho(\mathbf{A}) = m$. A recession direction $\bar{\mathbf{d}}_j \in \mathcal{S}$ is also an extreme direction if and only if $\bar{\mathbf{d}}_j$ is a positive scalar multiple of the n-vector

$$\begin{bmatrix}-\mathbf{B}^{-1}\mathbf{r}_j\\ \mathbf{e}_j\end{bmatrix} \geqq \mathbf{0},$$

where $\mathbf{B}^{-1}\mathbf{r}_j \leqq \mathbf{0}$ and $\mathbf{r}_j$ is the jth (nonbasic) column of $\mathbf{R}$.

PROOF. (sufficiency) Let $\mathbf{B}^{-1}\mathbf{r}_j \leqq \mathbf{0}$. Then $\mathbf{A}\bar{\mathbf{d}}_j = \mathbf{0}$, $\bar{\mathbf{d}}_j \geqq \mathbf{0}$ so that $\bar{\mathbf{d}}_j$ is a recession direction of $\mathcal{S}$. To show that $\bar{\mathbf{d}}_j$ is also extreme, construct

$$\bar{\mathbf{d}}_j = \lambda_1 \mathbf{d}_1 + \lambda_2 \mathbf{d}_2 = \bar{\lambda}_1 \begin{bmatrix} \mathbf{d}_{11} \\ \mathbf{e}_j \end{bmatrix} + \bar{\lambda}_2 \begin{bmatrix} \mathbf{d}_{21} \\ \mathbf{e}_j \end{bmatrix},$$

where $\lambda_1, \lambda_2, \bar{\lambda}_1, \bar{\lambda}_2 > 0$ and $\mathbf{d}_1, \mathbf{d}_2$ are recession directions of $\mathcal{S}$. With $\mathbf{A}\mathbf{d}_1 = \mathbf{A}\mathbf{d}_2 = \mathbf{0}$, we must have $\mathbf{d}_{11} = \mathbf{d}_{21} = -\mathbf{B}^{-1}\mathbf{r}_j$ so that $\mathbf{d}_1, \mathbf{d}_2$ are not distinct recession directions in $\mathcal{S}$. Hence $\bar{\mathbf{d}}_j$ must be an extreme direction of $\mathcal{S}$.

(necessity) Let $\bar{\mathbf{d}}'_j = (\bar{d}_1, \ldots, \bar{d}_k, \bar{d}_j \mathbf{e}'_j)$ be an extreme direction of $\mathcal{S}$ with $\bar{d}_i > 0, i=1,\ldots,k$ and $i=j$. Clearly the associated columns of $\mathbf{A}$, namely $\bar{\mathbf{a}}_1, \ldots, \bar{\mathbf{a}}_k$, must be linearly independent and, since $\rho(\mathrm{A}) = m$, we must have $k \leq m$. Let us assume to the contrary that these k vectors are linearly dependent, *i.e.*, there exist scalars $\mu_1, \ldots, \mu_k$ not all zero such that $\Sigma_{i=1}^k \mu_i \bar{\mathbf{a}}_i = \mathbf{0}$. Select $\boldsymbol{\mu}' = (\mu_1, \ldots, \mu_k, \mathbf{0}') \in \mathbf{R}^n$ and $0 < \alpha \in \mathbf{R}$ sufficiently small so that the recession directions $\mathbf{d}_1 = \bar{\mathbf{d}}_j + \alpha\boldsymbol{\mu}$, $\mathbf{d}_2 = \bar{\mathbf{d}}_j - \alpha\boldsymbol{\mu}$ are distinct, nonnegative, and satisfy $\mathbf{A}\mathbf{d}_1 = \mathbf{A}\mathbf{d}_2 = \mathbf{0}$ as required. But since $\bar{\mathbf{d}}_j = \frac{1}{2}\mathbf{d}_1 + \frac{1}{2}\mathbf{d}_2$, this contradicts the assumption that $\bar{\mathbf{d}}_j$ is an extreme direction. Hence the vectors $\bar{\mathbf{a}}_1, \ldots, \bar{\mathbf{a}}_k$ must be linearly independent.

Since $k \leq m$, the vectors $\bar{\mathbf{a}}_1, \ldots, \bar{\mathbf{a}}_k$, together with $m-k$ additional vectors from those remaining in the subset $\{\bar{\mathbf{a}}_i, i = k+1, \ldots, n;\ i \neq j\}$, form a basis for $\mathbf{R}^m$. For convenience, suppose these vectors correspond to the first m columns of $\mathbf{A}$ and constitute the columns of an mth order basis matrix B. Then $\mathbf{A}\bar{\mathbf{d}}_j = \mathbf{B}\hat{\mathbf{d}}_j + \bar{d}_j \mathbf{r}_j = \mathbf{0}$, where $\hat{\mathbf{d}}_j$ is a vector containing the first m components of $\bar{\mathbf{d}}_j$, and thus $\hat{\mathbf{d}}_j = -\bar{d}_j \mathbf{B}^{-1}\mathbf{r}_j$ so that

$$\bar{\mathbf{d}}_j = \begin{bmatrix} \hat{\mathbf{d}}_j \\ \bar{d}_j \mathbf{e}_j \end{bmatrix} = \bar{d}_j \begin{bmatrix} -\mathbf{B}^{-1}\mathbf{r}_j \\ \mathbf{e}_j \end{bmatrix} \geqq \mathbf{0}.$$

And since $\bar{\mathbf{d}}_j \geqq \mathbf{0}$ and $\bar{\mathrm{d}}_j > 0$, it follows that $\mathbf{B}^{-1}\mathbf{r}_j \leqq \mathbf{0}$. Q.E.D.

It is easily demonstrated that

8.5.2. COROLLARY. A convex polyhedron $\mathcal{S} = \{\mathbf{x}|\mathbf{A}\mathbf{x} = \mathbf{b}, \mathbf{x} \geqq \mathbf{0}, \mathbf{x} \in \mathbf{R}^n\}$ has a finite number of extreme directions.

PROOF. Let $\mathbf{A} = [\mathbf{B}, \mathbf{R}]$ be of order $(m \times n)$ with $\rho(\mathbf{A}) = m$. For each choice of an mth order basis matrix $\mathbf{B}$, there are $n-m$ possible ways to select a nonbasic vector $\mathbf{r}_j$ from $\mathbf{R}$ to form $\bar{\mathbf{d}}_j$ in (8.4). Hence an upper bound on the number of extreme directions which can be obtained in this fashion is $(n-m)[n!/m!(n-m)!] = n!/m!(n-m-1)!$ Q.E.D.

We next examine the conditions under which extreme directions exist. Specifically, we look to the following

8.5.3. EXISTENCE THEOREM. The convex polyhedron $\mathcal{S} = \{\mathbf{x}|\mathbf{A}\mathbf{x} = \mathbf{b}, \mathbf{x} \geqq \mathbf{0}, \mathbf{x} \in \mathbf{R}^n\}$, where $\mathbf{A}$ is $(m \times n)$ and $\rho(\mathbf{A}) = m$, has at least one extreme direction if and only if it is unbounded.

PROOF. (necessity) If $\mathcal{S}$ has an extreme direction, then clearly it must be unbounded. (sufficiency) Let $\mathcal{S}$ be unbounded and assume to the contrary that $\mathcal{S}$ has no extreme directions but that, from representation theorem 8.4.4, $\mathcal{S} = co(\mathcal{E})$, *i.e.*, for $\mathcal{E} = \{\bar{\mathbf{x}}_1, \ldots, \bar{\mathbf{x}}_r\}$ the set of extreme points of $\mathcal{S}$,

$$co\{\bar{\mathbf{x}}_1, \ldots, \bar{\mathbf{x}}_r\} = \left\{\mathbf{x} \,|\, \mathbf{x} = \sum_{i=1}^{r} \lambda_i \bar{\mathbf{x}}_i,\ 0 \leq \lambda_i \in \mathbf{R} \text{ for all } i,\ \sum_{i=1}^{r} \lambda_i = 1\right\}.$$

Then for any $\mathbf{x} \in \mathcal{S}$, we have, via the Cauchy-Schwarz inequality,

$$\|\mathbf{x}\| \leq \sum_{i=1}^{r} \lambda_i \|\bar{\mathbf{x}}_i\| \leq \sum_{i=1}^{r} \|\bar{\mathbf{x}}_i\|,$$

a violation of the assumption that $\mathcal{S}$ is unbounded. Thus $\mathcal{S}$ must have at least one extreme direction. Q.E.D.

We note briefly that if $\mathcal{S}$ possesses extreme directions, then they correspond to the extreme points of $0^+\mathcal{S}$.

We may relate the notion of an extreme direction of a polyhedral convex set $\mathcal{S}$ in $\mathbf{R}^n$ to the material on faces of polyhedra presented in section 8.3 above by offering

8.5.4. THEOREM. Given polyhedral $\mathcal{S} = \{\mathbf{x}|\mathbf{A}\mathbf{x} \leqq \mathbf{b}, \mathbf{x} \in \mathbf{R}^n\} \neq \phi$, a recession direction $\mathbf{d}$ $(\neq \mathbf{0}) \in \mathbf{R}^n$ is an extreme direction of $\mathcal{S}$ if and only if $\{\lambda\mathbf{d}|0 < \lambda \in \mathbf{R}\}$ is a one-dimensional face of the recession cone $0^+\mathcal{S} = \{\mathbf{d}|\mathbf{A}\mathbf{d} \leqq \mathbf{0}, \mathbf{d} \geqq \mathbf{0}\}$.

PROOF. (sufficiency) Let $\mathbf{A}^=$ be the submatrix of $\mathbf{A}$ containing those rows $\boldsymbol{\alpha}_i$ for which $\boldsymbol{\alpha}_i\mathbf{d} = 0$. If $\{\lambda\mathbf{d}|0 < \lambda \in \mathbf{R}\}$ is a one-dimensional face of $0^+\mathcal{S}$, then $\rho(\mathbf{A}^=) = n-1$. Hence all solutions of $\mathbf{A}^=\mathbf{y} = \mathbf{0}$ are of the form $\mathbf{y} = \lambda\mathbf{d} \in \mathbf{R}^n$, $0 < \lambda \in \mathbf{R}$. If $\mathbf{d}$ is expressible as a positive linear combination of two other distinct recession directions of $\mathcal{S}$, then, by definition, we obtain a contradiction. Thus $\mathbf{d}$ must be an extreme direction of $\mathcal{S}$.

(necessity) If $\mathbf{d} \in 0^+\mathcal{S}$ and $\rho(\mathbf{A}^=) < n-1$, there exists a direction $\bar{\mathbf{d}} \neq \lambda\mathbf{d}$, $0 < \lambda \in \mathbf{R}$, such that $\mathbf{A}^=\bar{\mathbf{d}} = \mathbf{0}$. Thus for δ sufficiently small, the recession directions $\mathbf{d}_1 = \mathbf{d} + \delta\bar{\mathbf{d}}$, $\mathbf{d}_2 = \mathbf{d} - \delta\bar{\mathbf{d}}$ in $\mathcal{S}$ allow us to write $\mathbf{d} = \frac{1}{2}\mathbf{d}_1 + \frac{1}{2}\mathbf{d}_2$. Obviously $\mathbf{d}$ is not an extreme direction. Q.E.D.

In some of the theorems to follow we shall have occasion to invoke the assumption that certain point sets are bounded. To appropriately characterize this concept let us now examine a couple of "unboundedness criteria." Specifically,

8.5.5. THEOREM. (a) An arbitrary convex set $\mathcal{S}$ in $\mathbf{R}^n$ is unbounded if and only if it contains a half-line. Hence a convex polyhedron $\mathcal{S} = \{\mathbf{x}|\mathbf{A}\mathbf{x} \leqq \mathbf{b}, \mathbf{x} \in \mathbf{R}^n\}$ is unbounded if and only if it admits a half-line or (b) if and only if the polyhedral convex cone $\mathcal{C} = \{\mathbf{x}|\mathbf{Ax} \leqq \mathbf{0}, \mathbf{x} \epsilon \mathbf{R}^n\}$ contains a non-null vector.

PROOF. (a) (sufficiency) If we assume to the contrary that $\mathcal{S}$ contains no half-line, then clearly $\mathcal{S}$ is bounded. (necessity) With $\mathcal{S}$ convex, $\mathbf{x} = \theta\mathbf{x}_1 + (1-\theta)\mathbf{x}_2 \epsilon \mathcal{S}$ for $\mathbf{x}_1, \mathbf{x}_2 \epsilon \mathcal{S}$, $0 \leq \theta \leq 1$. If a half-line generated by $\mathbf{x}$ does not lie in $\mathcal{S}$, then there exists a point $\mathbf{y} \notin \mathcal{S}$ on the said half-line such that $\|\mathbf{x}\| < \|\mathbf{y}\|$ and thus $\mathcal{S}$ is bounded.

(b) (necessity) If polyhedral $\mathcal{S}$ is unbounded, it contains a half-line, cone $\mathcal{C}$ does not have elements of both signs, and thus $\mathcal{C} \neq \{\mathbf{0}\}$. (sufficiency) If $\mathbf{x}_1 \epsilon \mathcal{S}$ and $\mathbf{x}_2$ $(\neq \mathbf{0}) \epsilon \mathcal{C}$, then $\mathbf{x}_1 + \lambda\mathbf{x}_2 \epsilon \mathcal{S}$ for all $0 < \lambda \epsilon \mathbf{R}$ so that $\mathcal{S}$ is an unbounded convex polyhedron. Q.E.D.

8.6. Combined Extreme Point and Extreme Direction Representation for Polyhedral Convex Sets

The principal result of this section (theorem 8.6.2) is a demonstration that a point within a convex polyhedron $\mathcal{S}$ in $\mathbf{R}^n$ can be represented as the sum of a convex combination of a finite number of extreme points from $\mathcal{S}$ and a nonnegative linear combination of a finite number of extreme directions for $\mathcal{S}$, *i.e.*, any $\mathbf{x} \epsilon \mathcal{S}$ is representable as the sum of the convex hull of a finite number of extreme points and the conical hull of a finite collection of extreme directions, all taken from $\mathcal{S}$. To set the stage for this result we state first

8.6.1. THEOREM. For polyhedral $\mathcal{S} = \{\mathbf{x}|\mathbf{Ax} \leqq \mathbf{b}, \mathbf{x} \epsilon \mathbf{R}^n\} \neq \phi$, let $\rho(\mathbf{A}) = n$.

(a) If $max\{\mathbf{u}'\mathbf{x}|\mathbf{x} \epsilon \mathcal{S}\} < +\infty$, then there exists an extreme point $\bar{\mathbf{x}} \epsilon \mathcal{S}$ such that $\mathbf{u}'\bar{\mathbf{x}} = max\{\mathbf{u}'\mathbf{x}|\mathbf{x} \epsilon \mathcal{S}\}$.

(b) If $max\{\mathbf{u}'\mathbf{x}|\mathbf{x} \epsilon \mathcal{S}\}$ is unbounded, then there exists an extreme direction $\bar{\mathbf{d}} \epsilon \mathcal{S}$ such that $\mathbf{u}'\bar{\mathbf{d}} > 0$.

PROOF. Relative to (a), the set of optimal solutions is the non-null face $\mathfrak{F} = \{\mathbf{x}|\mathbf{u}'\mathbf{x} = \bar{\mathrm{u}}, \mathbf{x} \in \mathcal{S}\}$. Then from 8.3.e.ii above, $\mathfrak{F}$ contains a face of dimension $n - \rho(\mathbf{A}) = 0$ and thus admits an extreme point $\bar{\mathbf{x}}$ so that $\mathbf{u}'\bar{\mathbf{x}} = \bar{\mathrm{u}}$.

Looking to (b), if $max\{\mathbf{u}'\mathbf{x}|\mathbf{A}\mathbf{x} \leqq \mathbf{b}, \mathbf{x} \in \mathbf{R}^n\}$ is unbounded, then there is no vector $\mathbf{y} \in \mathbf{R}^m$ which satisfies $\mathbf{A}'\mathbf{y} = \mathbf{u}, \mathbf{y} \geqq \mathbf{0}$ (since otherwise $\mathbf{u}'\mathbf{x} = \mathbf{y}'\mathbf{A}\mathbf{x} \leqq \mathbf{y}'\mathbf{b}$). Hence by Farkas' theorem 3.3.1.a, there exists a recession direction $\mathbf{d}$ such that $\mathbf{A}\mathbf{d} \leqq \mathbf{0}, \mathbf{u}'\mathbf{d} > 0$. Let $\mathbf{u}'\bar{\mathbf{d}} = max\{\mathbf{u}'\mathbf{d}|\mathbf{A}\mathbf{d} \leqq \mathbf{0}, \mathbf{u}'\mathbf{d} \leq \delta > 0\} = \delta$. Then by part (a) of this theorem, $\bar{\mathbf{d}}(\neq \mathbf{0})$ must be an extreme point of $0^+\mathcal{S} = \{\mathbf{d}|\mathbf{A}\mathbf{d} \leqq \mathbf{0}\}$ with $\rho(\mathbf{A}^=) = n-1$ and $\mathbf{u}'\bar{\mathbf{d}} > 0$. Then by theorem 8.5.4, $\bar{\mathbf{d}}$ must be an extreme direction of $\mathcal{S}$. Q.E.D.

We now consider

8.6.2. REPRESENTATION THEOREM [Minkowski, 1910].

Given a nonempty polyhedral convex set $\mathcal{S} = \{\mathbf{x}|\mathbf{A}\mathbf{x} \leqq \mathbf{b}, \mathbf{x} \in \mathbf{R}^n\}$, let $\mathbf{A}$ be of order $(m \times n)$ with $\rho(\mathbf{A}) = n$. In addition, let $\{\mathbf{x}_1, \ldots, \mathbf{x}_k\}$, $\{\mathbf{d}_1, \ldots, \mathbf{d}_l\}$ be finite sets of extreme points and extreme directions for $\mathcal{S}$ respectively. Then

$$\mathcal{S} = \left\{\mathbf{x}|\mathbf{x} = \sum_{i=1}^{k} \lambda_i \mathbf{x}_i + \sum_{j=1}^{l} \mu_j \mathbf{d}_j, \sum_{i=1}^{k} \lambda_i = 1, \right.$$
$$\left. 0 \leq \lambda_i \in \mathbf{R} \text{ for all } i;\ 0 \leq \mu_j \in \mathbf{R} \text{ for all } j \right\}$$
$$= co\{\mathbf{x}_1, \ldots, \mathbf{x}_k\} + coni\{\mathbf{d}_1, \ldots, \mathbf{d}_l\}.$$

PROOF. Let $\mathcal{K} = \{\mathbf{x}|\mathbf{x} = \Sigma_{i=1}^k \lambda_i \mathbf{x}_i + \Sigma_{j=1}^l \mu_j \mathbf{d}_j, \Sigma_{i=1}^k \lambda_i = 1, 0 \leq \lambda_i \in \mathbf{R}$ for all i; $0 \leq \mu_j \in \mathbf{R}$ for all $j\}$. Since $\mathbf{x}_i \in \mathcal{S}$ for all i and $\mathcal{S}$ is convex, it follows that $\bar{\mathbf{x}} = \Sigma_{i=1}^k \lambda_i \mathbf{x}_i \in \mathcal{S}$ for all λ_i satisfying $\Sigma_{i=1}^k \lambda_i = 1, 0 \leq \lambda_i \in \mathbf{R}$. And since the $\mathbf{d}_j$ are recession directions, $\bar{\mathbf{x}} + \Sigma_{j=1}^l \mu_j \mathbf{d}_j \in \mathcal{S}$ for all $0 \leq \mu_j \in \mathbf{R}$. Hence $\mathcal{K} \subseteq \mathcal{S}$.

To prove the reverse set inclusion, let us assume to the contrary that $\mathcal{K} \subset \mathcal{S}$ but $\mathcal{K} \neq \mathcal{S}$. Hence there exists a $\mathbf{y} \in \mathcal{S} \backslash \mathcal{K}$ such that $\mathbf{Ay} \leqq \mathbf{b}$. That is to say, there do not exist coefficients $\lambda_i \geq 0$, $i = 1,\ldots,k$; $\mu_j \geq 0$, $j=1,\ldots,l$, satisfying

$$\begin{aligned} \sum_{i=1}^{k} \lambda_i \mathbf{x}_i + \sum_{j=1}^{l} \mu_j \mathbf{d}_j &= \mathbf{y} \\ \sum_{i=1}^{k} \lambda_i \qquad\qquad &= 1. \end{aligned} \tag{8.5}$$

Then by Farkas' theorem 3.3.1.a, there exists a vector $(\mathbf{u}', \mathrm{u}_0) \in \mathbf{R}^{n+1}$ such that

$$\begin{aligned} &(a) \quad \mathbf{u}'\mathbf{x}_i - \mathrm{u}_0 \leq 0,\ i=1,\ldots,k, \\ &(b) \quad \mathbf{u}'\mathbf{d}_j \leq 0,\ j=1,\ldots,l, \\ &(c) \quad \mathbf{u}'\mathbf{y} - \mathrm{u}_0 > 0.^1 \end{aligned} \tag{8.6}$$

Let $max\{\mathbf{u}'\mathbf{x} | \mathbf{x} \in \mathcal{S}\} < +\infty$. Then according to part (a) of theorem 8.6.1, this finite maximum must occur at an extreme point of $\mathcal{S}$. But since $\mathbf{y} \in \mathcal{S}$ and, from (8.6.a,c), $\mathbf{u}'\mathbf{y} > \mathbf{u}'\mathbf{x}_i$ for all i, we obtain a contradiction. Additionally, if $max\{\mathbf{u}'\mathbf{x} | \mathbf{x} \in \mathcal{S}\}$ is unbounded, then from part (b) of theorem 8.6.1, there exists an extreme direction $\mathbf{d}_j$ with $\mathbf{u}'\mathbf{d}_j > 0$. But (8.5.b) yields another contradiction. Hence it must be true that $\mathcal{K} = \mathcal{S}$. Q.E.D.

[1] To see this write (8.5) as

$$\begin{bmatrix} \mathbf{A}_1 & \mathbf{A}_2 \\ \mathbf{1}' & \mathbf{0}' \end{bmatrix} \begin{bmatrix} \boldsymbol{\lambda} \\ \boldsymbol{\mu} \end{bmatrix} = \begin{bmatrix} \mathbf{y} \\ 1 \end{bmatrix},$$

where the columns of $\mathbf{A}_1, \mathbf{A}_2$ are the $\mathbf{x}_i$'s, $\mathbf{d}_j$'s respectively. Since (8.5) has no solution, Farkas' theorem renders $(\mathbf{u}', \mathrm{u}_0)$ such that

$$\begin{bmatrix} \mathbf{A}_1' & 1 \\ \mathbf{A}_2' & 0 \end{bmatrix} \begin{bmatrix} \mathbf{u} \\ \mathrm{u}_0 \end{bmatrix} \leqq \mathbf{0}, \begin{bmatrix} \mathbf{y} \\ 1 \end{bmatrix}' \begin{bmatrix} \mathbf{u} \\ \mathrm{u}_0 \end{bmatrix} > 0$$

and consequently (8.6).

We note briefly that the converse of this theorem is also true, *i.e.*, every $co\{\mathbf{x}_1,\ldots,\mathbf{x}_k\}+coni\{\mathbf{d}_1,\ldots,\mathbf{d}_l\}$ for finite sets $\{\mathbf{x}_1,\ldots,\mathbf{x}_k\}$, $\{\mathbf{d}_1,\ldots,\mathbf{d}_l\}$ is a convex polyhedron.

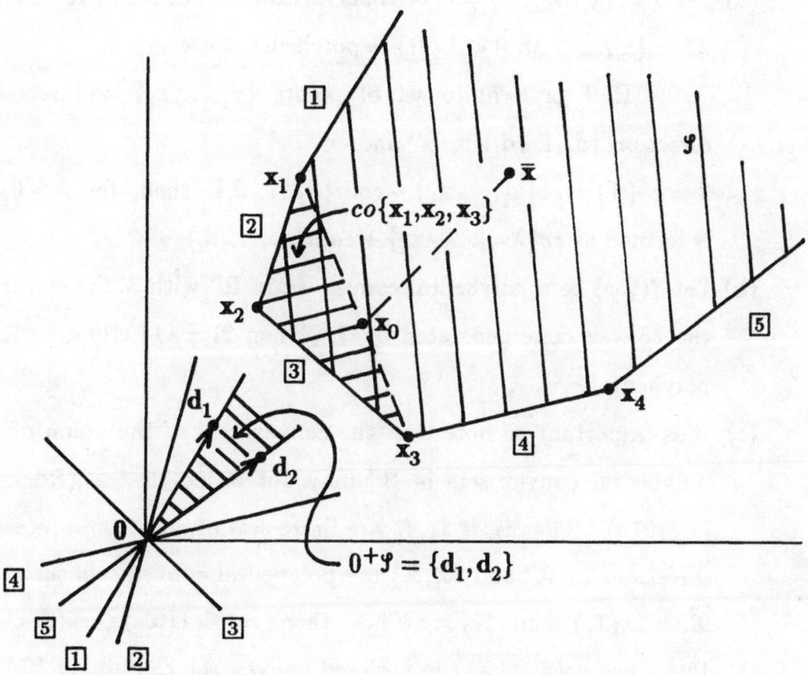

Figure 8.4

Looking to the geometry of this theorem (Figure 8.4), let us assume that $\mathcal{S}$ is unbounded so that $coni\{\mathbf{d}_1,\ldots,\mathbf{d}_l\}\neq\phi$. If the defining hyperplanes of $\mathcal{S}$ are translated parallel to themselves so that they all pass through the origin, then we obtain the recession cone $0^+\mathcal{S}$. By virtue of theorem 8.6.2, any $\mathbf{x}\in\mathcal{S}$ can be expressed as a convex combination of extreme points plus a conical combination of extreme directions of $\mathcal{S}$, *e.g.*, a point $\bar{\mathbf{x}}$ can be written as $\mathbf{x}_0 + \mu_1\mathbf{d}_1 + \mu_2\mathbf{d}_2$. Since $\mathbf{x}_0 = \lambda_1\mathbf{x}_1 + \lambda_3\mathbf{x}_3$, we have

$$\bar{\mathbf{x}} = \lambda_1\mathbf{x}_1 + \lambda_3\mathbf{x}_3 + \mu_1\mathbf{d}_1 + \mu_2\mathbf{d}_2,\ \lambda_1 + \lambda_3 = 1; \qquad \lambda_1, \lambda_3, \mu_1, \mu_2 \geq 0.$$

We state further that:

(a) Let $\mathcal{S} = \{\mathbf{x}|\mathbf{A}\mathbf{x} \leqq \mathbf{b}\} \neq \phi$ be a polyhedral convex set in $\mathbf{R}^n$. Then $\lambda\mathcal{S} = \{\mathbf{x}|\mathbf{A}\mathbf{x} \leqq \lambda\mathbf{b}, 0 < \lambda \in \mathbf{R}\}$ is polyhedral for every λ.

If $\mathcal{T}$ is a finite set of points $\{\mathbf{x}_1,\ldots,\mathbf{x}_k\}$ and recession direction $\{\mathbf{d}_1,\ldots,\mathbf{d}_l\}$ in $\mathbf{R}^n$ and $\mathcal{S} = co(\mathcal{T}) = co\{\mathbf{x}_1,\ldots,\mathbf{x}_k\} + coni\{\mathbf{d}_1,\ldots,\mathbf{d}_l\}$, then, for $\lambda > 0$, $\lambda\mathcal{S}$ is formed as $co\{\lambda\mathbf{x}_1,\ldots,\lambda\mathbf{x}_k\} + coni\{\mathbf{d}_1,\ldots,\mathbf{d}_l\}$.

(b) Let $\mathcal{S}(\neq\phi)$ be a polyhedral convex set in $\mathbf{R}^n$ with $\mathcal{K}$ the closure of the convex cone generated by $\mathcal{S}$. Then $\mathcal{K} = \cup\{\lambda\mathcal{S}|0 < \lambda \in \mathbf{R}\}$ is polyhedral.

(c) It is important to note that the convex hull of the union of two polyhedral convex sets in $\mathbf{R}^n$ may not be polyhedral (Rockafellar, 1970). That is, if $\mathcal{S}_1, \mathcal{S}_2$ are finite sets of points and recession directions in $\mathbf{R}^n$ and $\mathcal{K}_1, \mathcal{K}_2$ are polyhedral convex sets such that $\mathcal{K}_1 = co(\mathcal{S}_1)$ and $\mathcal{K}_2 = co(\mathcal{S}_2)$, then $co(\mathcal{K}_1 \cup \mathcal{K}_2) \subset co(\mathcal{S}_1 \cup \mathcal{S}_2)$. But since $\overline{co(\mathcal{K}_1 \cup \mathcal{K}_2)}$ is a closed convex set containing $\mathcal{K}_1$ and $\mathcal{K}_2$, it must recede in all the directions in which both $\mathcal{K}_1, \mathcal{K}_2$ recede so that $co(\mathcal{S}_1 \cup \mathcal{S}_2) = \overline{co(\mathcal{K}_1 \cup \mathcal{K}_2)}$ is polyhedral (since $co(\mathcal{S}_1 \cup \mathcal{S}_2)$ is finitely generated).

8.7 Resolution of Convex Polyhedra

In this section we shall demonstrate that every polyhedral convex set $\mathcal{S}$ in $\mathbf{R}^n$ is decomposable into the sum of a bounded convex polyhedron and a polyhedral convex cone. To set the stage for this "resolution of polyhedra," let us first consider a supporting definition. Specifically, a pair of finite point sets $\mathcal{K}_1 = \{\mathbf{x}_1,\ldots,\mathbf{x}_k\}$, $\mathcal{K}_2 = \{\mathbf{x}_{k+1},\ldots,\mathbf{x}_m\}$ is termed a ***finite basis*** for a polyhedron $\mathcal{S}$ in $\mathbf{R}^n$ if $\mathcal{S} = co(\mathcal{K}_1) + coni(\mathcal{K}_2)$, *i.e.*, if $\mathbf{x} \in \mathcal{S}$, then $\mathbf{x}$ has the representation $\mathbf{x} = \Sigma_{i=1}^k \lambda_i \mathbf{x}_i + \Sigma_{i=k+1}^m \lambda_i \mathbf{x}_i$, $\Sigma_{i=1}^k \lambda_i = 1$, $0 \leq \lambda_i \in \mathbf{R}$, $i=1,\ldots,m$.

The resolution theorem for polyhedra to be presented below involves a statement of the so-called

8.7.1. FINITE BASIS THEOREM. Every convex polyhedron in $\mathbf{R}^n$ has a finite basis.

The converse of this finite basis theorem is essentially an extension of Weyl's theorem 4.6.1 (every finitely generated cone in $\mathbf{R}^n$ is polyhedral) to general convex polyhedra. It can also be viewed as the converse of Minkowski's theorem 8.6.2 (a set in $\mathbf{R}^n$ obtained as the convex hull of a finite set of vectors plus the conical combination of some other finite set of vectors is polyhedral).

In the proof underlying the aforementioned decomposition we shall have occasion to make a transition from a polyhedral convex set $\mathcal{S}$ involving vectors $\mathbf{x} \in \mathbf{R}^n$ to a polyhedral convex cone $\hat{\mathcal{C}}$ containing vectors $\binom{\mathbf{x}}{t} \in \mathbf{R}^{n+1}$, where $\hat{\mathcal{C}}$ is contained in the half-space $\mathcal{G} = \left\{\binom{\mathbf{x}}{t} \middle| 0 \leq t \in \mathbf{R}\right\}$ and intersects the hyperplane $\mathcal{H} = \left\{\binom{\mathbf{x}}{t} \,\middle|\, t = 1\right\}$ (Goldman, 1956). Specifically, there exists a one-to-one inclusion-preserving correspondence between the set $\mathcal{S} = \{\mathbf{x} | \mathbf{A}\mathbf{x} \leqq \mathbf{b}\}$ in $\mathbf{R}^n$ and the polyhedral convex cone $\hat{\mathcal{C}} = \left\{\binom{\mathbf{x}}{t} \middle| \mathbf{A}\mathbf{x} - \mathbf{b}t \leqq \mathbf{0}, 0 \leq t \in \mathbf{R}\right\}$ in $\mathbf{R}^{n+1}$. For instance, if $\mathbf{x}_0 \in \{\mathbf{x} | \mathbf{A}\mathbf{x} \leqq \mathbf{b}\}$, then $\binom{\mathbf{x}_0}{1} \in \left\{\binom{\mathbf{x}}{t} \middle| \mathbf{A}\mathbf{x} - \mathbf{b}t \leqq \mathbf{0}, t \geq 0\right\}$. Conversely, any polyhedral convex cone $\mathcal{D}$ in $\binom{\mathbf{x}}{t}$ - space such that $\mathcal{D} \subset \mathcal{G}$ can be written as $\mathcal{D} = \left\{\binom{\mathbf{x}}{t} \middle| \mathbf{A}\mathbf{x} - \mathbf{b}t \leqq \mathbf{0}, 0 \leq t \in \mathbf{R}\right\}$. Then corresponding to $\mathcal{D}$ there exists in $\mathbf{R}^n$ a convex polyhedron $\left\{\mathbf{x} \middle| \binom{\mathbf{x}}{1} \in \mathcal{D} \cap \mathcal{H} \neq \phi\right\} = \{\mathbf{x} | \mathbf{A}\mathbf{x} \leqq \mathbf{b}\}$.

On the basis of these considerations we now look to the following

8.7.2. RESOLUTION THEOREM FOR CONVEX POLYHEDRA [Motzkin, 1936; Goldman, 1956]. A polyhedral convex set $\mathcal{S} = \{\mathbf{x}|\mathbf{A}\mathbf{x} \leqq \mathbf{b}, \mathbf{x} \in \mathbf{R}^n\} \neq \phi$ can be expressed as the (vector) sum

$$co\{\mathbf{x}_1,\ldots,\mathbf{x}_k\} + coni\{\mathbf{x}_{k+1},\ldots,\mathbf{x}_m\} =$$

$$\{\mathbf{x}|\mathbf{x}=\mathbf{y}+\mathbf{z},\ \mathbf{y}\in co\{\mathbf{x}_1,\ldots,\mathbf{x}_k\},\ \mathbf{z}\in coni\{\mathbf{x}_{k+1},\ldots,\mathbf{x}_m\}\} =$$

$$\Big\{\mathbf{x}|\mathbf{x} = \Sigma_{i=1}^k\lambda_i\mathbf{x}_i + \Sigma_{i=k+1}^m\lambda_i\mathbf{x}_i,\ \Sigma_{i=1}^k\lambda_i = 1,$$

$$0 \leq \lambda_i \in \mathbf{R}, i=1,\ldots,m\Big\} \tag{8.7}$$

of a polytope (bounded convex polyhedron) $co\{\mathbf{x}_1,\ldots,\mathbf{x}_k\}$ and a polyhedral convex cone $coni\{\mathbf{x}_{k+1},\ldots,\mathbf{x}_m\}$. Conversely, any set $\mathcal{S} \neq \phi$ of the form $co\{\mathbf{x}_1,\ldots,\mathbf{x}_k\} + coni\{\mathbf{x}_{k+1},\ldots,\mathbf{x}_m\}$ is a polyhedral convex set.

PROOF. Let $\mathcal{S} = \{\mathbf{x} \mid \mathbf{A}\mathbf{x} \leqq \mathbf{b}, \mathbf{x}\in\mathbf{R}^n\}$ be a convex polyhedron

By theorem 4.7.2, the polyhedral convex cone $\hat{\mathcal{C}} = \Big\{\binom{\mathbf{x}}{t} \Big| \mathbf{A}\mathbf{x} - \mathbf{b}t \leqq \mathbf{0}, \mathbf{x}\in\mathbf{R}^n, 0 \leq t \in \mathbf{R}\Big\} \subset \mathbf{R}^{n+1}$ is finitely generated by the set of $(n+1)$-vectors $\Big\{\binom{\mathbf{x}_1}{t_1},\ldots,\binom{\mathbf{x}_m}{t_m}\Big\}$, where $t_i = 0$ or 1, $i=1,\ldots,m$. In this regard: for $t_i = 1$, the $\mathbf{x}_i$, $i=1,\ldots,k$, form a convex hull; for $t_i = 0$, the $\mathbf{x}_i$, $i = k+1,\ldots,m$, form a polyhedral convex cone. Thus $\mathbf{x}\in\mathcal{S}$ if and only if $\binom{\mathbf{x}}{1}\in\hat{\mathcal{C}}$ or $\mathbf{x}\in\mathcal{S}$ if and only if $\binom{\mathbf{x}}{1}\in coni\Big\{\binom{\mathbf{x}_1}{t_1},\ldots,\binom{\mathbf{x}_m}{t_m}\Big\}$, where $t_i = 1, i=1,\ldots,k$; $t_i=0$, $i=k+1,\ldots,m$. Thus $\mathcal{S} = \Big\{\mathbf{x} \Big| \binom{\mathbf{x}}{1}\in\hat{\mathcal{C}}\Big\} = co\{\mathbf{x}_1,\ldots,\mathbf{x}_k\} + coni\{\mathbf{x}_{k+1},\ldots,\mathbf{x}_m\}$.

Looking to the converse, let $\mathcal{S}$ be finitely generated, *i.e.*, $\mathcal{S} = co\{\mathbf{x}_1,\ldots,\mathbf{x}_k\} + coni\{\mathbf{x}_{k+1},\ldots,\mathbf{x}_m\}$. Then $\mathbf{x}_0\in\mathcal{S}$ if and only if

$$\begin{pmatrix}\mathbf{x}_0\\1\end{pmatrix}\epsilon\, coni\left\{\begin{pmatrix}\mathbf{x}_1\\1\end{pmatrix},\ldots,\begin{pmatrix}\mathbf{x}_k\\1\end{pmatrix},\begin{pmatrix}\mathbf{x}_{k+1}\\0\end{pmatrix},\ldots,\begin{pmatrix}\mathbf{x}_m\\0\end{pmatrix}\right\}.$$

Again invoking theorem 4.7.2, the preceding statement is equivalent to $\mathbf{x}_0 \epsilon \mathscr{S}$ if and only if $\begin{pmatrix}\mathbf{x}_0\\1\end{pmatrix}\epsilon\hat{\mathscr{C}}$, *i.e.*, $\mathbf{Ax}_0 - \mathbf{b}(1) \leqq \mathbf{0}$ or $\mathbf{Ax}_0 \leqq \mathbf{b}$. Thus $\mathscr{S}$ is a convex polyhedron. Q.E.D.

In sum, a set $\mathscr{S}$ of vectors in $\mathbf{R}^n$ is a convex polyhedron if and only if $\mathscr{S} = \mathscr{S}_1 + \mathscr{S}_2$, where $\mathscr{S}_1$ is a convex polytope and $\mathscr{S}_2$ is a polyhedral convex cone.

It is important to note that in any decomposition such as (8.7), $coni\{\mathbf{x}_{k+1},\ldots,\mathbf{x}_m\} = \{\mathbf{x}|\mathbf{Ax} \leqq \mathbf{0}, \mathbf{x}\epsilon\mathbf{R}^n\}$. We can easily verify this by remembering that $\mathscr{S} = co\{\mathbf{x}_1,\ldots,\mathbf{x}_k\} + coni\{\mathbf{x}_{k+1},\ldots,\mathbf{x}_m\} = \left\{\mathbf{x}\middle|\begin{pmatrix}\mathbf{x}\\1\end{pmatrix}\epsilon\hat{\mathscr{C}}\right\}$, where $\hat{\mathscr{C}} = \left\{\begin{pmatrix}\mathbf{x}\\t\end{pmatrix}\middle| \mathbf{Ax} - \mathbf{b}t \leqq \mathbf{0}, t \geqq 0\right\}$. Then $coni\{\mathbf{x}_{k+1},\ldots,\mathbf{x}_m\} = \left\{\mathbf{x}\middle|\begin{pmatrix}\mathbf{x}\\0\end{pmatrix}\epsilon\hat{\mathscr{C}}\right\} = \{\mathbf{x}|\mathbf{Ax} \leqq \mathbf{0}\}$.

Two immediate consequences of theorem 8.7.2 are:

8.7.3. COROLLARY [Goldman, 1956]. Let $\mathscr{S} = \{\mathbf{x}|\mathbf{Ax} \leqq \mathbf{b}, \mathbf{x}\epsilon\mathbf{R}^n\} \neq \phi$ be a polyhedral convex set with $\rho(\mathbf{A}) = n$. Then $\mathscr{S}$ has a minimal basis $\{\mathbf{x}_1,\ldots,\mathbf{x}_k;\ \mathbf{x}_{k+1},\ldots,\mathbf{x}_m\}$ which is unique (up to positive multiples of the $\mathbf{x}_i$ for $i = k+1,\ldots,m$). Here $\{\mathbf{x}_1,\ldots,\mathbf{x}_k\}$ is the set of extreme vectors of $\mathscr{S}$.

8.7.4. COROLLARY [Goldman, 1956]. A polyhedral convex set $\mathscr{S} = \{\mathbf{x}|\mathbf{Ax} \leqq \mathbf{b}, \mathbf{x}\epsilon\mathbf{R}^n\} \neq \phi$ is bounded if and only if $coni\{\mathbf{x}_{k+1},\ldots,\mathbf{x}_m\} = \{\mathbf{x} \mid \mathbf{Ax} \leqq \mathbf{0}\} = \{\mathbf{0}\}$.

PROOF. (sufficiency) By virtue of theorem 8.7.2, if $coni\{\mathbf{x}_{k+1},\ldots,\mathbf{x}_m\} = \{\mathbf{0}\}$, then $\mathscr{S} = co\{\mathbf{x}_1,\ldots,\mathbf{x}_k\}$ is a bounded polyhedral convex set. (necessity) If $\mathbf{x}_0(\neq\mathbf{0})\epsilon\{\mathbf{x}|\mathbf{Ax} \leqq \mathbf{0}\}$, then $\mathbf{x}_0$ can serve as a recession direction of $\mathscr{S}$ in that the half-line $\{\mathbf{x}|\mathbf{x}_1 + \lambda\mathbf{x}_0, 0 < \lambda\epsilon\mathbf{R}\}$ is a subset of $\mathscr{S}$ (for fixed $\mathbf{x}_1 \epsilon \mathscr{S}$) and thus $\mathscr{S}$ is unbounded. Q.E.D.

Theorem 8.7.2 also provides the foundation for the assertion of a "finite basis theorem for polytopes." That is,

8.7.5. THEOREM. A polyhedral convex set $\mathcal{I} = \{\mathbf{x}|\mathbf{A}\mathbf{x} \leqq \mathbf{b}, \mathbf{x} \in \mathbf{R}^n\} \neq \mathbf{0}$ is a polytope if and only if $\mathcal{I}$ is bounded.

PROOF. (sufficiency) If $\{\mathbf{x}|\mathbf{A}\mathbf{x} \leqq \mathbf{0}\} = \{\mathbf{0}\}$, then, by corollary 8.7.4, $\mathcal{I}$ is bounded and thus $\mathcal{I} = co\{\mathbf{x}_1, \ldots, \mathbf{x}_k\}$ is a polytope. (necessity) Let $\{\mathbf{x}|\mathbf{A}\mathbf{x} \leqq \mathbf{0}\} = \{\mathbf{0}\}$ so that $\mathcal{I} = co\{\mathbf{x}_1, \ldots, \mathbf{x}_k\}$ is a polytope. Additionally, $\{\mathbf{x}|\mathbf{A}\mathbf{x} \leqq \mathbf{0}\} = \{\mathbf{0}\}$ implies that $\rho(\mathbf{A}) = n$. By corollary 8.7.3, $\{\mathbf{x}_1, \ldots, \mathbf{x}_k\}$ is the set of extreme vectors of $\mathcal{I}$. If $\mathcal{K}$ is a finite set of vectors with $\mathcal{I} = co\{\mathbf{x}_1, \ldots, \mathbf{x}_k\}$ as its convex hull, then $\mathcal{K}$ is a basis for $\mathcal{I}$ and thus, by corollary 8.7.3, $\mathcal{K}$ contains all extreme vectors of $\mathcal{I}$. With $\mathcal{I}$ the convex hull of a finite set of extreme vectors, it follows that $\mathcal{I}$ must be bounded. Q.E.D.

An additional application of the aforementioned transformation from a polyhedral convex set $\mathcal{I}$ in $\mathbf{R}^n$ to a polyhedral cone $\hat{\mathcal{C}}$ containing vectors $\begin{pmatrix}\mathbf{x}\\ t\end{pmatrix}$ in $\mathbf{R}^{n+1}$ is found in

8.7.6. THEOREM. (a) Every polyhedral convex cone $\mathcal{C}$ in $\mathbf{R}^n$ and (b) every polytope $\mathcal{K}$ in $\mathbf{R}^n$ is a convex polyhedron.

PROOF. (a) For $\mathcal{C} = \{\mathbf{x}|\mathbf{x} = \mathbf{A}'\boldsymbol{\lambda}, \boldsymbol{\lambda} \geqq \mathbf{0}, \mathbf{x} \in \mathbf{R}^n\}$, let $\mathcal{C}^* = \{\mathbf{y}|\mathbf{A}\mathbf{y} \leqq \mathbf{0}, \mathbf{y} \in \mathbf{R}^n\}$. Since $\mathcal{C}^*$ is a polyhedral convex cone, it has the alternative representation $\{\mathbf{y}|\mathbf{y} = \mathbf{B}'\mathbf{v}, \mathbf{v} \geqq \mathbf{0}\}$ and thus $\mathcal{C}^{**} = \mathcal{C} = \{\mathbf{u}|\mathbf{B}\mathbf{u} \leqq \mathbf{0}, \mathbf{u} \in \mathbf{R}^n\}$ is polyhedral.

(b) To verify this assertion we must demonstrate that every convex hull of a finite set of vectors is representable as the intersection of a finite number of closed half-spaces. Given $\mathcal{I} = \{\mathbf{x}|\mathbf{A}\mathbf{x} \leqq \mathbf{b}\}$, $\hat{\mathcal{C}}^*$ is a polyhedral convex cone and thus, by part (a), $\hat{\mathcal{C}}^* = \{\mathbf{w}|\mathbf{w} = \mathbf{D}'\boldsymbol{\mu}, \boldsymbol{\mu} \geqq \mathbf{0}, \mathbf{w} \in \mathbf{R}^{n+1}\}$. Then $\hat{\mathcal{C}} = \hat{\mathcal{C}}^{**} = \{\mathbf{r}|\mathbf{D}\mathbf{r} \leqq \mathbf{0}, \mathbf{r} \in \mathbf{R}^{n+1}\} \cap \mathcal{H} = \left\{\begin{pmatrix}\mathbf{x}\\ \mathrm{t}\end{pmatrix} \middle| \ t=1\right\}$ is a representation of $\mathcal{I}$ as the intersection of a finite number of closed half-spaces. Q.E.D.

It is also true that:

(a) If $\mathcal{S}_1, \mathcal{S}_2$ are polyhedral convex sets in $\mathbf{R}^n$, then their sum $\mathcal{S}_1 + \mathcal{S}_2$ and difference $\mathcal{S}_1 - \mathcal{S}_2$ is polyhedral.

(b) Let $\mathcal{G}_1, \mathcal{H}_1$ and $\mathcal{G}_2, \mathcal{H}_2$ be finite bases for convex polyhedra $\mathcal{S}_1, \mathcal{S}_2$ respectively in $\mathbf{R}^n$, *i.e.*, from theorems 8.7.1, 2

$$\mathcal{S}_1 = co(\mathcal{G}_1) + coni(\mathcal{H}_1),$$
$$\mathcal{S}_2 = co(\mathcal{G}_2) + coni(\mathcal{H}_2).$$

Then

$$\mathcal{S}_1 + \mathcal{S}_2 = \big(co(\mathcal{G}_1) + co(\mathcal{G}_2)\big) + \big(coni(\mathcal{H}_1) + coni(\mathcal{H}_2)\big)$$
$$= co(\mathcal{G}_1 + \mathcal{G}_2) + coni(\mathcal{H}_1 \cup \mathcal{H}_2).$$

Similarly,

$$\mathcal{S}_1 - \mathcal{S}_2 = co(\mathcal{G}_1 - \mathcal{G}_2) + coni(\mathcal{H}_1 \cup (-\mathcal{H}_2)).$$

Thus both $\mathcal{S}_1 + \mathcal{S}_2$, $\mathcal{S}_1 - \mathcal{S}_2$ have finite bases.

8.8. Separation of Convex Polyhedra

In this section we extend some of the basic separation results for arbitrary convex sets encountered earlier to the case where at least one of the sets being separated by a hyperplane is a convex polyhedron. We start with the strong separation of a polytope and a polyhedral convex cone and then consider the strong separation of two polyhedral convex sets.

8.8.1. STRONG SEPARATION THEOREM [Goldman, 1956]. Let $\mathcal{S} = \{\mathbf{x} | \mathbf{x} = \mathbf{B}'\boldsymbol{\mu}, \boldsymbol{\mu} \geqq \mathbf{0}, \mathbf{1}'\boldsymbol{\mu} = 1\}$ be a convex polytope (bounded convex polyhedron) and $\mathcal{C} = \{\mathbf{x} | \mathbf{x} = \mathbf{A}'\boldsymbol{\lambda}, \boldsymbol{\lambda} \geqq \mathbf{0}\}$ a polyhedral convex cone in $\mathbf{R}^n$ with $\mathcal{S} \cap \mathcal{C} = \phi$. Then there exists a hyperplane $\mathcal{H} = \{\mathbf{x} | \mathbf{C}'\mathbf{x} = 0, \mathbf{C} \neq \mathbf{0}, \mathbf{x} \in \mathbf{R}^n\}$ passing through the origin which separates $\mathcal{S}$, $\mathcal{C}$ in the sense that $\mathcal{S}$ lies in the open half-space $(\mathcal{H}^+) = \{\mathbf{x} | \mathbf{C}'\mathbf{x} > 0\}$ and $\mathcal{C}$ lies in the closed half-space $[\mathcal{H}^-] = \{\mathbf{x} | \mathbf{C}'\mathbf{x} \leq 0\}$.

PROOF. Let $\mathbf{B}' = [\mathbf{b}_1, \ldots, \mathbf{b}_p]$ be of order $(n \times p), p > 1$. Since $\mathcal{S} \cap \mathcal{C} = \phi, \mathbf{b}_i$, $1 \le i \le p$, does not lie in the polyhedral convex cone generated by the set

$$\{-\mathbf{b}_1, \ldots, -\mathbf{b}_{i-1}, -\mathbf{b}_{i+1}, \ldots, -\mathbf{b}_p; \mathbf{a}_1, \ldots, \mathbf{a}_q\}, \tag{8.8}$$

where $\mathbf{A}' = [\mathbf{a}_1, \ldots, \mathbf{a}_q]$ is of order $(n \times q)$. For if

$$\mathbf{b}_i = \sum_{k \neq i}^{p} \bar{\mu}_k(-\mathbf{b}_k) + \sum_{j=1}^{q} \bar{\lambda}_j \mathbf{a}_j, \; 0 \le \bar{\mu}_k \epsilon \mathbf{R}, \; 0 \le \bar{\lambda}_j \epsilon \mathbf{R},$$

then

$$\sum_{k \neq i}^{p} \bar{\mu}_k \mathbf{b}_k + \mathbf{b}_i = \sum_{j=1}^{q} \bar{\lambda}_j \mathbf{a}_j.$$

If we now set

$$\mu = 1 + \sum_{k \neq i}^{p} \bar{\mu}_k > 0, \; \mu_k = \bar{\mu}_k / \mu \;\text{ for } k \neq i,$$

$$\mu_i = \mu^{-1}, \text{ and } \lambda_j = \bar{\lambda}_j / \mu,$$

then we obtain

$$\sum_{k=1}^{p} \mu_k \mathbf{b}_k = \sum_{j=1}^{q} \lambda_j \mathbf{a}_j, \; \sum_{k=1}^{p} \mu_k = 1, \; \mu_k \ge 0, \; \lambda_j \ge 0.$$

Since this contradicts the assertion that $\mathcal{S} \cap \mathcal{C} = \phi$, $\mathbf{b}_i$ is not an element of (8.8).

By proposition 4.8.b, for each i, $1 \le i \le p$, there exists a vector $\mathbf{y}_i$ such that $\mathbf{y}_i'\mathbf{b}_i > 0$ and $\mathbf{y}_i'\mathbf{x} \le 0$ for each vector $\mathbf{x}$ within the spanning set (8.8). Thus $\mathbf{y}_i'\mathbf{b}_i > 0$, $\mathbf{y}_i'\mathbf{b}_k \ge 0$ for $k \neq i$, and $\mathbf{y}_i'\mathbf{x} \le 0$ for all $\mathbf{x} \epsilon \mathcal{C}$. If $\mathbf{C} = \mathbf{y}_1 + \cdots + \mathbf{y}_p$, then $\mathbf{C}'\mathbf{b}_i > 0$ and $\mathbf{C}'\mathbf{x} \le 0$ for all $\mathbf{x} \epsilon \mathcal{C}$. Hence $\mathcal{H}$ separates $\mathcal{S}$, $\mathcal{C}$ strongly. Q.E.D.

Next,

8.8.2. STRONG SEPARATION THEOREM. Let $\mathcal{S}_1, \mathcal{S}_2$ be non-null convex polyhedra in $\mathbf{R}^n$ with $\mathcal{S}_1 \cap \mathcal{S}_2 = \phi$. Then $\mathcal{S}_1, \mathcal{S}_2$ can be strongly separated by a hyperplane $\mathcal{H} = \{\mathbf{x}|\mathbf{C}'\mathbf{x} = \alpha, \mathbf{C} \neq \mathbf{0}, \mathbf{x} \in \mathbf{R}^n\}$ such that $\mathcal{H} \cap \mathcal{S}_1 = \mathcal{H} \cap \mathcal{S}_2 = \phi$.

PROOF. By proposition 8.7.b, $\mathcal{S}_1 - \mathcal{S}_2$ has a finite basis and thus represents a convex polyhedron. Since $\mathbf{0} \notin \mathcal{S}_1 - \mathcal{S}_2$, there exists a supporting hyperplane $\mathcal{H} = \{\mathbf{x}|\mathbf{C}'\mathbf{x} = \alpha\}$ not containing 0 which strongly separates $\mathcal{S}_1 - \mathcal{S}_2$ and $\mathbf{0}$. For every $\mathbf{x}_1 \in \mathcal{S}_1, \mathbf{x}_2 \in \mathcal{S}_2$ we have $\mathbf{C}'(\mathbf{x}_1 - \mathbf{x}_2) \leq \alpha < 0$ and thus $\mathbf{C}'\mathbf{x}_1 < \mathbf{C}'\mathbf{x}_2$. Let $\mathbf{C}'\mathbf{x}$ be bounded above on $\mathcal{S}_1$ so that $\alpha_1 = sup\{\mathbf{C}'\mathbf{x}|\mathbf{x} \in \mathcal{S}_1\}$. Similarly, for $\mathbf{C}'\mathbf{x}$ bounded below on $\mathcal{S}_2$, take $\alpha_2 = inf\{\mathbf{C}'\mathbf{x}|\mathbf{x} \in \mathcal{S}_2\}$. Hence $\mathbf{C}'\mathbf{x}_1 < \mathbf{C}'\mathbf{x}_2$ implies $\alpha_1 < \alpha_2$ and thus $\alpha_1 < \alpha < \alpha_2$. Clearly $\mathcal{H}$ strongly separates $\mathcal{S}_1, \mathcal{S}_2$. Q.E.D.

An alternative strong separation theorem for convex polyhedra $\mathcal{S}_1, \mathcal{S}_2$ which explicitly utilizes their linear inequality structure to obtain a separating hyperplane is provided by

8.8.3. STRONG SEPARATION THEOREM. Let $\mathcal{S}_1 = \{\mathbf{x}|\mathbf{A}_1\mathbf{x} \leqq \mathbf{b}_1\}$, $\mathcal{S}_2 = \{\mathbf{x}|\mathbf{A}_2\mathbf{x} \leqq \mathbf{b}_2\}$ be non-null polyhedral convex sets in $\mathbf{R}^n$ with $\mathcal{S}_1 \cap \mathcal{S}_2 = \phi$. Then there exists a pair of disjoint half-spaces $[\mathcal{H}_1^-]$, $[\mathcal{H}_2^+]$ in $\mathbf{R}^n$ such that $\mathcal{S}_1 \subseteq [\mathcal{H}_1^-], \mathcal{S}_2 \subseteq [\mathcal{H}_2^+]$.

PROOF. With $\mathcal{S}_1 \cap \mathcal{S}_2 = \phi$, the combined systems $\mathbf{A}_1\mathbf{x} \leqq \mathbf{b}_1$, $\mathbf{A}_2\mathbf{x} \leqq \mathbf{b}_2$ are inconsistent. Thus, by theorem 7.2.5, there exist vectors $\mathbf{0} \leqq \mathbf{y}_1, \mathbf{y}_2 \in \mathbf{R}^m$ such that $\mathbf{y}_1'\mathbf{A}_1 + \mathbf{y}_2'\mathbf{A}_2 = \mathbf{0}$, $\mathbf{y}_1'\mathbf{b}_1 + \mathbf{y}_2'\mathbf{b}_2 < 0$. Let $\mathbf{y}_1'\mathbf{b}_1 < 0$. Hence $\mathbf{y}_1'\mathbf{A}_1 \neq \mathbf{0}$ can be used to define a set of half-spaces, *i.e.*, $(\mathbf{y}_1'\mathbf{A}_1)\mathbf{x} \leq \mathbf{y}_1'\mathbf{b}_1$; and $(\mathbf{y}_2'\mathbf{A}_2)\mathbf{x} \leq \mathbf{y}_2'\mathbf{b}_2, \mathbf{y}_2'\mathbf{A}_2 = -\mathbf{y}_1'\mathbf{A}_1$ taken together imply that $(\mathbf{y}_1'\mathbf{A}_1)\mathbf{x} \geq -\mathbf{y}_2'\mathbf{b}_2$. Hence

$$[\mathcal{H}_1^-] = \{\mathbf{x} \,|\, (\mathbf{y}_1'\mathbf{A}_1)\mathbf{x} \leq \mathbf{y}_1'\mathbf{b}_1\},$$

$$[\mathcal{H}_2^+] = \{\mathbf{x} \,|\, (\mathbf{y}_1'\mathbf{A}_1)\mathbf{x} \geq -\mathbf{y}_2'\mathbf{b}_2\}.$$

And since $\mathbf{y}_1'\mathbf{b}_1 < -\mathbf{y}_2'\mathbf{b}_2$, it follows that $[\mathcal{H}_1^-] \cap [\mathcal{H}_2^+] = \phi$. Since every $\mathbf{x} \epsilon \mathcal{S}_1$ is also in $[\mathcal{H}_1^-]$, it follows that $\mathcal{S}_1 \subseteq [\mathcal{H}_1^-]$. With $\mathbf{y}_1'\mathbf{A}_1 = -\mathbf{y}_2'\mathbf{A}_2$, every $\mathbf{x} \epsilon \mathcal{S}_2$ also satisfies $(\mathbf{y}_2'\mathbf{A}_2)\mathbf{x} \leq \mathbf{y}_2'\mathbf{b}_2$ so that $\mathbf{x} \epsilon [\mathcal{H}_2^+]$. Thus $\mathcal{S}_2 \subseteq [\mathcal{H}_2^+]$. Q.E.D.

Additional separation propositions are:

(a) Let $\mathcal{S}_1, \mathcal{S}_2$ be non-null convex sets in $\mathbf{R}^n$ with $\mathcal{S}_1$ polyhedral. A necessary and sufficient condition for the existence of a hyperplane separating $\mathcal{S}_1, \mathcal{S}_2$ properly and not containing $\mathcal{S}_2$ is that $\mathcal{S}_1 \cap ri(\mathcal{S}_2) = \phi$.

(b) Let $\mathcal{S}_1, \mathcal{S}_2$ be non-null convex sets in $\mathbf{R}^n$ with $\mathcal{S}_1 \cap \mathcal{S}_2 = \phi$. If $\mathcal{S}_1$ is a convex polyhedron, $\mathcal{S}_2$ is closed, and $\mathcal{S}_1, \mathcal{S}_2$ have no common recession directions (except possibly those directions in which $\mathcal{S}_2$ is linear), then there exists a hyperplane separating $\mathcal{S}_1, \mathcal{S}_2$ strongly.

Let us now review some of the salient features of the preceding few sections of this chapter and, at the same time, relate them to our earlier discussions on convex sets and finite cones. We note first that the property of a set $\mathcal{S}$ in $\mathbf{R}^n$ being ***polyhedral*** reflects the notion that $\mathcal{S}$ is the intersection of finitely many closed half-spaces; it is the solution set for a finite system of inequalities $\mathbf{a}_i'\mathbf{x} \leq \mathbf{b}_i$, $i=1,\ldots,m$. Moreover, the said polyhedral convex set is a ***finite cone*** $\mathcal{C}$ if the closed half-spaces all have bounding hyperplanes which pass through the origin. Hence $\mathcal{C}$ is the solution set to $\mathbf{a}_i'\mathbf{x} \leq 0$, $i=1,\ldots,m$.

In this regard, the quality of being polyhedral imposes a "finiteness condition" on the ***external representation*** of a convex set (*i.e.*, from theorem 3.2.4.a, an n-dimensional closed convex set is the intersection of its set of supporting closed half-spaces) and, dually, on the ***internal representation*** of a convex set (according to theorem 8.6.2, a polyhedral convex set can be represented as the convex hull of its set of extreme points plus the conical combination of its set of extreme directions) (Rockafellar, 1970).

In general, a ***finitely generated convex set*** $\mathcal{S}$ is the convex hull of a finite collection of points and recession directions, *i.e.*, $\mathbf{x} \in co(\mathcal{S})$ if and only if $\mathbf{x} = co\{\text{set of points}\} + coni\{\text{set of recession directions}\}$; and a ***finitely generated convex cone*** $\mathfrak{C}$ is the convex hull of the null vector and finitely many recession directions.

8.9. Exercises

1. For $\mathcal{S} = \{\mathbf{x} \mid 3x_1 + 4x_2 \leq 12, x_1 \geq 0, x_2 \geq 0\}$, express $\mathcal{S}$ as the convex hull of its collection of extreme points.

2. Which of the following convex polyhedra is also a convex polytope?
$\mathcal{S}_1 = \{\mathbf{x} \mid 3x_1 + 4x_2 \leq 12, -x_1 + x_2 \leq 1, x_1 \geq 0, x_2 \geq 0\}$;
$\mathcal{S}_2 = \{\mathbf{x} \mid -x_1 + x_2 \leq 1, x_1 \geq 3, x_2 \geq 0\}$.

3. For set $\mathcal{S}_2$ of the preceding problem, determine the sets of extreme points and extreme directions and express a point $\mathbf{x} \in \mathcal{S}_2$ as a convex combination of extreme points plus a conical combination of extreme directions. Do the same for
$$\mathcal{S} = \{\mathbf{x} \mid -2x_1 + \tfrac{1}{2}x_2 \leq 6, \; -x_1 + x_2 \leq 8, \; x_1 \geq 0, \; x_2 \geq 0\}.$$

4. Identify the extreme points and extreme directions for
$$\mathcal{S} = \{\mathbf{x} \mid -x_1 + 2x_2 \leq 4, \; x_1 + x_2 \leq 4, \; x_2 \leq 6, \; x_1 \geq 0, \; x_2 \geq 0\}.$$
Express the point $\bar{\mathbf{x}}' = (6,2)$ as a convex combination of extreme points plus a conical combination of extreme directions of $\mathcal{S}$.

5. Determine which of the following equations defining
$$\mathcal{S} = \{\mathbf{x} \mid 2x_1 + 4x_2 + x_3 = 20, \; x_1 + 2x_2 + x_3 = 20, \; 3x_1 + 6x_2 + 2x_3 = 40, \; x_1 \geq 0, \; x_2 \geq 0, \; x_3 \geq 0\}$$
is redundant.

6. Find the dimension of $\mathcal{S} = \{\mathbf{x} \mid \mathbf{A}\mathbf{x} = \mathbf{b}, \mathbf{x} \geqq \mathbf{0}\}$, where
$$\mathbf{A} = \begin{bmatrix} 1 & 2 & 3 \\ 4 & 2 & -1 \end{bmatrix}, \qquad \mathbf{b} = \begin{bmatrix} 1 \\ 1 \end{bmatrix}.$$

CHAPTER 9
SIMPLICIAL TOPOLOGY AND FIXED POINT THEOREMS

9.1 Simplexes

We noted above that a set $\mathscr{S}$ in $\mathbf{R}^n$ is termed a convex polytope if it is expressible as the convex hull of finitely many points $\{\mathbf{x}_1,\cdots,\mathbf{x}_k\}$ or $\mathscr{S} = \mathrm{co}\{\mathbf{x}_1,\cdots,\mathbf{x}_k\}$. It was also mentioned that a set of $k+1$ vectors $\{\mathbf{x}_0,\mathbf{x}_1,\cdots,\mathbf{x}_k\}$ in $\mathbf{R}^n$ is affinely independent if and only if the translated set $\{\mathbf{x}_1-\mathbf{x}_0,\cdots,\mathbf{x}_k-\mathbf{x}_0\}$ is linearly independent or $\rho[\mathbf{x}_1-\mathbf{x}_0,\cdots,\mathbf{x}_k-\mathbf{x}_0]=k$ (for a simple geometric version of this characterization see Figures 9.1.a, b).

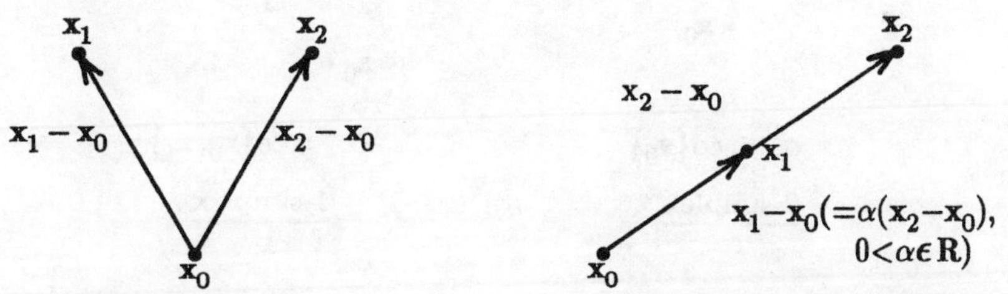

$\{\mathbf{x}_1-\mathbf{x}_0,\mathbf{x}_2-\mathbf{x}_0\}$ linearly independent; $\{\mathbf{x}_0,\mathbf{x}_1,\mathbf{x}_2\}$ affinely independent

a.

$\{\mathbf{x}_1-\mathbf{x}_0,\mathbf{x}_2-\mathbf{x}_0\}$ not linearly independent; $\{\mathbf{x}_0,\mathbf{x}_1,\mathbf{x}_2\}$ not affinely independent.

b.

Figure 9.1

Upon combining these definitions we may now state that a convex polytope $\mathscr{S}$ in $\mathbf{R}^n$ is a ***k-dimensional simplex*** (or simply a ***k-simplex***) if and only if it is the convex hull of a set of $k+1$ affinely independent vectors $\{\mathbf{x}_0,\mathbf{x}_1,\cdots,\mathbf{x}_k\}$, $k\leq n+1$. A k-simplex spanned by the points $\{\mathbf{x}_0,\mathbf{x}_1,\cdots,\mathbf{x}_k\}$ in $\mathbf{R}^n$ will be denoted as

$$\begin{aligned}\sigma^k &= \text{co}\{\mathbf{x}_0, \mathbf{x}_1, \cdots, \mathbf{x}_k\}\\ &= \left\{\mathbf{x} \mid \mathbf{x} = \sum_{i=0}^{k} \lambda_i \mathbf{x}_i,\ \sum_{i=0}^{k} \lambda_i = 1,\ 0 \le \lambda_i \epsilon \mathbf{R} \text{ for all } i\right\}.^{1}\end{aligned} \tag{9.1}$$

In general, $dim(\sigma^k)$ = number of vertices minus $1 = k$ since the vertices of σ^k determine a ***k-plane*** or affine subspace of dimension k, *i.e.*, the vertices lie on no plane of dimension $k-1$. Examples of k-simplexes for $k = 0, 1, 2, 3$ are illustrated in Figures 9.2.a, b, c, d. Clearly the $k+1$ vertices of a k-simplex are extreme points of the same. Hence we may view a k-simplex as a convex polytope having exactly $k+1$ extreme points. We note briefly that a subset of σ^k is said to be a ***subsimplex*** of σ^k if it is itself a simplex.

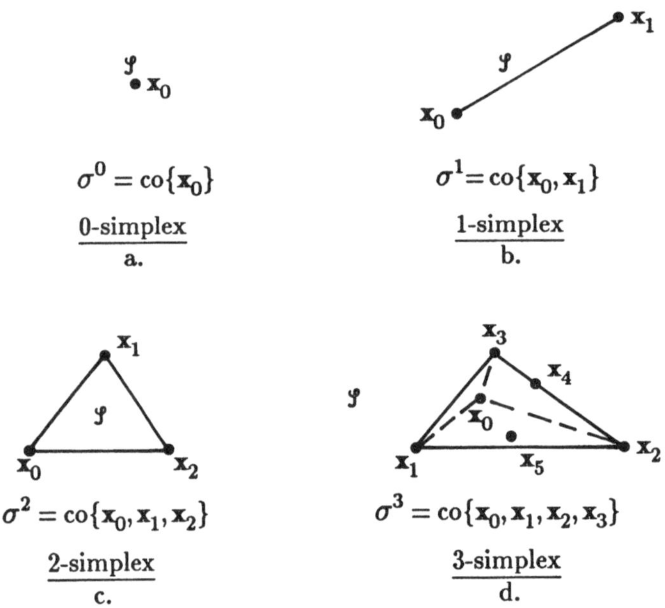

Figure 9.2

[1]Some authors define a k-simplex as the set of all "strictly positive" ($\lambda_i > 0$) convex combinations of a collection of $k+1$ affinely independent points. Clearly any such set must be open. Equation (9.1) is then used to define the "closure of an (open) k-simplex" and subsequently called a "closed k-simplex."

Based upon the preceding definition of a k-simplex we may now state that if the set of vectors $\{\mathbf{x}_1,\cdots,\mathbf{x}_n\}$ in $\mathbf{R}^n$ is linearly independent, then the convex hull of $\{\mathbf{0},\mathbf{x}_1,\cdots,\mathbf{x}_n\}$ is an ***n*-simplex**. In particular, the ***standard n-simplex*** in $\mathbf{R}^n$ is the convex hull of $\{\mathbf{0},\mathbf{e}_1,\cdots,\mathbf{e}_n\}$, where $\mathbf{e}_i, i=1,\cdots,n$, is the ith unit column vector. Moreover, an n-dimensional ***unit simplex*** in $\mathbf{R}^{n+1}$ is formed as the convex hull of the set of $n+1$ unit column vectors $\{\mathbf{e}_1,\cdots,\mathbf{e}_{n+1}\}$ and appears as

$$\sigma^n = \left\{\mathbf{x}\,\middle|\,\sum_{i=1}^{n+1}\mathbf{x}_i = 1,\ 0\le \mathbf{x}_i \in \mathbf{R} \text{ for all } i\right\}.$$

The ***boundary*** of a k-dimensional simplex contains simplexes of lower dimension which are generally called ***simplicial faces*** or just ***faces***. The number of such faces of dimension ℓ is

$$\binom{k+1}{\ell+1} = \frac{(k+1)!}{(\ell+1)!(k-\ell)!}.$$

As evidenced by Figure 9.2.c, σ^2 has $\binom{2+1}{0+1}=3$ zero-dimensional faces (the vertices), $\binom{2+1}{1+1}=3$ one-dimensional faces (the sides of the triangle), and $\binom{2+1}{2+1}=1$ two-dimensional face (namely σ^2 itself). And from Figure 9.2.d, it is easily shown that σ^3 has four zero-dimensional faces, six one-dimensional faces, and four two-dimensional faces. Clearly the union of all the faces of σ^k is its ***closure***.

More specifically, an ℓ-simplex σ^ℓ is an ℓ-face of a k-simplex σ^k if all vertices of σ^ℓ are also vertices of σ^k. Here the vertices of σ^ℓ form a subset of the vertices of σ^k and the convex hull of this subset is the resulting face. If $\ell=0$, $\sigma^\ell=\sigma^0$ is a vertex of σ^k; and if $\ell=k-1$, σ^ℓ is a specialized face which is called a ***facet*** of σ^k. In this latter case there is exactly one vertex in σ^k which is not in σ^ℓ. If this vertex is denoted as $\mathbf{x}_i$ for some $0 \le i \le k$, then we may call σ^ℓ the ***face (facet) of σ^k opposite the vertex*** $\mathbf{x}_i$ and refer to it as a $(k-1)$-simplex. For $\mathbf{x} = \Sigma_{i=0}^{k}\lambda_i\mathbf{x}_i \in \mathrm{co}\{\mathbf{x}_0, \mathbf{x}_1, \cdots, \mathbf{x}_k\}$, let index set $\mathfrak{I}(\mathbf{x}) = \{i \mid \lambda_i > 0,\ 0 \leq i \leq k\} = \{i_0, \cdots, i_\ell\}$. Then $\mathbf{x} \in \sigma^\ell$, where the face of σ^ℓ is termed the ***carrier face*** of $\mathbf{x}$ and is formed as $\mathrm{co}\{\mathbf{x}_{i_0}, \cdots, \mathbf{x}_{i_l}\}$. In fact, $\mathbf{x}$ belongs to exactly one face of σ^k and amounts to the face of lowest dimension of σ^k containing $\mathbf{x}$. As exhibited in Figure 9.2.d, the carrier of $\mathbf{x}_4$ is the face spanned by $\mathbf{x}_2, \mathbf{x}_3$ (*i.e.*, $\mathbf{x}_4 \in \mathrm{co}\{\mathbf{x}_2, \mathbf{x}_3\}$) while the carrier of $\mathbf{x}_5$ is the face generated by $\mathbf{x}_0, \mathbf{x}_1$, and $\mathbf{x}_2$ ($\mathbf{x}_5 \in \mathrm{co}\{\mathbf{x}_0, \mathbf{x}_1, \mathbf{x}_2\}$). The carrier face of any point interior to σ^3 is σ^3 itself.

Additional considerations concerning simplexes are the following:

(a) (Equivalent definitions of affine independence and a k-simplex.) A set of vectors $\{\mathbf{x}_0, \mathbf{x}_1, \cdots, \mathbf{x}_k\}$ in $\mathbf{R}^n$ is affinely independent if

$$\sum_{i=0}^{k} \lambda_i \mathbf{x}_i = \mathbf{0}$$

$$\sum_{i=0}^{k} \lambda_i = 0$$

imply that $\lambda_i = 0, i = 0, 1, \cdots, k$. If we let $\hat{\mathbf{x}}_i = \begin{bmatrix}\mathbf{x}_i\\ 1\end{bmatrix}$, $i = 0, 1, \cdots, k$, so that the preceding two equations appear as

$$\sum_{i=0}^{k} \lambda_i \begin{bmatrix}\mathbf{x}_i\\ 1\end{bmatrix} = \sum_{i=0}^{k} \lambda_i \hat{\mathbf{x}}_i = \mathbf{0},$$

then the set $\{\mathbf{x}_0, \mathbf{x}_1, \cdots, \mathbf{x}_k\}$ in $\mathbf{R}^n$ is ***affinely independent*** if and only if the set $\{\hat{\mathbf{x}}_0, \hat{\mathbf{x}}_1, \cdots, \hat{\mathbf{x}}_k\}$ in $\mathbf{R}^{n+1}$ is linearly independent. In this regard, a k-simplex can be considered as the conical hull of the $k+1$ linearly independent vectors $\{\hat{\mathbf{x}}_0, \hat{\mathbf{x}}_1, \cdots, \hat{\mathbf{x}}_k\}$ in $\mathbf{R}^{n+1}$, *i.e.*, $\sigma^k = \mathrm{coni}\{\hat{\mathbf{x}}_0, \hat{\mathbf{x}}_1, \cdots, \hat{\mathbf{x}}_k\} = \{\hat{\mathbf{x}} \mid \hat{\mathbf{x}} = \Sigma_{i=0}^{k} \lambda_i \hat{\mathbf{x}}_i,\ 0 \leq \lambda_i \in \mathbf{R}$ for all $i\}$, where $\hat{\mathbf{x}} = \begin{bmatrix}\mathbf{x}\\ 1\end{bmatrix}$. Again the vectors $\{\mathbf{x}_0, \mathbf{x}_1, \cdots, \mathbf{x}_k\}$ are the vertices of the k-simplex.

(b) For $\{\mathbf{x}_0, \mathbf{x}_1, \cdots, \mathbf{x}_k\}$ in $\mathbf{R}^n$, the following are equivalent:

(i) $\mathbf{x}_0, \mathbf{x}_1, \cdots, \mathbf{x}_k$ are affinely independent;

(ii) the matrix

$$[\hat{\mathbf{x}}_0, \hat{\mathbf{x}}_1, \cdots, \hat{\mathbf{x}}_k] = \begin{bmatrix} \mathbf{x}_0 & \mathbf{x}_1 & \cdots & \mathbf{x}_k \\ 1 & 1 & \cdots & 1 \end{bmatrix}$$

has rank $k+1$;

(iii) the vectors $\mathbf{y}_i = \mathbf{x}_i - \mathbf{x}_{i-1}$, $1 \leq i \leq k$, are linearly independent; and

(iv) $\dim\left(\{\mathbf{x}_0, \mathbf{x}_1, \cdots, \mathbf{x}_k\}\right) = k$.

(c) Simplexes are convex sets since the convex hull of any set in $\mathbf{R}^n$ is convex. In fact, simplexes are the most elementary of convex sets.

(d) If $\sigma^k = \text{co}\{\mathbf{x}_0, \mathbf{x}_1, \cdots, \mathbf{x}_k\}$ is a k-simplex in $\mathbf{R}^n$, then $ri(\sigma^k) \neq \phi$, *i.e.*, σ^k has interior points.

(e) $\mathcal{S} = \{\phi\}$ is termed a (−1) - ***dimensional empty simplex*** or just a (−1) - ***simplex*** and denoted as σ^{-1}.

(f) Given a simplex σ^k in $\mathbf{R}^n$, the set of all points $\mathbf{x} = \Sigma_{i=0}^k \lambda_i \mathbf{x}_i$, $\Sigma_{i=0}^k \lambda_i = 1$, $0 \leq \lambda_i \in \mathbf{R}$ defines the ***closure of*** σ^k. The points of σ^k for which $\lambda_i > 0$, $i = 0, 1, \cdots, k$, are ***interior points*** of the simplex.

(g) Let σ^k be a k-dimensional simplex in $\mathbf{R}^n$ and let $\cup_i \sigma^{k_i}$ represent the union of all its faces. Then $\cup_i \sigma^{k_i} = \sigma^k$. Moreover, the simplexes σ^{k_i} are disjoint.

9.2. Simplicial Decomposition and Subdivision

As we shall now demonstrate, simplexes may be thought of as the fundamental components of more complicated constructs such as complexes and convex polytopes. First, let $\mathcal{S} = \{\sigma^{k_1}, \cdots, \sigma^{k_r}\}$ depict a finite set of simplexes in $\mathbf{R}^n$. Then $\mathcal{S}$ is termed a ***finite simplicial complex*** $\mathcal{G}$ if the following conditions hold:

(a) if simplex $\sigma^{k_j} \in \mathcal{G}$, then each of its faces is also a member of $\mathcal{G}$;

(b) if simplexes σ^{k_j}, $\sigma^{k_\ell} \in \mathcal{G}$, then $\sigma^{k_j} \cap \sigma^{k_\ell} = \phi$.

Clearly a complex in $\mathbf{R}^n$ amounts to a collection of mutually disjoint simplexes. As this definition implies, if $\sigma^2 = \mathrm{co}\{\mathbf{x}_0, \mathbf{x}_1, \mathbf{x}_2\}$ is a 2-simplex, then its associated 2-dimensional complex has the form

$$\begin{aligned}\mathcal{G} &= \sigma^{-1} \cup \{\sigma_i^0\}_{i=1}^3 \cup \{\sigma_i^1\}_{i=1}^3 \cup \sigma^2 \\ &= \left\{\sigma^{-1}, \mathbf{x}_0, \mathbf{x}_1, \mathbf{x}_2, \ \mathrm{co}\{\mathbf{x}_0, \mathbf{x}_1\}, \ \mathrm{co}\{\mathbf{x}_0, \mathbf{x}_2\}, \ \mathrm{co}\{\ \mathbf{x}_1, \mathbf{x}_2\}, \ \sigma^2\right\},\end{aligned}$$

i.e., $\mathcal{G}$ contains the empty simplex, all vertices and edges of σ^2, and the 2-simplex σ^2 itself. In all, a complex in $\mathbf{R}^n$ will admit 2^{k+1} mutually disjoint simplexes. We note further that a subset of $\mathcal{G}$ is called a ***subcomplex of*** $\mathcal{G}$ if it is a complex in its own right while $\dim(\mathcal{G}) = \max\left\{\dim(\sigma^{k_j}), \mathrm{j} = 1, \cdots, r\right\}$.

Based upon the preceding discussion we state that for $\mathcal{S} = \left\{\sigma^{k_1}, \cdots, \sigma^{k_r}\right\}$ again taken to be a finite set of simplexes in $\mathbf{R}^n$, their union $\mathcal{P} = \bigcup_{i=1}^{r} \sigma^{k_i}$ is a ***convex polytope.*** If $\mathcal{P}$ happens to be the union of all simplexes within a simplicial complex $\mathcal{G}$ in $\mathbf{R}^n$, then complex $\mathcal{G}$ is said to be a ***(simplicial) decomposition*** of the convex polytope $\mathcal{P}$ into "disjoint simplexes." (This decomposition is alternatively termed a ***triangulation*** of $\mathcal{P}$.) So for $\mathcal{G}$ the 2-dimensional complex specified in the preceding paragraph, it is obvious that $\mathcal{G}$ serves as the decomposition or triangulation of σ^2. Moreover, the union of all simplexes in $\mathcal{G}$ above is the polytope

$$\mathcal{P} = \sigma_0^{-1} \cup \mathbf{x}_0 \cup \mathbf{x}_1 \cup \mathbf{x}_2 \cup \mathrm{co}\{\mathbf{x}_0, \mathbf{x}_1\} \ \cup \mathrm{co}\{\mathbf{x}_0, \mathbf{x}_2\} \cup \mathrm{co}\{\mathbf{x}_1, \mathbf{x}_2\} \cup \sigma^2.$$

If $\mathcal{P}$ has the simplicial decomposition $\mathcal{G}$ and $\mathbf{x} \in \mathcal{P}$, then the subsimplex of smallest dimension of $\mathcal{G}$ which contains $\mathbf{x}$ is called the ***carrier simplex*** of $\mathbf{x}$.

A ***(simplicial) subdivision*** $\mathcal{D}$ ***of a k-simplex*** in $\mathbf{R}^n$ is a decomposition or finite collection of simplexes $\{\sigma^{k_j} \mid j \in \mathcal{J}\}$ such that:

(a) $\bigcup_{j \in \mathcal{J}} \sigma^{k_j} = \sigma^k$, *i.e.*, the subsimplexes cover σ^k;

(b) for any $\ell, m \in \mathcal{J}$, either $\sigma^{k_\ell} \cap \sigma^{k_m} = \phi$ or the intersection is a face common to both subsimplexes;

(c) the faces of any σ^{k_i} are members of the subdivision $\mathcal{D}$.

In this regard, any facet or $(k-1)$ - face of a subsimplex of the subdivision of σ^k is either interior to σ^k (and consequently the face of exactly two simplexes of the subdivision) or is on the boundary of σ^k (and thus a face of exactly one subsimplex). The ***mesh*** of a subdivision is the diameter of the largest subsimplex. Moreover, two different subsimplexes $\sigma^{k}{}_{\ell}, \sigma^{k}{}_{m}$ of a given subdivision $\mathfrak{D}$ are ***adjacent*** if they share a common facet or $(k-1)$-dimensional face.

If σ^n is a unit simplex, a ***restricted (simplicial) subdivision*** of σ^k is a subdivision which contains no vertices on the faces of σ^n other that the unit column vectors $\mathbf{e}_1, \cdots, \mathbf{e}_{n+1}$.

In general, any (sub)simplex obtained as the result of the subdivision of a simplex σ^k will be called a ***derived (sub)simplex***.

An important specialized type of subdivision of a k-simplex in $\mathbf{R}^n$ is called a ***barycentric subdivision.*** To execute this type of subdivision let us first note that for any simplex σ^k in $\mathbf{R}^n$, the ***barycenter*** of σ^k is the point $\mathbf{y}_k$ having the ***barycentric weights*** $\lambda_0 = \lambda_1 = \cdots = \lambda_k = \frac{1}{(k+1)}$. Then from (9.1),

$$\mathbf{y}_k = \frac{1}{k+1}\sum_{i=0}^{k} \mathbf{x}_i. \tag{9.2}$$

If we determine the barycenters of all the faces of dimensions $0,1,2,\ldots$ of a simplex σ^k, then we obtain a family $\mathfrak{D}^{(1)}$ of k-simplexes called the ***barycentric subdivision of σ^k of order 1.*** Its properties are:

(a) each k-simplex of $\mathfrak{D}^{(1)}$ contains the barycenter of a face of dimension 0, the barycenter of a face of dimension 1, etc.;

(b) if a k-simplex of $\mathfrak{D}^{(1)}$ contains the barycenters of $\sigma^{kp} = \text{co}\{\mathbf{x}_{i_0}, \cdots, \mathbf{x}_{i_p}\}$ and $\sigma^{kq} = \text{co}\{\mathbf{x}_{j_0}, \cdots, \mathbf{x}_{j_q}\}$, with $p < q$, then $\{i_0, \cdots, i_p\} \subset \{j_0, \cdots, j_q\}$.

If we next divide each k-simplex of $\mathfrak{D}^{(1)}$ in a similar fashion, we generate a new family of k-simplexes called the ***barycentric subdivision of σ^k of order 2*** (denoted $\mathfrak{D}^{(2)}$). In general, we define recursively the vth order barycentric subdivision of a k-simplex as: the ***barycentric subdivision of σ^k of order v***, $\mathfrak{D}^{(v)}$, is the first-order barycentric subdivision of $\mathfrak{D}^{(v-1)}$, where the latter family of simplexes is the $(v-1)^{th}$ order barycentric subdivision of σ^k. And as v increases, we obtain a collection of barycentric subdivisions of arbitrarily small mesh.

For instance, the barycentric subdivision of a zero-dimensional simplex $\sigma^0 = \{\mathbf{x}_0\}$ (a single vertex) is itself, trivially, a subdivided simplex. Hence from (9.2), $\mathbf{y}_0 = \mathbf{x}_0$. The barycentric subdivision of a one-dimensional simplex $\sigma^1 = \text{co}\{\mathbf{x}_0, \mathbf{x}_1\}$ consists of two subsimplexes of the same dimension $\text{co}\{\mathbf{x}_0, \mathbf{y}_1\}$, $\text{co}\{\mathbf{y}_1, \mathbf{x}_1\}$, where $\mathbf{y}_1 = \frac{1}{2}(\mathbf{x}_0 + \mathbf{x}_1)$ is the barycenter of σ^1 (Figure 9.3.a). In general, a simplex of dimension k is barycentricially subdivided into $(k+1)!$ subsimplexes of the same dimension, *e.g.*, as Figure 9.3.b reveals, $\sigma^2 = \text{co}\{\mathbf{x}_0, \mathbf{x}_1, \mathbf{x}_2\}$ is subdivided into $3! = 6$ subsimplexes of dimension two. If we focus on the shaded portion $\text{co}\{\mathbf{y}_0, \mathbf{y}_1, \mathbf{y}_2\}$ (here $\mathbf{y}_0$ is the barycenter of a face of dimension zero, $\mathbf{y}_1$ is the barycenter of a face of dimension one, and $\mathbf{y}_2$ is the barycenter of a face of dimension two), then it can be seen that under barycentric subdivision this derived subsimplex has associated with it a unique hierarchical sequence $\mathscr{V}_0 \subset \mathscr{V}_1 \subset \mathscr{V}_2$ of proper subsets of $\{\mathbf{x}_0, \mathbf{x}_1, \mathbf{x}_2\}$.

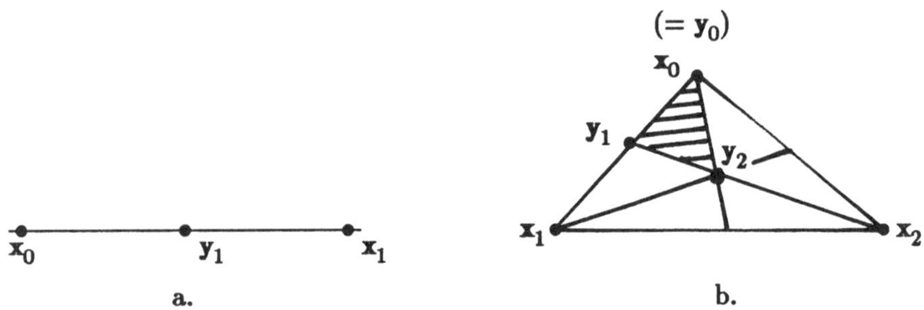

Figure 9.3

In general, any s-dimensional derived subsimplex σ^s is in one-to-one correspondence with a unique hierarchical sequence $\mathscr{V}_0(\neq\phi) \subset \mathscr{V}_1 \subset \cdots$

$\subset \mathcal{V}_r \subset \cdots \subset \mathcal{V}_s$ of proper subsets of $\{\mathbf{x}_0, \mathbf{x}_1, \cdots, \mathbf{x}_k\}$ and such that σ^s is uniquely expressible as $\text{co}\{\mathbf{y}_0, \mathbf{y}_1, \cdots, \mathbf{y}_r, \cdots, \mathbf{y}_s\}$, where $\mathbf{y}_r$ is the barycenter of the subsimplex spanned by $\mathcal{V}_r$.

Based upon this discussion we may observe that:

(a) A $(k-1)$-dimensional derived subsimplex σ^{k-1} in the barycentric subdivision of σ^k is either:

(i) a face of exactly one k-dimensional derived subsimplex if σ^{k-1} completely lies on the boundary of σ^k or;

(ii) a face common to exactly two k-dimensional derived subsimplexes.

(b) Let $\mathfrak{D}^{(v)}$ represent a vth order barycentric subdivision of the k-dimensional simplex σ^k. Then a $(k-1)$-dimensional derived subsimplex $\sigma^{(k-1,v)}$ of order v is either:

(i) a face of exactly one k-dimensional derived subsimplex of order v if $\sigma^{(k-1,v)}$ completely lies on the boundary of σ^k; or

(ii) a face common to exactly two k-dimensional derived subsimplexes of order v.

(c) Let $\sigma^{(v)}$ be any derived subsimplex of order v in the vth barycentric subdivision of the k-dimensional simplex σ^k with $d(\sigma^{(v)})$, $d(\sigma^k)$ their respective diameters. Then

$$d\left(\sigma^{(v)}\right) \leq \left(\frac{k}{k+1}\right)^v d(\sigma^k).$$

At this point it is helpful to introduce a systematic computational procedure which renders a concise description of the barycentric subdivision of a k-simplex (Shapley (1973), Scarf (1973)). Given that the structure of σ^k is provided by (9.1), the first-order barycentric subdivision of σ^k is a simplical subdivision consisting of $(k+1)!$ subsimplexes, each one associated with a particular permutation $i_0, \cdots, i_k$ of the integers $0, \cdots, k$. For each such permutation the derived k-dimensional subsimplex $\sigma^{(k,1)}_{i_0,\cdots,i_k}$ is defined as consisting of all vertices with $\lambda_{i_0} \geq \lambda_{i_1} \geq \cdots \geq \lambda_{i_k} \geq 0$. In general, the vertices of the derived subsimplexes $\sigma^{(k,1)}_{i_0,\ldots,i_k}$ of a first-order barycentric subdivision of σ^k are:

$$\begin{aligned} \mathbf{y}_{i_0} &= \mathbf{x}_{i_0} \\ \mathbf{y}_{i_1} &= \tfrac{1}{2}\left(\mathbf{x}_{i_0}+\mathbf{x}_{i_1}\right) \\ &\vdots \\ \mathbf{y}_{i_k} &= \frac{1}{k+1}\left(\mathbf{x}_{i_0}+\mathbf{x}_{i_1}+\cdots+\mathbf{x}_{i_k}\right). \end{aligned} \tag{9.3}$$

Additionally, any higher order barycentric subdivision of σ^k may be obtained recursively from its immediately preceding one.

EXAMPLE 9.1. For $k=2$, let $\sigma^2 = \text{co}\{\mathbf{x}_0,\mathbf{x}_1,\mathbf{x}_2\}$. Hence there are $(k+1)! = 3! = 6$ derived subsimplexes constituting the barycentric subdivision of σ^2 (Figure 9.4). Clearly the six permutations of the subscripts "0, 1, 2" are

$$0,1,2;\quad 1,0,2;\quad 2,1,0;\quad 0,2,1;\quad 2,0,1;\quad 1,2,0.$$

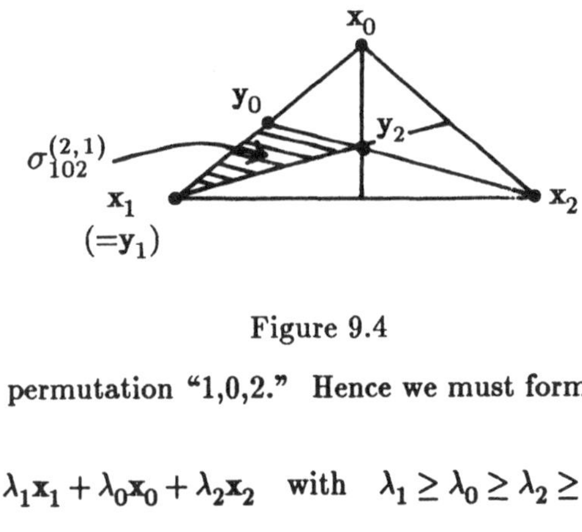

Figure 9.4

Let us select the permutation "1,0,2." Hence we must form

$$\lambda_1\mathbf{x}_1+\lambda_0\mathbf{x}_0+\lambda_2\mathbf{x}_2 \quad \text{with} \quad \lambda_1 \geq \lambda_0 \geq \lambda_2 \geq 0.$$

Using (9.3), the vertices of $\sigma^{(2,1)}_{102} = \text{co}\{\mathbf{y}_1,\mathbf{y}_0,\mathbf{y}_2\}$ are determined as

$$\begin{aligned} \mathbf{y}_1 &= \mathbf{x}_1 \\ \mathbf{y}_0 &= \tfrac{1}{2}\left(\mathbf{x}_1+\mathbf{x}_0\right) \\ \mathbf{y}_2 &= \tfrac{1}{3}\left(\mathbf{x}_1+\mathbf{x}_0+\mathbf{x}_2\right), \end{aligned}$$

i.e., $\sigma_{102}^{(2,1)}$ consists of all vectors of the form

$$\mathbf{y}=\theta_1\mathbf{y}_1+\theta_0\mathbf{y}_0+\theta_2\mathbf{y}_2 \quad \text{with} \quad \theta_1,\theta_0,\theta_2\geq 0.$$

Substituting from above we have

$$\begin{aligned}\mathbf{y}&=\theta_1\mathbf{x}_1+\theta_0\left(\tfrac{1}{2}(\mathbf{x}_1+\mathbf{x}_0)\right)+\theta_2\left(\tfrac{1}{3}(\mathbf{x}_1+\mathbf{x}_0+\mathbf{x}_2)\right)\\&=(\theta_1+\tfrac{1}{2}\theta_0+\tfrac{1}{3}\theta_2)\,\mathbf{x}_1+\left(\tfrac{1}{2}\theta_0+\tfrac{1}{3}\theta_2\right)\mathbf{x}_0+\tfrac{1}{3}\theta_2\mathbf{x}_2\\&=\lambda_1\mathbf{x}_1+\lambda_0\mathbf{x}_0+\lambda_2\mathbf{x}_2.\end{aligned}$$

As required,

$$\lambda_1=\theta_1+\tfrac{1}{2}\theta_0+\tfrac{1}{3}\theta_2>\lambda_0=\tfrac{1}{2}\theta_0+\tfrac{1}{3}\theta_2>\lambda_2=\tfrac{1}{3}\theta_2.$$

We have just verified that any vector of the form $\theta_1\mathbf{y}_1+\theta_0\mathbf{y}_0+\theta_2\mathbf{y}_2$ with $\theta_i\geq 0$ for $i=0,1,2$ can be expressed as $\lambda_1\mathbf{x}_1+\lambda_0\mathbf{x}_0+\lambda_2\mathbf{x}_2$ with $\lambda_1\geq\lambda_0\geq\lambda_2\geq 0$ and conversely. The remaining five derived subsimplexes of the first-order barycentric subdivision of σ^2 are determined in a similar fashion. We note further that the second-order barycentric subdivision of σ^2 is obtained by subjecting each of the six derived subsimplexes to its own first-order barycentric subdivision. Here there will be a total of $((k+1)!)^2$ derived subsimplexes in the barycentric subdivision of σ^2 of order two.

Finally:

(a) The standard n-simplex $\text{co}\{\mathbf{0}, \mathbf{e}_1,\cdots,\mathbf{e}_n\}$ in $\mathbf{R}^n$ has the property that the vector of barycentric coordinates of a point $\mathbf{x}\in\text{co}\{\cdot\}$ is $\mathbf{x}$ itself.

(b) For any integer m, the set

$$\mathcal{Y}=\left\{\mathbf{y}\mid y_i=\frac{\lambda_i}{m}, i=0,1,\cdots,n;\ \sum_{i=0}^{n}\lambda_i=m\right\}$$

in $\mathbf{R}^{n+1}$ is the set of vertices of a simplicial subdivision of σ^n called an ***equilateral subdivision***. This subdivision has m^n n-simplexes of diameter $\frac{\sqrt{2}}{m}$.

9.3. Simplicial Mappings and Labeling

Let us now define a mapping from one simplex or convex polytope to another given that an appropriate simplicial decomposition has been defined on each of them, *i.e.*, simplexes $\sigma^{k_1}, \sigma^{k_2}$ can be thought of as the union of all disjoint subsimplexes within their respective simplicial complexes $\mathcal{G}_1$, $\mathcal{G}_2$ in $\mathbf{R}^n$. Specifically, $f:\sigma^{k_1}\rightarrow\sigma^{k_2}$ is a single-valued ***simplicial mapping*** from the complex $\mathcal{G}_1$ into the complex $\mathcal{G}_2$ if:

(a) the vertices of subsimplex $\sigma_i^{k_1}$ of σ^{k_1} are always mapped by f into the vertices of subsimplex $\sigma_j^{k_2}$ of σ^{k_2}, *i.e.*, f assigns to each vertex $\mathbf{x}_r$, $r=0,1,\ldots,k_1$, of $\sigma_i^{k_1}$ a vertex $\mathbf{y}_t=f(\mathbf{x}_r)$, $t=0,1,\cdots,k_2$, of $\sigma_j^{k_2}$;

(b) interior points of subsimplex $\sigma_i^{k_1}$ of σ^{k_1} are always mapped by f into interior points of subsimplex $\sigma_j^{k_2}$ of σ^{k_2}, *i.e.*, f assigns to each point

$$\begin{aligned}\mathbf{x}\in\sigma_i^{k_1} &= \mathrm{co}\{\mathbf{x}_0,\mathbf{x}_1,\cdots,\mathbf{x}_{k_1}\}\\ &=\left\{\mathbf{x}\,|\,\mathbf{x}=\sum_p\lambda_p\mathbf{x}_p,\ \sum_p\lambda_p=1,\ 0<\lambda_p\in\mathbf{R},\ p=0,1,\cdots,k_1\right\}\end{aligned}$$

a point $\mathbf{y}=f(\mathbf{x})=\sum_p\lambda_p f(\mathbf{x}_p)\in\sigma^{k_2}=\mathrm{co}\{\mathbf{y}_0,\mathbf{y}_1,\cdots,\mathbf{y}_{k_2}\}$.

Based upon this definition we now have

9.3.1. THEOREM. For $\mathcal{G}'$ a subdivision of a simplicial complex $\mathcal{G}$ in $\mathbf{R}^n$, let the mapping $f:\mathcal{G}'\rightarrow\mathcal{G}$ assign to each vertex $\mathbf{x}_i\in\mathcal{G}'$ an arbitrary vertex $\mathbf{y}_j=f(\mathbf{x}_i)$ of the carrier face of $\mathbf{x}_i$ in $\mathcal{G}$. Then f is a simplicial mapping from $\mathcal{G}'$ into $\mathcal{G}$.

PROOF. Let $\mathbf{x}_i$ be a vertex of an arbitrary derived k-simplex $\sigma^{k'}$ of $\mathcal{G}'$. Then there exists a k-simplex σ^k of $\mathcal{G}$ such that $\sigma^{k'}\subset\sigma^k$ and thus $\mathbf{x}_i\in\sigma^k$. Since a carrier simplex of vertex $\mathbf{x}_i$ is a face simplex (of lowest dimension) of σ^k, the image $\mathbf{y}_j=f(\mathbf{x}_i)$ must be a vertex of σ^k and thus f is a simplicial mapping. Q.E.D.

Before we see exactly how the concept of a simplicial mapping may be used, let us examine the notion of an "integer label." Let the k-simplex

$\sigma^k = \text{co}\{\mathbf{x}_0, \mathbf{x}_1, \cdots, \mathbf{x}_k\}$ be simplicially subdivided and let $\mathcal{V}$ represent the collection of all the vertices of all the derived subsimplexes. A ***labeling function*** ℓ is a mapping from the set of vertices $\mathcal{V}$ into the set of integer labels $\mathcal{I} = \{0, 1, \cdots, k\}$. Here $\ell(\mathbf{x}_i)$ is called the ***label*** of $\mathbf{x}_i \in \mathcal{V}$ and the label assignment must be such that $\mathbf{x}_i \in \text{co}\{\mathbf{x}_{i_0}, \mathbf{x}_{i_1}, \cdots, \mathbf{x}_{i_m}\}$ implies $\ell(\mathbf{x}_i) \in \{i_0, i_1, \cdots, i_m\}$. In particular, if $\mathbf{x}_{i_1}, \cdots, \mathbf{x}_{i_p}$ are the vertices of some simplex σ^{kp} in $\text{co}\{\mathbf{x}_0, \mathbf{x}_1 \cdots, \mathbf{x}_k\}$, then σ^{kp} is characterized as labeled by the set $\{\ell(\mathbf{x}_{i_1}), \cdots, \ell(\mathbf{x}_{i_p})\}$. If for all $\mathbf{x} \in \mathcal{V}$ we have $\ell(\mathbf{x}) \in \mathcal{I}$, then the mapping ℓ is called a ***proper labeling*** of the subdivision. Moreover, a subsimplex is termed ***completely labeled*** if $\ell(\mathbf{x})$ assumes all the values in $\mathcal{I}$ on its set of vertices.

A derived subsimplex $\text{co}\{\mathbf{y}_0, \mathbf{y}_1, \cdots, \mathbf{y}_k\}$ is said to be ***regular*** if $\ell(\mathbf{y}_i) \neq \ell(\mathbf{y}_j)$, $i \neq j$, *i.e.*, different vertices in $\mathcal{V}$ have different labels under ℓ. Clearly any regular derived subsimplex must be completely labeled. In addition, a $(k-1)$-dimensional derived subsimplex $\text{co}\{\mathbf{v}_0, \mathbf{v}_1, \cdots, \mathbf{v}_{k-1}\}$ is regular if $\ell(\mathbf{v}_i) \neq \ell(\mathbf{v}_j)$, $i \neq j$, and $\ell(\mathbf{v}_i) \neq k$ for $i=0, 1, \cdots, k-1$.

The next three theorems (Sperner's lemma, the Knaster-Kuratowski-Mazurkiewicz (K-K-M) theorem, and Brouwer's theorem) are all mathematically equivalent and enable us to address the question of the existence of a fixed point for a continuous point-to-point mapping. As we shall now see, Sperner's lemma ensures the existence of at least one completely labeled derived subsimplex in a simplicial subdivision. The K-K-M theorem provides a set of assumptions which, via Sperner's lemma, guarantee that the intersection of a finite collection of sets on a simplex is not vacuous, and, by also invoking Sperner's lemma, Brouwer's theorem demonstrates that a continuous mapping of a simplex into itself admits a fixed point.

9.3.2. SPERNER'S LEMMA [Sperner (1928); Nikaido (1968); Kuhn (1968); Stoer, Witzgall (1970)]. Let $\mathcal{D}$ be a simplicial subdivision of the k-simplex $\sigma^k = \text{co}\{\mathbf{x}_0, \mathbf{x}_1, \cdots, \mathbf{x}_k\}$. If σ^k is properly labeled by the labeling function ℓ, then there exists an odd number of completely labeled subsimplexes in $\mathcal{D}$.

PROOF. (By induction on k.) For $k=0$, σ^0 has but a single point $\mathbf{x}_0$ which bears the label "0" and thus there is one completely labeled subsimplex, namely $\mathbf{x}_0$ itself.

Assume the theorem true for $k-1$ and prove it for k. Then under this assumption the number $\Delta^{(k-1)}$ of $(k-1)$-subsimplexes with label set $\{0,1,\cdots,k-1\}$ on the facet $\sigma^{k-1} = \text{co}\{\mathbf{x}_0,\mathbf{x}_1,\cdots,\mathbf{x}_{k-1}\}$ of σ^k is odd. For each k-simplex $\sigma^{(k)} = \text{co}\{\mathbf{y}_0,\mathbf{y}_1,\cdots,\mathbf{y}_k\}$ in $\mathfrak{D}$, let $\delta(\sigma^{(k)})$ denote the number of facets of $\sigma^{(k)}$ with label set $\{0,1,\cdots,k-1\}$. Then

$$\delta\left(\sigma^{(k)}\right) = \begin{cases} 1 \text{ if } \ell(\sigma^{(k)}) = \{0,1,\cdots,k\}; \\ 0 \text{ or } 2 \text{ otherwise}. \end{cases} \tag{9.4}$$

To see this let $\sigma^{(k)}$ be a regular derived subsimplex. Then exactly k vertices of $\sigma^{(k)}$, namely $\mathbf{y}_0,\mathbf{y}_1,\cdots,\mathbf{y}_{k-1}$, are mapped by ℓ into $\{0,1,\cdots,k-1\}$ while the remaining vertex $\mathbf{y}_k$ is mapped by ℓ to the integer label k. Thus exactly one facet of $\sigma^{(k)}$, $\text{co}\{\mathbf{y}_0,\mathbf{y}_1, \cdots,\mathbf{y}_{k-1}\}$, is regular. Other facets have a common vertex $\mathbf{y}_k$ mapped to k and hence are not regular. Thus $\delta(\sigma^{(k)}) = 1$ for all regular $\sigma^{(k)}$.

If $\sigma^{(k)}$ is not regular or completely labeled (some of the integer labels $0,1,\cdots,k$ are not the image of any vertices of $\sigma^{(k)}$ under ℓ), then:

(a) some of the labels $0,1,\cdots,k-1$ are not the image of any vertices of $\sigma^{(k)}$; or

(b) each of the labels $0,1,\cdots,k-1$ is the image of some vertex of $\sigma^{(k)}$ while k is not.

So with $\sigma^{(k)}$ having no regular facets, $\delta(\sigma^{(k)}) = 0$ if (a) holds. If (b) obtains, the set of all vertices of $\sigma^{(k)}$ is mapped onto $\{0,1,\cdots,k-1\}$. For $\sigma^{(k)} = \text{co}\{\mathbf{y}_0,\mathbf{y}_1,\cdots,\mathbf{y}_k\}$, $\ell(\mathbf{y}_i) = i$, $i=0,1,\cdots,k-1$; $\ell(\mathbf{y}_k) = j$ for $0 \leq j \leq k-1$. Thus $\sigma^{(k)}$ has exactly two regular facets: one spanned by the k vertices of $\sigma^{(k)}$ other than $\mathbf{y}_k$; and one spanned by the k vertices of $\sigma^{(k)}$ other than $\mathbf{y}_j$. Hence $\ell(\sigma^{(k)}) = 2$.

Under (9.4), if $\Delta^{(k)}$ represents the number of k-simplexes in $\mathfrak{D}$ with label set $\{0, 1, \cdots, k-1\}$, then we must have

$$\Delta^{(k)} \equiv \sum_{\sigma^{(k)} \subseteq \mathfrak{D}} \delta(\sigma^{(k)}) \pmod 2.^2$$

Looking to the term $\Sigma\delta(\sigma^{(k)})$ (which depicts the number of regular derived subsimplexes counted at least once), we see, from the preceding section, that all interior $(k-1)$ - subsimplexes labeled $\{0, 1, \cdots, k-1\}$ are counted twice while those on the boundary of σ^k are counted once. However, all $(k-1)$-subsimplexes labeled $\{0, 1, \cdots, k-1\}$ and on the boundary of σ^k must be on the facet $\sigma^{k-1} = \text{co}\{\mathbf{x}_0, \mathbf{x}_1, \cdots, \mathbf{x}_{k-1}\}$. Thus $\Delta^{(k-1)}$ is the number of all $(k-1)$ - subsimplexes labeled $\{0, 1, \cdots, k-1\}$ which are not counted twice in $\Sigma\delta(\sigma^{(k)})$ so that

$$\Delta^{(k-1)} \equiv \sum_{\sigma^{(k)} \subseteq \mathfrak{D}} \delta(\sigma^{(k)}) \pmod 2.$$

This congruence relation coupled with the preceding one renders $\Delta^{(k)} \equiv \Delta^{(k-1)} \pmod 2$. And since $\Delta^{(k-1)}$ is odd under the induction hypothesis, it must be true that $\Delta^{(k)}$ is odd. Q.E.D.

In essence, Sperner's lemma informs us that if a k-simplex σ^k is (barycentrically) subdivided and a labeling function ℓ assigns to a vertex $\mathbf{y}_i$ in the subdivision a label $\ell(\mathbf{y}_i)$ of a vertex $\mathbf{x}_{\ell(\mathbf{y}_i)}$ of a carrier of $\mathbf{y}_i$, then there is some k-dimensional derived simplex $\sigma^{(k)} = \text{co}\{\mathbf{y}_0, \mathbf{y}_1, \cdots, \mathbf{y}_k\}$ for which $\ell(\mathbf{y}_i) \neq \ell(\mathbf{y}_j)$, $i \neq j$, *i.e.*, $\sigma^{(k)}$ is regular and thus completely labeled. Moreover, the number of such simplexes is odd.

The preceding version of Sperner's lemma is known as the "strong form." There is also a "weak form" which asserts that

[2] This expression is a particularization of the general ***congruence relation*** $a \equiv b \pmod m$. Here we may state that "integers a, b are congruent modulo m" if and only if their difference is divisible by m, *i.e.*, if and only if there exists an integer r such that $\frac{a-b}{m} = r$. In short, $a \equiv b \pmod m \leftrightarrow \frac{a-b}{m} = r$.

9.3.2.1. (WEAK) SPERNER'S LEMMA. Let $\mathfrak{D}$ be a simplicial subdivision of the k-simplex $\sigma^k = \text{co}\{\mathbf{x}_0, \mathbf{x}_1, \cdots, \mathbf{x}_k\}$. If σ^k is properly labeled by the labeling function ℓ, then there exists at least one completely labeled subsimplex in $\mathfrak{D}$.

For a generalization of Sperner's lemma see Sperner (1980), Forster (1980), Kuhn (1980), Shapley (1973), and Kannai (1988).

Our next theorem provides conditions which guarantee that the intersection of $k+1$ closed sets $\mathcal{A}_0, \mathcal{A}_1, \cdots, \mathcal{A}_k$ defined on a simplex σ^k is not empty. Specifically, we have the

9.3.3. KNASTER - KURATOWSKI - MAZURKIEWICZ THEOREM [Knaster, Kuratowski, Mazurkiewicz (1929); Berge (1963); Border (1985)]. Let $\sigma^k = \text{co}\{\mathbf{x}_0, \mathbf{x}_1, \cdots, \mathbf{x}_k\}$ be a k-simplex in $\mathbf{R}^n$ and let $\{\mathcal{A}_0, \mathcal{A}_1, \cdots, \mathcal{A}_k\}$ be a family of closed subsets of σ^k. If for each index set $\{i_0, i_1, \cdots, i_m\} \subset \{0, 1, \cdots, k\}$ we have $\text{co}\{\mathbf{x}_{i_0}, \mathbf{x}_{i_1}, \cdots, \mathbf{x}_{i_m}\} \subset \cup_{j=0}^{m} \mathcal{A}_{i_j}$, then $\cap_{i=0}^{k} \mathcal{A}_i \neq \phi$ is compact.

PROOF. Since σ^k is compact and $\sigma^k \subset \cup_{i=0}^{k} \mathcal{A}_i$, the $\mathcal{A}_i$ constitute a closed covering of σ^k. By Lebesgue's theorem 1.3.1, there exists an $\epsilon > 0$ such that each subsimplex in a subdivision $\mathfrak{D}$ of σ^k has diameter $\leq \epsilon$. Let ℓ be a labeling function which assigns to each vertex $\mathbf{x}_i$ of $\mathfrak{D}$ an integer label $\ell(\mathbf{x}_i) \in \{0, 1, \cdots, k\}$. In this regard, if $\mathbf{x}_i \in \mathfrak{D}$ belongs to the face $\text{co}\{\mathbf{x}_{i_0}, \cdots, \mathbf{x}_{i_m}\} \subset \cup_{j=0}^{m} \mathcal{A}_{i_j}$, then there is some index $i \in \{i_0, \cdots, i_m\}$ with $\mathbf{x}_i \in \mathcal{A}_i$. If all vertices are labeled in this fashion, then according to Sprener's lemma, there is a completely labeled subsimplex $\text{co}\{\mathbf{x}_{\ell_0}, \cdots, \mathbf{x}_{\ell_k}\}$ with $\mathbf{x}_{\ell_i} \in \mathcal{A}_i$, $i = 0, 1, \cdots, k$. Since this subsimplex has: (1) a point in common with each $\mathcal{A}_i$, $i = 0, 1, \cdots, k$; and (2) a diameter $\leq \epsilon$, it must be true that $\cap_{i=0}^{k} \mathcal{A}_i \neq \phi$. And since this intersection is a closed subset of compact σ^k, it too must be compact. Q.E.D.

As this theorem reveals, if ***each*** face simplex of σ^k is a proper subset of a subcollection of closed sets $\mathcal{A}_0, \mathcal{A}_1, \cdots, \mathcal{A}_k$ in $\mathbf{R}^n$, then the intersection of all $k+1$ of these sets is nonempty.

Two consequences of the K-K-M theorem are:

9.3.4. COROLLARY. Let $\text{co}\{\mathbf{x}_0, \mathbf{x}_1, \cdots, \mathbf{x}_k\}$ be a subset of $\mathbf{R}^n$ with $\{\mathcal{A}_0, \mathcal{A}_1, \cdots, \mathcal{A}_k\}$ a family of closed sets such that for every $\{i_0, \cdots, i_\ell\} \subset \{0, 1, \cdots, k\}$ we have $\text{co}\{\mathbf{x}_{i_0}, \cdots, \mathbf{x}_{i_\ell}\} \subset \cup_{j=0}^{\ell} \mathcal{A}_{i_j}$. Then $\cap_{i=0}^{k} \mathcal{A}_i \neq \phi$.

9.3.5. COROLLARY. Let $\{\mathcal{A}_0, \mathcal{A}_1, \cdots, \mathcal{A}_k\}$ in $\mathbf{R}^n$ be a collection of closed sets covering the k-simplex $\sigma^k = \text{co}\{\mathbf{x}_0, \mathbf{x}_1, \cdots, \mathbf{x}_k\}$ in a fashion such that: (1) $\mathcal{A}_i$ contains $\mathbf{x}_i$; and (2) $\mathcal{A}_i$ does not meet the face opposite to $\mathbf{x}_i$, $i=1, \cdots, k$. Then $\cap_{i=0}^{k} \mathcal{A}_i \neq \phi$.

A generalization of the K-K-M theorem has been provided by Fan (1961) and Shapley (1973). See also Border (1985) and Shapley and Vohra (1991).

9.4. The Existence of Fixed Points[3]

We may now demonstrate that a continuous point-to-point mapping of a simplex σ^k into itself has at least one ***fixed point***, *i.e.*, a point which is transformed into itself by the mapping. The accompanying proof is the classical one which utilizes Sperner's lemma.

9.4.1. BROUWER'S (FIXED POINT) THEOREM [Brouwer (1912); Stoer, Witzgall (1970); K-K-M (1929); Border (1985); Scarf (1973)]. Let $f: \sigma^k \to \sigma^k$ be a continuous point-to-point mapping. Then f has a fixed point $\bar{\mathbf{x}} = f(\bar{\mathbf{x}})$.

PROOF. Let $\sigma^k = \text{co}\{\mathbf{x}_0, \mathbf{x}_1, \cdots, \mathbf{x}_k\}$. Then for each $\mathbf{x} \in \sigma^k$,

[3] Those readers not familiar with the concept of a continuous point-to-point function are encouraged to read section A.1 of the appendix to this chapter.

$$\mathbf{x} = \sum_{i=0}^{k} \lambda_i \mathbf{x}_i, \ \sum_{i=0}^{k} \lambda_i = 1, \ 0 \le \lambda_i \in \mathbf{R}, \ i = 0, 1, \cdots, k.$$

Correspondingly, the image of $\mathbf{x}$ under f or $f(\mathbf{x})$ (also within σ^k) is expressible as

$$f(\mathbf{x}) = \sum_{i=0}^{k} \hat{\lambda}_i \mathbf{x}_i, \ \sum_{i=0}^{k} \hat{\lambda}_i = 1, \ 0 \le \hat{\lambda}_i \in \mathbf{R}, \ i=0, 1, \cdots, k.$$

Since we may view the coefficients $\lambda_i, \hat{\lambda}_i$ in these convex combinations as continuous functions $\lambda_i(\mathbf{x}), \hat{\lambda}_i(f(\mathbf{x}))$ respectively on σ^k, it follows that the sets

$$\mathcal{F}_i = \left\{ \mathbf{x} \mid \hat{\lambda}_i \left(f(\mathbf{x}) \right) \le \lambda_i(\mathbf{x}), \mathbf{x} \in \sigma^k \right\}, \ i = 0, 1, \cdots, k,$$

are closed. Moreover, for every subset $\{i_0, i_1, \cdots, i_\ell\}$ of $\{0, 1, \cdots, k\}$ they satisfy

$$\text{co} \left\{ \mathbf{x}_{i_0}, \mathbf{x}_{i_1}, \cdots, \mathbf{x}_{i_\ell} \right\} \subseteq F_{i_0} \cup F_{i_1} \cup \cdots \cup F_{i_\ell}. \tag{9.5}$$

(If a point $\mathbf{x} \in \sigma^k$ is not in $\cup_{j=0}^{\ell} F_{i_j}$, then $\hat{\lambda}_{i_h}(f(\mathbf{x})) > \lambda_{i_h}$ for $0 \le h \le k$ and thus

$$1 \ge \hat{\lambda}_{i_0} \left(f(\mathbf{x}) \right) + \cdots + \hat{\lambda}_{i_\ell} \left(\mathrm{f}(\mathbf{x}) \right) > \lambda_{i_0} (\mathbf{x}) + \cdots + \lambda_{i_\ell}(\mathbf{x})$$

so that $\mathbf{x} \notin \text{co}\{\mathbf{x}_{i_0}, \mathbf{x}_{i_1}, \cdots, \mathbf{x}_{i_\ell}\}$.)

Let $\mathfrak{D}$ be a simplicial subdivision of σ^k and let $\mathbf{u}$ be an arbitrary vertex in $\mathfrak{D}$. There exists a smallest face $\text{co}\{\mathbf{x}_{i_0}, \mathbf{x}_{i_1}, \cdots, \mathbf{x}_{i_\ell}\}$ containing $\mathbf{u}$ and, by (9.5), there exists an integer h such that $\mathbf{u} \in \mathcal{F}_{i_h}$. For $\ell(\mathbf{u}) = i_h$ a labeling of $\mathfrak{D}$, Sperner's lemma 9.3.2 yields the existence of a k-simplex $\sigma_m^k \in \mathfrak{D}$ with label set $\{0, 1, \cdots, k-1\}$. Under this labeling $\mathbf{u} \in \mathcal{F}_{\ell(\mathbf{u})}$ and thus σ_m^k meets each of the closed sets $\mathcal{F}_i, i = 0, 1, \cdots, k$.

Consider a sequence of simplicial subdivisions $\{\mathfrak{D}^{(v)}\}_{v=1,2,\ldots}$ of σ^k such that the mesh of $\mathfrak{D}^{(v)}$ tends to zero as $\mathrm{v} \to \infty$. In each $\mathfrak{D}^{(v)}$ there exists a k-simplex $\sigma_m^{(k,v)}$ with vertices $\mathbf{x}_0^{(v)}, \mathbf{x}_1^{(v)}, \cdots, \mathbf{x}_k^{(v)}$ such that $\mathbf{x}_i^{(v)} \in \mathcal{F}_i$, $i = 0, 1, \cdots, k$. Since σ^k is compact, $\{\mathfrak{D}^{(v)}\}_{v=1,2,\ldots}$ has a subsequence which admits a convergent sequence of vertices $\mathbf{x}_i^{(v)} \to \mathbf{x}_i^{(\infty)}$, $i = 0, 1, \cdots, k$. Since $\mathcal{F}_i$

is closed, $\mathbf{x}_i^{(\infty)} \in \mathfrak{F}_i, i = 0, 1, \cdots, k$. Let $\sigma_m^{(k,v)} = \text{co}\{\mathbf{x}_0^{(v)}, \mathbf{x}_1^{(v)}, \cdots, \mathbf{x}_k^{(v)}\}$. Since $d(\sigma_m^{(k,v)}) \rightarrow 0$ as $v \rightarrow \infty$, all vertices converge to a single point $\bar{\mathbf{x}} \in \mathfrak{F}_i$. Since $\bar{\mathbf{x}}$ satisfies $\hat{\lambda}_i(f(\bar{\mathbf{x}})) \leq \lambda_i(\bar{\mathbf{x}})$ for all i and $\Sigma_{i=0}^k \hat{\lambda}_i(f(\bar{\mathbf{x}})) = \Sigma_{i=0}^k \lambda_i(\bar{\mathbf{x}}) = 1$, it follows that $\hat{\lambda}_i(f(\bar{\mathbf{x}})) = \lambda_i(\bar{\mathbf{x}})$ for all i so that $\bar{\mathbf{x}} = f(\bar{\mathbf{x}})$ is a fixed point of σ^k. Q.E.D.

A couple of points of clarification concerning this theorem are now in order. First, when we state that f is a continuous mapping of σ^k into itself we mean that there exists a rule or law of correspondence which associates with each $\mathbf{x} \in \sigma^k$ an image $f(\mathbf{x})$ also in σ^k which depends in a continuous way on $\mathbf{x}$. Note that there is no requirement that every $\mathbf{x} \in \sigma^k$ be an image of some point of σ^k, *i.e.*, f may only map σ^k into a small subset of itself. Second, f is completely arbitrary; all that is required is that the image of each $\mathbf{x}$ be an element of σ^k. That is, if $\mathbf{x} \in \sigma^k$, then $\mathbf{y} = f(\mathbf{x})$ or $y_i = f_i(\mathbf{x}) \geq 0$ for all i and $\sum_i y_i = \sum_i f_i(\mathbf{x}) = 1$.

It should be intuitively clear that the motivation underlying the proof of Brouwer's fixed point theorem was to first employ Sperner's lemma to establish the existence of a completely labeled subsimplex of the (arbitrary) decomposition $\mathfrak{D}$. By taking progressively finer decompositions we obtain a sequence of arbitrarily small completely labeled subsimplexes $\sigma^{(k,v)} = \text{co}\{\mathbf{x}_1^{(v)}, \mathbf{x}_2^{(v)}, \cdots, \mathbf{x}_k^{(v)}\}$. From the compactness of σ^k itself the $\sigma^{(k,v)}$ get smaller and smaller and consequently converge (on a subsequence) in the limit to a point $\bar{\mathbf{x}}$, *i.e.*, all vertices of $\sigma^{(k,v)}$ converge to $\bar{\mathbf{x}}$ so that $\mathbf{x}_i^{(v)} \rightarrow \bar{\mathbf{x}}$ as $v \rightarrow \infty$, $i = 0, 1, \cdots, k$.

In order to extend Brouwer's theorem to general convex sets let us observe first that two convex sets $\mathcal{S}_1, \mathcal{S}_2$ are said to be ***homeomorphic*** if there exists a one-to-one and onto correspondence f which is continuous in both directions between them, *i.e.*, both the point-to-point mapping $f: \mathcal{S}_1 \rightarrow \mathcal{S}_2$ and its inverse f^{-1} are continuous. Here f is called a ***homeomorphism*** between

$\mathcal{S}_1, \mathcal{S}_2$. Moreover, two compact convex sets $\mathcal{S}_1, \mathcal{S}_2$ in $\mathbf{R}^n$ are homeomorphic if dim $(\mathcal{S}_1)$ = dim$(\mathcal{S}_2) = k$ while a compact convex set in $\mathbf{R}^n$ is homeomorphic to the unit ball $\mathcal{B} = \{\mathbf{x} \mid \|\mathbf{x}\| \leq 1\}$ of the same dimension and thus to a simplex of the same dimension. Be virtue of this observation we see that Brouwer's theorem is applicable to $\mathcal{B}$ and thus, in general, to any set homeomorphic to σ^k. In this regard, we may now state as corollary to theorem 9.4.1

9.4.1.1 BROUWER'S (FIXED POINT) THEOREM. Let $\mathcal{S}$ in $\mathbf{R}^n$ be homeomorphic to σ^k with $f: \mathcal{S} \to \mathcal{S}$ a continuous point-to-point mapping. Then f has a fixed point $\bar{\mathbf{x}} = f(\bar{\mathbf{x}})$.

PROOF. Let $f: \mathcal{S} \to \mathcal{S}$ be continuous. Given a homeomorphism $\zeta: \sigma^k \to \mathcal{S}$, the composition mapping $g = \zeta^{-1} \circ f \circ \zeta: \sigma^k \to \sigma^k$ is continuous with fixed point $\hat{\mathbf{x}} \epsilon \sigma^k$. Since $\hat{\mathbf{x}} = \zeta^{-1} \circ f \circ \zeta(\hat{\mathbf{x}})$, it follows that $\bar{\mathbf{x}} = \zeta(\hat{\mathbf{x}}) = f(\bar{\mathbf{x}})$ is a fixed point of f. Q.E.D.

An even stronger version of this corollary is provided by

9.4.1.2. BROUWER'S (FIXED POINT) THEOREM. Let $\mathcal{S}$ in $\mathbf{R}^n$ be a nonempty closed, bounded, and convex set with $f: \mathcal{S} \to \mathcal{S}$ a continuous point-to-point mapping. Then f has a fixed point $\bar{\mathbf{x}} = f(\bar{\mathbf{x}})$.

PROOF. Since $\mathcal{S}$ is bounded, it is contained within some simplex σ^k in $\mathbf{R}^n$ and hence is a retraction of σ^k. Let g represent the retraction of σ^konto $\mathcal{S}$. Since $f: \mathcal{S} \to \mathcal{S}$ is continuous, f may be extended to a continuous mapping $h = f \circ g: \sigma^k \to \sigma^k$. Hence any fixed point $\bar{\mathbf{x}}$ of h must be a fixed point of f. Q.E.D.

We note briefly that:

(a) A simplex σ^k is the smallest compact convex set which contains its vertices.

(b) If f represents a point-to-point mapping of σ^k into its boundary, then f does not have a fixed point.

(c) Given a set $\mathcal{S}$ in $\mathbf{R}^n$ with $f:\mathcal{S}\to\mathbf{R}^n$ continuous, the set of fixed points $\mathcal{F} = \{\mathbf{x} \mid \mathbf{x} = f(\mathbf{x}) \epsilon \mathbf{R}^n\}$ is a closed (possibly empty) subset of $\mathcal{S}$. If $\mathcal{S}$ is compact, then $\mathcal{F}$ is also compact.

(d) Let set $\mathcal{S}(\neq\phi)$ in $\mathbf{R}^n$ be compact and convex with $f:\mathcal{S}\to\mathbf{R}^n$ continuous. Then there exists a point $\bar{\mathbf{x}} \epsilon \mathcal{S}$ such that $\|\bar{\mathbf{x}} - f(\bar{\mathbf{x}})\| \leq \|\mathbf{x} - f(\bar{\mathbf{x}})\|$ for all $\mathbf{x} \epsilon \mathcal{S}$. Moreover, if $f(\mathcal{S}) \subset \mathcal{S}$, then $\bar{\mathbf{x}}$ is a fixed point of f.

(e) Let set $\mathcal{S}(\neq\phi)$ in $\mathbf{R}^n$ be compact and convex with $f:\mathcal{S}\to\mathcal{S}$ continuous and let $\mathcal{F} = \{\mathbf{x} \mid \mathbf{x} = f(\mathbf{x}) \epsilon \mathbf{R}^n\}$ depict the set of fixed points of f. Then for every $\in > 0$ there exists a $\delta > 0$ such that $\|\mathbf{x} - f(\mathbf{x})\| < \delta$ implies $\mathbf{x} \epsilon \bigcup_{\mathbf{x}\epsilon\mathcal{F}} B(\mathbf{x}, \in)$.

(f) Let $\{\mathcal{A}_0, \mathcal{A}_1, \cdots, \mathcal{A}_k\}$ in $\mathbf{R}^n$ be a family of closed subsets of $\sigma^k = \mathrm{co}\{\mathbf{x}_0, \mathbf{x}_1, \cdots, \mathbf{x}_k\}$ such that for each index set $\{i_0, i_1, \cdots, i_m\} \subset \{0, 1, \cdots, k\}$ we have $\mathrm{co}\{\mathbf{x}_{i_0}, \mathbf{x}_{i_1}, \cdots, \mathbf{x}_{i_m}\} \subset \bigcup_{j=0}^{m} \mathcal{A}_{i_j}$ (the hypothesis of the K-K-M theorem). Furthermore, let σ^k be simplicially subdivided and labeled (in a fashion such that if $\mathbf{x}_i \epsilon \mathfrak{D}$ belongs to the face $\mathrm{co}\{\mathbf{x}_{i_0}, \cdots, \mathbf{x}_{i_m}\}$, then there is some index $i \epsilon \{i_0, \cdots, i_m\}$ with $\mathbf{x}_i \epsilon \mathcal{A}_i$) and let $\mathcal{A} = \bigcap_{i=0}^{m} \mathcal{A}_i$. Then for every $\in > 0$ there exists a $\delta > 0$ such that if the mesh of subdivision $\mathfrak{D}$ is less than δ, then every completely labeled subsimplex lies in $\bigcup_{\mathbf{x}\epsilon\mathcal{A}} B(\mathbf{x}, \in)$.

(g) Let $f:\sigma^k \to \sigma^k$ with $\mathcal{F} = \{\mathbf{x} \mid \mathbf{x} = f(\mathbf{x}) \epsilon \mathbf{R}^n\}$ the set of fixed points of f. In addition, let σ^k be subdivided and labeled (in a fashion such that for $\mathbf{u}$ a vertex of $\mathfrak{D}$ and $\mathbf{u} \epsilon \mathrm{co}\{\mathbf{x}_{i_0}, \cdots, \mathbf{x}_{i_\ell}\}$ choose $\ell(\mathbf{u}) \epsilon \{i_0, \cdots, i_\ell\}$ such that (9.5) holds). (ℓ thus satisfies the hypothesis of Sperner's lemma.) Then for every $\in > 0$ there exists a $\delta > 0$ such that if the mesh of the subdivision $\mathfrak{D}$ is less than δ, then every completely labeled subsimplex lies in $\bigcup_{\mathbf{x}\epsilon\mathcal{F}} B(\mathbf{x}, \in)$.

We noted above that Sperner's lemma, the K-K-M theorem, and Brouwer's (fixed point) theorem are mathematically equivalent. Moreover, we used Sperner's lemma to verify the K-K-M and Brouwer results (theorems 9.3.3, 9.4.1). Let us now rationalize the notion that the K-K-M theorem implies Brouwer's theorem. To see this let the point-to-point mapping $f:\sigma^k \rightarrow \sigma^k$ be continuous and let $\mathcal{A}_i = \{\mathbf{x} \mid \hat{\lambda}_i(f(\mathbf{x})) \leq \lambda_i(\mathbf{x}),\quad \mathbf{x} \epsilon \sigma^k\}$, $i=0,1,\cdots,k$ (see theorem 9.4.1). Here $\mathcal{A}_i$ is a closed subset of σ^k for all i. Since $\mathbf{x}$ is a point on the face $\text{co}\{\mathbf{x}_{i_0}, \mathbf{x}_{i_1}, \cdots, \mathbf{x}_{i_\ell}\}$ it follows that

$$\lambda_{i_0}(\mathbf{x}) + \lambda_{i_1}(\mathbf{x}) + \cdots + \lambda_{i_\ell}(\mathbf{x}) \;=\; 1 \;\geq\; \hat{\lambda}_{i_0}(f(\mathbf{x})) + \hat{\lambda}_{i_1}(f(\mathbf{x})) + \cdots + \hat{\lambda}_{i_\ell}(f(\mathbf{x}))$$

so that $\hat{\lambda}_i(f(\mathbf{x})) \leq \lambda_i(\mathbf{x})$ for at least one $i \epsilon \{i_{i_0}, i_1, \cdots, i_\ell\}$. Hence $\mathbf{x} \epsilon \bigcup_{j=1}^{\ell} \mathcal{A}_{i_j}$ and thus $\text{co}\{\mathbf{x}_{i_0}, \mathbf{x}_{i_1}, \cdots, \mathbf{x}_{i_\ell}\} \subset \bigcup_{j=1}^{\ell} \mathcal{A}_{i_j}$. Hence the K-K-M assumptions hold, $\bigcap_{i=0}^{k} \mathcal{A}_i \neq \phi$, and thus there is an $\mathbf{x} \epsilon \bigcap_{i=0}^{k} \mathcal{A}_i$ such that $\hat{\lambda}_i(f(\mathbf{x})) \leq \lambda_i(\mathbf{x})$ for all i. In this regard, there exists a point $\bar{\mathbf{x}}$ between $\mathbf{x}$ and its image $f(\mathbf{x})$ for which $\hat{\lambda}_i(f(\bar{\mathbf{x}})) \leq \lambda_i(\bar{\mathbf{x}})$, $i=0,1,\cdots,k$. But since $\Sigma_{i=0}^{k} \hat{\lambda}_i(f(\bar{\mathbf{x}})) = \Sigma_{i=0}^{k} \lambda_i(\bar{\mathbf{x}}) = 1$, the latter set of inequalities hold only if $\hat{\lambda}_i(f(\bar{\mathbf{x}})) = \lambda_i(\bar{\mathbf{x}})$ for all i or $\bar{\mathbf{x}} = f(\bar{\mathbf{x}})$.

That Brouwer's theorem implies the K-K-M result can be demonstrated as follows. Let us assume to the contrary of the K-K-M conclusion that $\bigcap_{i=0}^{k} \mathcal{A}_i = \phi$. Since $\{\mathcal{A}_0, \mathcal{A}_1, \cdots, \mathcal{A}_k\}$ is a family of closed subsets of $\sigma^k = \text{co}\{\mathbf{x}_0, \mathbf{x}_0, \cdots, \mathbf{x}_k\}$, the collection $\{\mathcal{A}_0^c, \mathcal{A}_1^c, \cdots, \mathcal{A}_k^c\}$ constitutes an open cover of σ^k (*i.e.*, each $\mathcal{A}_i^c, i=0,1,\cdots,k$, is open and $\sigma^k \subset \bigcup_{i=0}^{k} \mathcal{A}_i^c$). Since σ^k is compact, $\{\mathcal{A}_i^c,\ i=0,1,\cdots,k\}$ has a finite subcover $\{\mathcal{A}_{i_0}^c, \cdots, \mathcal{A}_{i_m}^c\}$. Let us define, via theorem A.1.4, the finite set of continuous functions $g_i: \sigma^k \rightarrow \{\mathbf{x} \mid \mathbf{x} \geqq \mathbf{0},\ \mathbf{x} \epsilon R^n\}$ as

$$g_i(\mathbf{x}) \;=\; \begin{cases} d(\mathbf{x},\mathbf{y}),\ \mathbf{y} \epsilon \mathcal{A}_{i_j}^c,\ j=0,1,\ldots,m; \\ 0,\ \ \mathbf{y} \notin \mathcal{A}_{i_j}^c,\ j=0,1,\ldots,m. \end{cases}$$

If we set $f_i(\mathbf{x}) = g_i(\mathbf{x}) \Big/ \Sigma_{j=0}^{m} g_i(\mathbf{x})$, then clearly $\Sigma_{i=0}^{m} f_i(\mathbf{x}) \equiv 1$. Next, let $h{:}\,\sigma^k \to \sigma^k$ with $h(\mathbf{x}) = \Sigma_{i=0}^{m} f_i(\mathbf{x})\, \mathbf{x}_i$. Clearly h is continuous and, with σ^k a compact convex set, Brouwer's theorem informs us that h has a fixed point $\bar{\mathbf{x}}$. For $\{f_0, f_1, \cdots, f_m\}$ define index set $\mathcal{I} = \{i \mid f_i(\bar{\mathbf{x}}) > 0\}$. Then $\bar{\mathbf{x}} \in co\{\mathbf{x}_i, i \in \mathcal{I}\}$ and $\bar{\mathbf{x}} \notin \mathcal{A}_i$ for $i \in \mathcal{I}$. But this contradicts the hypothesis of K-K-M that $co\{\mathbf{x}_i, i \in \mathcal{I}\} \subset \bigcup_{i \in \mathcal{I}} \mathcal{A}_i$.

To verify that Brouwer's fixed point theorem implies Sperner's lemma let us start with a barycentric subdivision $\mathcal{D}$ of the k-simplex $\sigma^k = co\{\mathbf{x}_0, \mathbf{x}_1, \ldots, \mathbf{x}_k\}$, where $\mathcal{V} = \{\mathbf{x}_1, \ldots, \mathbf{x}_r\}$ is the collection of all the vertices of all the derived subsimplexes, each $\mathbf{x}_i$ has barycentric coordinates $\mathbf{x}_i' = (\mathbf{x}_{i0}, x_{i1}, \ldots, x_{ik})$, $i=1,\ldots,r$, and $\ell : \mathcal{V} \to \{0, 1, \ldots, k\}$ is a proper labeling of the subdivision. Following Yoseloff (1974), let $f : \mathcal{V} \to \sigma^k$ be defined as:

$$f(\mathbf{x}_i) = (\hat{x}_{i0}, \hat{x}_{i1}, \ldots, \hat{x}_{ik}) = \hat{\mathbf{x}}_i,$$

$$\hat{x}_{ij} = \begin{cases} x_{ij} - \epsilon, & j = \ell(\mathbf{x}_i); \\ x_{ij} + \frac{\epsilon}{k}, & j \neq \ell(\mathbf{x}_i), \end{cases} \tag{9.6}$$

where $\epsilon = \min_i \{x_{i\ell(\mathbf{x}_i)}\}$. Clearly $f(\mathbf{x}_i) \in \sigma^k$ for all $\mathbf{x}_i \in \mathcal{V}$.

Having defined f on $\mathcal{V}$, let us extend it to all of σ^k by extending it to each of the subsimplexes in $\mathcal{D}$ (a more formal definition of a linear extension is given in section 9.6). That is, let $\sigma_i^k \in \mathcal{D}$ with vertices $\mathbf{x}_{i_0}, \mathbf{x}_{i_1}, \ldots, \mathbf{x}_{i_k}$. Then

$$\sigma_i^k = \left\{ \mathbf{x} \in \sigma^k \;\middle|\; \mathbf{x} = \sum_{j=0}^{k} \lambda_j \mathbf{x}_{i_j}, \sum_{i=0}^{k} \lambda_j = 1, 0 \leq \lambda_j \in \mathcal{R} \text{ for all } j \right\}.$$

So if $\mathbf{x} = \sum_{j=0}^{k} \lambda_j \mathbf{x}_{i_j}$, then

$$f(\mathbf{x}) = \sum_{j=0}^{k} \lambda_j f(\mathbf{x}_{i_j}). \tag{9.7}$$

Here $f(\mathbf{x})$ is continuous for all points in σ^k and maps σ^k into itself. Since f thus satisfies the hypotheses of Brouwer's theorem, there must exist at least one point $\mathbf{x}\epsilon\sigma^k$ such that $\mathbf{x}=f(\mathbf{x})$. Yoseloff then demonstrates (see his lemma 1) that $\mathbf{x}$ is contained in a subsimplex σ_i^k of $\mathfrak{D}$ which has a complete set of labels on its vertices. This is actually verified by proving the equivalent statement: if $\sigma_i^k\epsilon\mathfrak{D}$ is not completely labeled, then f does not admit a fixed point in σ_i^k. To see this let σ_i^k have vertices $\mathbf{x}_{i_0},\mathbf{x}_{i_1},\ldots,\mathbf{x}_{i_k}$, where $\mathbf{x}'_{i_j}=(x_{j_0},x_{j_1},\ldots,x_{j_k})$. Assume that no vertex of σ_i^k carries the label p. From (9.6), if $f(\mathbf{x}_{i_j})=(\hat{x}_{j_0},\hat{x}_{j_1},\ldots,\hat{x}_{j_k})$, then component $\hat{x}_{jp}=x_{jp}+\frac{\epsilon}{k}, j=0,1,\ldots,k$. Let $\mathbf{x}\epsilon\sigma_i^k$. From (9.7), the pth component of $f(\mathbf{x})$ is $\Sigma_{j=0}^k\lambda_j\hat{x}_{jk}=\Sigma_{j=0}^k\lambda_j x_{jk}+\frac{\epsilon}{k}$. Since this is just the pth component of $\mathbf{x}$ plus $\frac{\epsilon}{k}$, it follows that $f(\mathbf{x})\neq\mathbf{x}$. So if to the contrary f has a fixed point, a completely labeled subsimplex σ_i^k must exist in $\mathfrak{D}$ and thus Sperner's lemma holds. (In fact, the only points which remain fixed under f are the barycenters of simplexes in $\mathfrak{D}$ which carry a complete set of labels.)

9.5 Fixed Points of Compact Point-to-Point Functions

A useful generalization of Brouwer's fixed point theorem 9.4.1 has been developed by Schauder (1930). Here a continuous point-to-point mapping is replaced by one that is either completely continuous or compact. To define these latter concepts let $\mathcal{S}(\neq\phi)$ be a subset of $\mathbf{R}^n$ with f a continuous point-to-point mapping of $\mathcal{S}$ into itself. Then $f:\mathcal{S}\rightarrow\mathcal{S}$ is ***completely continuous*** if the image of each bounded subset $\mathcal{A}$ of $\mathcal{S}$ is contained within a compact set; and $f:\mathcal{S}\rightarrow\mathcal{S}$ is said to be ***compact*** if $f(\mathcal{S})$ is compact.

Two additional concepts merit our attention. First, for $\mathcal{S}$ a compact subset of $\mathbf{R}^n$ and $0<\epsilon\in\mathbf{R}$, the set $\{\mathbf{x}_1,\cdots,\mathbf{x}_p\}$ is called an ***ϵ-net*** of $\mathcal{S}$ if for $\mathbf{x}\epsilon\mathcal{S}$ there is an $\mathbf{x}_i$ such that $\|\mathbf{x}-\mathbf{x}_i\|<\epsilon, i=1,\cdots,p$. Next, let $\mathcal{S}$ be a compact subset of $\mathbf{R}^n$ with $\{\mathbf{x}_1,\cdots,\mathbf{x}_p\}$ an ϵ-net of $\bar{\mathcal{S}}$. For $\mathbf{x}\in\bar{\mathcal{S}}$, define

$$h_\epsilon(\mathbf{x}) = \frac{\sum_{i=1}^n g_i(\mathbf{x})\mathbf{x}_i}{\sum_{i=1}^n g_i(\mathbf{x})}, \tag{9.8}$$

where

$$g_i(\mathbf{x}) = \begin{cases} \epsilon - \|\mathbf{x}-\mathbf{x}_i\| & \text{if } \|\mathbf{x}-\mathbf{x}_i\| \le \epsilon; \\ 0 & \text{if } \|\mathbf{x}-\mathbf{x}_i\| > \epsilon. \end{cases}$$

Given these considerations we have

9.5.1. THEOREM [Cronin (1964)]. For $\mathcal{S}$ a compact subset of $\mathbf{R}^n$, let f be a compact point-to-point mapping with domain $\mathcal{A}$, a bounded subset of $\mathbf{R}^n$, and let $f(\mathcal{A}) \subset \mathcal{S}$. If h_ϵ is defined on $\bar{\mathcal{S}}$ and $\mathbf{x}\in\mathcal{A}$, then

$$\|f(\mathbf{x}) - h_\epsilon(f(\mathbf{x}))\| < \epsilon.$$

We now turn to the main result of this section, namely the

9.5.2. SCHAUDER (FIXED POINT) THEOREM [Cronin (1964), Schauder (1930)]. Let $\mathcal{S}$ be a closed convex set in $\mathbf{R}^n$ with $f:\mathcal{S}\to\mathbf{R}^n$ a compact point-to-point mapping such that $f(\mathcal{S}) \subset \mathcal{S}$. Then f has a fixed point, *i.e.*, there exists an $\bar{\mathbf{x}}\in\mathcal{S}$ such that $\bar{\mathbf{x}} = f(\bar{\mathbf{x}})$.

PROOF. With $\mathcal{S}$ closed $\overline{f(\mathcal{S})} \subset \mathcal{S}$. Let $\{\epsilon_n\}$ be a monotonic decreasing sequence such that $\lim_{n\to\infty}\{\epsilon_n\} = 0$. From (9.8), let $f_n = h_{\epsilon_n}(f)$ be defined on $\mathcal{S}$. With $\mathcal{S}$ convex, $f_n(\mathcal{S}) \subset \mathcal{S}$ since

$$h_{\epsilon_n}(\mathbf{x}) = \frac{\sum_{i=1}^{p_n} g_i(\mathbf{x})\mathbf{x}_i}{\sum_{i=1}^{p_n} g_i(\mathbf{x})},$$

where $\{\mathbf{x}_1,\cdots,\mathbf{x}_{p_n}\}$ serves as an ϵ-net of $\overline{f(\mathcal{S})} \subset \mathcal{S}$.

Let $\mathcal{V}_n$ be a subspace of $\mathbf{R}^n$ spanned by $\{\mathbf{x}_1,\cdots,\mathbf{x}_{p_n}\}$ with $\mathcal{S}_n = \mathcal{S}\cap\mathcal{V}_n$. Clearly $\mathcal{S}_n$ is a closed convex subset of $\mathcal{V}_n$ and the mapping

f_n is defined on $\mathcal{S}_n$ with $f_n(\mathcal{S}_n) \subset \mathcal{S}_n$. By corollary 9.4.1.1 to Brouwer's theorem, there exists a point $\mathbf{x}_n \in \mathcal{S}_n$ such that $\mathbf{x}_n = f_n(\mathbf{x}_n)$. The set $f(\mathbf{x}_n)$ is contained in the compact set $f(\mathcal{S})$ and thus $f(\mathbf{x}_n) \subset \mathcal{S}$. Hence $f(\mathbf{x}_n)$ has a limit point $\bar{\mathbf{x}} \in \mathcal{S}$. Assume that the sequence $\{f(\mathbf{x}_n)\}$ converges to $\bar{\mathbf{x}}$, *i.e.*, $\|f(\mathbf{x}_n) - \bar{\mathbf{x}}\| < \epsilon$ for $n > n_\epsilon$. From the definition of f_n, $\|f_n(\mathbf{x}_n) - f(\mathbf{x}_n)\| < \epsilon_n$. Upon adding these two inequalities we obtain $\|f_n(\mathbf{x}_n) - \bar{\mathbf{x}}\| < \epsilon + \epsilon_n$. But since $f_n(\mathbf{x}_n) = \mathbf{x}_n$ (the fixed point result), the preceding inequality becomes $\|\mathbf{x}_n - \bar{\mathbf{x}}\| < \epsilon + \epsilon_n$. With f continuous, if $\epsilon + \epsilon_n < \delta(\epsilon^{(1)})$, then $\|f(\mathbf{x}_n) - f(\bar{\mathbf{x}})\| < \epsilon^{(1)}$. Thus the sequence $\{f(\mathbf{x}_n)\}$ converges to $f(\bar{\mathbf{x}})$. And since the limit of the sequence $\{f(\mathbf{x}_n)\}$ is unique, $\bar{\mathbf{x}} = f(\bar{\mathbf{x}})$. Q.E.D.

It is important to mention that the preceding fixed point theorem 9.5.2 has the equivalent representations:

9.5.2.1. SCHAUDER'S (FIXED POINT) THEOREM. For $\mathcal{S}$ a subset of $\mathbf{R}^n$ let $f : \mathcal{S} \to \mathcal{S}$ be a completely continuous point-to-point mapping with $f(\mathcal{S})$ bounded. Then f has a fixed point $\bar{\mathbf{x}} = f(\bar{\mathbf{x}})$.

9.5.2.2 SCHAUDER'S (FIXED POINT) THEOREM. If $\mathcal{S}$ is a compact convex set in $\mathbf{R}^n$, then every continuous point-to-point mapping $f : \mathcal{S} \to \mathcal{S}$ has a fixed point $\bar{\mathbf{x}} = f(\bar{\mathbf{x}})$.

We next state as a corollary to theorem 9.5.2

9.5.2.3. THEOREM [Cronin, 1964]. Let $\mathcal{S}$ be homeomorphic to a closed convex set in $\mathbf{R}^n$ with $f : \mathcal{S} \to \mathbf{R}^n$ a compact point-to-point mapping such that $f(\mathcal{S}) \subset \mathcal{S}$. Then f has a fixed point $\bar{\mathbf{x}} = f(\bar{\mathbf{x}})$.

Further extensions and generalizations of Schauder's fixed point theorem 9.5.2 can be found in Tychonoff (1935), Klee (1960), Browder (1959), 1965a, and 1965b), and Fan (1969). While not strictly a fixed point theorem, Fan (1961) provides a theorem which generalizes Tychonoff's fixed point theorem.

9.6. Fixed Points of Point-to-Set Functions[4]

Brouwer's theorem established the existence of a fixed point for a continuous point-to-point mapping f of a compact convex set $\mathcal{S}$ into itself. If f is replaced by an upper hemicontinuous point-to-set mapping of $\mathcal{S}$ into the set of all compact convex subsets of $\mathcal{S}$, then a generalization of Brouwer's theorem is provided by

9.6.1 KAKUTANI'S (FIXED POINT) THEOREM [Kakutani (1941); Nikaido (1968); Berge (1963); Klein (1973); Stoer, Witzgall (1970)]. Let $\mathcal{S}$ be a nonempty compact convex set in $\mathbf{R}^n$ and let $F: \mathcal{S} \to \mathcal{S}$ be an upper hemicontinuous set-valued function with nonempty compact convex image sets $F(\mathbf{x})$ for each $\mathbf{x} \in \mathcal{S}$. Then F has a fixed point $\overline{\mathbf{x}} \in F(\overline{\mathbf{x}})$.

PROOF. We first consider a particularization of $\mathcal{S}$ to an n-simplex. The proof is then modified by having $\mathcal{S}$ represent an arbitrary compact convex subset of an n-simplex σ^n. A retraction of σ^n onto $\mathcal{S}$ then generalizes the proof.

(1) Let $\mathcal{S}$ be an n-simplex σ^n with the point-to-set function $F: \sigma^n \to \sigma^n$ upper hemicontinuous. Consider a sequence of simplicial subdivisions $\{\mathfrak{D}^{(v)}\}_{v=1,2,\dots}$ of σ^n such that the mesh of $\mathfrak{D}^{(v)}$ tends to zero as $v \to \infty$. For a given v let $\mathbf{x}^{(v)}$ represent a vertex of an n-simplex $\sigma^{(n,v)}$ of $\mathfrak{D}^{(v)}$. For each $\mathbf{x}^{(v)}$ let $\mathbf{y}^{(v)}$ correspond to an arbitrary point of the image set $F(\mathbf{x}^{(v)})$ and write $\mathbf{y}^{(v)} = f^{(v)}(\mathbf{x}^{(v)})$. Here $f^{(v)}$, along with its extension within each simplex $\sigma^{(n,v)}$ of $\mathfrak{D}^{(v)}$, forms a continuous point-to-point mapping from σ^n into itself or $f^{(v)}: \sigma^n \to \sigma^n$.[5]

[4]Readers not familiar with the concept of a hemicontinuous point-to-set function are urged to review section A.2 of the appendix to this chapter.

[5]For each $\mathbf{x} \in \sigma^n$ consider the n-simplex $\sigma^{(n,v)} = \text{co}\{\mathbf{x}_0^{(v)}, \mathbf{x}_1^{(v)}, \dots, \mathbf{x}_n^{(v)}\}$ of $\mathfrak{D}^{(v)}$ which contains it. Then a ***linear extension*** on $\sigma^{(n,v)}$ can be constructed as

$$f^{(v)}(\mathbf{x}) = f^{(v)}\Big(\sum_{i=0}^{n} \theta_i^{(v)} \mathbf{x}_i^{(v)}\Big) = \sum_{i=0}^{n} \theta_i^{(v)} f^{(v)}(\mathbf{x}_i^{(v)}).$$

Here the funciton $f^{(v)}$ is uniquely determined and, being linear in the interior of $\mathfrak{D}^{(v)}$, is continuous in σ^n.

By Brouwer's theorem 9.4.1, $f^{(v)}$ has a fixed point $\bar{\mathbf{x}}^{(v)} = f^{(v)}(\bar{\mathbf{x}}^{(v)})$. Consider the sequence $\{\bar{\mathbf{x}}^{(v)}\}_{v=1,2,\ldots}$ generated under $f^{(v)}$ for $v = 1,2,\ldots$. With σ^n compact, this sequence has a subsequence which converges to a point $\bar{\mathbf{x}} \epsilon \sigma^n$, the fixed point of F. It thus remains to demonstrate that $\bar{\mathbf{x}} \epsilon F(\bar{\mathbf{x}})$. To this end let the set of $n+1$ vertices of the simplex $\sigma^{(n,v)}$ containing $\bar{\mathbf{x}}^{(v)}$ be represented on $\{\mathbf{x}_0^{(v)}, \mathbf{x}_1^{(v)}, \ldots, \mathbf{x}_n^{(v)}\}$. Then

$$\bar{\mathbf{x}}^{(v)} = \sum_{i=0}^{n} \theta_i^{(v)} \mathbf{x}_i^{(v)}, \quad \sum_{i=0}^{n} \theta_i^{(v)} = 1, \quad 0 \leq \theta_i^{(v)} \epsilon \mathbf{R}, \ i=0,1,\ldots,n.$$

With $f^{(v)}$ a linear extension on $\sigma^{(n,v)}$,

$$\bar{\mathbf{x}}^{(v)} = f^{v}(\bar{\mathbf{x}}^{(v)}) = \sum_{i=0}^{n} \theta_i^{(v)} f^{(v)}(\mathbf{x}_i^{(v)}),$$

where $f^{(v)}(\mathbf{x}_i^{(v)}) \epsilon F(\mathbf{x}_i^{(v)})$, $i=0,1,\ldots,n$.

Since σ^n is compact, $\{\mathfrak{D}^{(v)}\}_{v=1,2,\ldots}$ has a subsequence which admits a convergent sequence of vectors and scalars

$$\begin{aligned} &\bar{\mathbf{x}}^{(v)} \rightarrow \bar{\mathbf{x}}, \\ &f^{(v)}\left(\mathbf{x}_i^{(v)}\right) \rightarrow f(\mathbf{x}_i), \\ &\theta_i^{(v)} \rightarrow \theta_i \end{aligned}$$

which satisfy, in the limit,

$$\bar{\mathbf{x}} = \sum_{i=0}^{n} \theta_i f(\mathbf{x}_i), \ \sum_{i=0}^{n} \theta_i = 1, 0 \leq \theta_i \epsilon \mathbf{R}, \ i=0,1,\ldots,n.$$

Since $\bar{\mathbf{x}}^{(v)} \epsilon \sigma^{(n,v)}$ of $\mathfrak{D}^{(v)}$ and the mesh of $\mathfrak{D}^{(v)}$ tends to zero, we have $f^{(v)}(\mathbf{x}_i^{(v)}) \rightarrow \bar{\mathbf{x}}$ for $i=0,1,\ldots,n$. And with $f^{(v)}(\mathbf{x}_i^{(v)}) \epsilon F(\mathbf{x}_i^{(v)})$ and $f^{(v)}(\mathbf{x}_i^{(v)}) \rightarrow f(\mathbf{x}_i)$, $i=0,1,\ldots,n$, it follows from the upper hemicontinuity of F that $f(\mathbf{x}_i) \epsilon F(\bar{\mathbf{x}})$ for all i. And since $F(\bar{\mathbf{x}})$ is convex, we have,

$$\bar{\mathbf{x}} = \sum_{i=0}^{n} \theta_i f(\mathbf{x}_i) \in F(\bar{\mathbf{x}})$$

and thus $\bar{\mathbf{x}}$ is a fixed point of F.

(2) Next, let $\mathcal{S}$ be any compact convex set in $\mathbf{R}^n$ and let σ^n be an n-simplex containing $\mathcal{S}$. For g a retraction of σ^n onto $\mathcal{S}$ (*i.e.*, $g: \sigma^n \rightarrow \mathcal{S}$ is a continuous point-to-point mapping such that $g(\mathbf{x}) = \mathbf{x}$ for all $\mathbf{x} \in \mathcal{S}$), $F(g)$ is upper hemicontinuous in $\mathcal{S}$. Then by the preceding proof, there exists a point $\bar{\mathbf{x}}$ such that $\bar{\mathbf{x}} \in F(g(\bar{\mathbf{x}}))$. Since $F(g(\bar{\mathbf{x}})) \subset \mathcal{S}$, it follows that $\bar{\mathbf{x}} \in \mathcal{S}$ and thus $g(\bar{\mathbf{x}}) = \bar{\mathbf{x}}$. Hence $\bar{\mathbf{x}} \in F(g(\bar{\mathbf{x}})) = F(\bar{\mathbf{x}})$. Q.E.D.

Mentioned in the appendix to this chapter is the observation that if F is a closed-valued point-to-set mapping of set $\mathcal{X}$ into a compact set $\mathcal{Y}$, then the notions of a closed correspondence and an upper hemicontinuous correspondence coincide. This equivalence leads us to state Kakutani's theorem in the alternative form

9.6.1.1. KAKUTANI'S (FIXED POINT) THEOREM. Let $\mathcal{S}$ be a nonempty compact convex set in $\mathbf{R}^n$ and let $F: \mathcal{S} \rightarrow \mathcal{S}$ be a closed set-valued function with nonempty convex image sets $F(\mathbf{x})$ for each $\mathbf{x} \in \mathcal{S}$. Then F has a fixed point $\bar{\mathbf{x}} \in F(\bar{\mathbf{x}})$.

Two additional variants of Kakutani's theorem 9.6.1, one employing directly the notion of lower hemicontinuity of a set-valued mapping and the other utilizing the inverse image of the same, are stated respectively as

9.6.1.2. KAKUTANI'S (FIXED POINT) THEOREM [Hildenbrand, Kirman (1976)]. Let $\mathcal{S}$ be a nonempty compact convex set in $\mathbf{R}^n$ and let $F: \mathcal{S} \rightarrow \mathcal{S}$ be a lower hemicontinuous set-valued mapping with nonempty closed convex image sets $F(\mathbf{x})$ for each $\mathbf{x} \in \mathcal{S}$. Then F has a fixed point $\bar{\mathbf{x}} \in F(\bar{\mathbf{x}})$.

9.6.1.3. KAKUTANI'S (FIXED POINT) THEOREM [Browder (1968)]. Let $\mathcal{S}$ be a nonempty compact convex set in $\mathbf{R}^n$ and let $F: \mathcal{S} \rightarrow \mathcal{S}$ have nonempty convex image sets $F(\mathbf{x})$

and satisfy $F^{-1}(\mathbf{x})$ is open for all $\mathbf{x} \in \mathcal{S}$. Then F has a fixed point $\bar{\mathbf{x}} \in F(\bar{\mathbf{x}})$.

Kakutani's fixed point theorem 9.6.1 has been generalized by Begle (1950) and Eilenberg and Montgomery (1946). Further results concerning extensions of fixed point theorems of set-valued functions are provided by Fan (1969).

9.7. Exercises

1. For simplex $\sigma^3 = co\{\mathbf{x}_0, \mathbf{x}_1, \mathbf{x}_2, \mathbf{x}_3\}$, find:
   (a) the three dimensional complex associated with σ^3; and
   (b) the first barycentric subdivision of σ^3.

2. Does the set $\mathcal{S} = \{\mathbf{x} \mid x_1 + x_2 \leq 10, x_1 \geq 0, x_2 \geq 0\}$ in $\mathbf{R}^2$ constitute a two-dimensional simplex? Express $\mathcal{S}$ as the convex hull of its set of extreme points or vertices. Are the vertices affinely independent?

3. Let $\mathcal{S} = co\{\mathbf{x}_1, \mathbf{x}_2, \mathbf{x}_3\}$ be a subset of $\mathbf{R}^2$. Form a finite collection of 2-simplices which constitutes a subdivision of $\mathcal{S}$. Form one which does not.

4. Let $f(x)$ be a continuous point-to-point mapping of the closed bounded interval $[-1,1]$ into itself. Demonstrate that there must exist a point $\bar{x} \in [-1,1]$ such that $\bar{x} = f(\bar{x})$. (Hint: consider a continuous function $F(x) = f(x) - x$ defined on $[-1,1]$. Clearly $F(-1) \geq 0, F(1) \leq 0$. Since f is continuous on $[-1,1]$ there must exist a point $\bar{x} \in [-1,1]$ such that $F(\bar{x}) = 0$.) Graphically illustrate the attainment of the fixed point of f.

5. Does the function $f(x) = x^2 - \frac{1}{4}x + \frac{1}{4}$ map the closed interval $[0,1]$ into itself?

6. Which of the following are point-to-set functions:
   (a) $F(x) = \{u \mid x \leq u \leq x + 5\}$;
   (b) $F(x) = \{x\}$;
   (c) $F(\mathbf{x}) = \{\mathbf{u} \mid x_i \leq u_i, i=1,\ldots,n\}$, where $\mathbf{x}, \mathbf{u} \in \mathbf{R}^n$.

7. Demonstrate that the point-to-set function $F(x) = \{u \mid -x \leq u \leq x\}$ is upper hemicontinuous.

8. Let a point-to-set function F be defined as $\mathcal{S} = [-1, 1]$ as

$$F(x) = \begin{cases} \frac{1}{2} + \frac{1}{2}x \leq u \leq 1 & \text{for} \quad 0 < x \leq 1; \\ -1 \leq u \leq -\frac{1}{2} + \frac{1}{2}x & \text{for} \quad -1 \leq x \leq 0; \\ -1 \leq u \leq 1 & \text{for} \quad x = 0. \end{cases}$$

Does F admit a fixed point?

APPENDIX

CONTINUOUS AND HEMICONTINUOUS FUNCTIONS

A.1. Continuous Point-to-Point Mappings

Given sets $\mathfrak{X}$ in $\mathbf{R}^n$ and $\mathfrak{Y}$ in $\mathbf{R}^m$, a ***single-valued function*** or ***point-to-point mapping*** $f:\mathfrak{X}\rightarrow\mathfrak{Y}$ is a rule which associates with each point (vector) $\mathbf{x}\in\mathfrak{X}$ a unique point (vector) $\mathbf{y}\in\mathfrak{Y}$. Here $\mathbf{y}=f(\mathbf{x})$ is the ***image*** of $\mathbf{x}$ under rule f. While $\mathfrak{X}$ is called the ***domain of f*** (denoted $\mathfrak{D}_f$) the collection of those $\mathbf{y}$'s in $\mathfrak{Y}$ which are the image of at least one $\mathbf{x}\in\mathfrak{X}$ is called the ***range of f*** and denoted $\mathfrak{R}_f$. Clearly the range of f is a subset of $\mathfrak{Y}$. If $\mathfrak{R}_f\subset\mathfrak{Y}$, then f is an ***into mapping***. And if $\mathfrak{R}_f=\mathfrak{Y}$ (*i.e.*, every $\mathbf{y}\in\mathfrak{Y}$ is the image of at least one $\mathbf{x}\in\mathfrak{X}$), then f is termed ***surjective*** or ***onto***. Moreover, f is said to be ***injective*** or ***one-to-one*** if no $\mathbf{y}\in\mathfrak{Y}$ is the image of more than one $\mathbf{x}\in\mathfrak{X}$ (*i.e.*, $\mathbf{x}_1\neq\mathbf{x}_2$ implies $f(\mathbf{x}_1)\neq f(\mathbf{x}_2)$). Finally, f is called ***bijective*** if it is both surjective and injective, *i.e.*, both onto and one-to-one.

If $\mathfrak{X}$ is a set in $\mathbf{R}^n$ and $\mathfrak{Y}=\mathbf{R}$, then $f:\mathfrak{X}\rightarrow\mathfrak{Y}$ is termed a ***scalar-valued function or mapping*** of a point or vector into a unique real number. Here the image of each vector $\mathbf{x}\in\mathfrak{X}$ is a scalar $y=f(\mathbf{x})\in\mathbf{R}$.

Given $\mathfrak{X}$ in $\mathbf{R}^n$ and $\mathfrak{Y}$ in $\mathbf{R}^m$ with $\mathcal{A}\subset\mathfrak{X}$, let $f_1:\mathcal{A}\rightarrow\mathfrak{Y}$ be a point-to-point mapping of $\mathcal{A}$ into $\mathfrak{Y}$ and $f_2:\mathfrak{X}\rightarrow\mathfrak{Y}$ a point-to-point mapping of $\mathfrak{X}$ into $\mathfrak{Y}$. Then f_1 is said to be a ***restriction*** of f_2 and f_2 is termed an ***extension*** of f_1 if and only if for each $\mathbf{x}\in\mathcal{A}$, $f_1(\mathbf{x})=f_2(\mathbf{x})$.

For f a point-to-point mapping of $\mathfrak{X}$ into $\mathfrak{Y}$, the subset $\mathfrak{G}_f=\{(\mathbf{x},\mathbf{y})\mid\mathbf{x}\in\mathfrak{X},\ \mathbf{y}=f(\mathbf{x})\in\mathfrak{Y}\}$ of $\mathfrak{X}\times\mathfrak{Y}$ is called the ***graph of f***. If the point-to-point mapping is bijective, then its ***single-valued inverse mapping*** $f^{-1}:\mathfrak{Y}\rightarrow\mathfrak{X}$ exists. In this instance, to each point $\mathbf{y}\in\mathfrak{Y}$ there corresponds a unique inverse image point $\mathbf{x}\in\mathfrak{X}$ such that $\mathbf{x}=f^{-1}(\mathbf{y})=f^{-1}(f(\mathbf{x}))$. Here the domain $\mathfrak{D}_{f^{-1}}$ of f^{-1} is $\mathfrak{Y}$ and its range $\mathfrak{R}_{f^{-1}}$ is $\mathfrak{X}$. Clearly f^{-1} must also be bijective (Figure A.1.1).

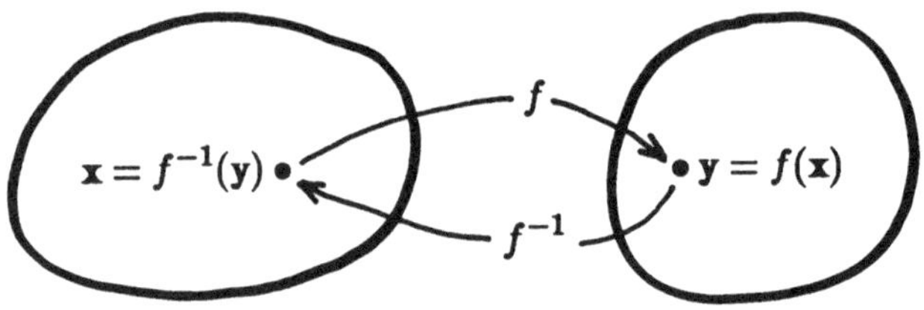

Figure A.1.1

Next, for $\mathfrak{X}, \mathfrak{Y}$ subsets of $\mathbf{R}^n$, $\mathbf{R}^m$ respectively, let $f: \mathfrak{X} \rightarrow \mathfrak{Y}$ be a point-to-point mapping of $\mathfrak{X}$ into $\mathfrak{Y}$. f is said to be ***continuous at a point*** $\mathbf{x}_0 \in \mathfrak{X}$ if either of the following definitions hold:

(a) for any $\in > 0$ there exists a $\delta_\in > 0$ such that $\|\mathbf{x} - \mathbf{x}_0\| < \delta_\in$ implies $\|f(\mathbf{x}) - f(\mathbf{x}_0)\| < \in$ (here the $\in$ subscript on δ means that "δ depends upon the $\in$ chosen"); or

(b) for each open ball $B(f(\mathbf{x}_0), \in)$ centered on $f(\mathbf{x}_0)$ there exists an open ball $B(\mathbf{x}_0, \delta_\in)$ centered on $\mathbf{x}_0$ such that $f(B(\mathbf{X}_0, \delta_\in)) \subseteq B(f(\mathbf{x}_0), \in)$, *i.e.*, points "near" $\mathbf{x}_0$ are mapped by f into points "near" $f(\mathbf{x}_0)$.

In general, the point-to-set mapping $f: \mathfrak{X} \rightarrow \mathfrak{Y}$ is said to be ***continuous on*** $\mathfrak{X}$ if it is continuous at each point of $\mathfrak{X}$.

The following two theorems provide us with a set of necessary and sufficient conditions for the continuity of a point-to-point mapping at a given point $\mathbf{x}_0$ and at any arbitrary point $\mathbf{x}$ of $\mathfrak{X}$ respectively. Specifically,

A.1.1. THEOREM (continuity in terms of convergent sequences). For sets $\mathfrak{X}$ in $\mathbf{R}^n$ and $\mathfrak{Y}$ in $\mathbf{R}^m$, the point-to-point mapping f of $\mathfrak{X}$ into $\mathfrak{Y}$ is continuous at $\mathbf{x}_0 \in \mathfrak{X}$ if and only if $\mathbf{x}_k \rightarrow \mathbf{x}_0$ implies $f(\mathbf{x}_k) \rightarrow f(\mathbf{x}_0)$ for every sequence $\{\mathbf{x}_k\}_{k \in K}$ in $\mathfrak{X}$.

Clearly f is a continuous mapping of $\mathfrak{X}$ into $\mathfrak{Y}$ if it "sends convergent sequences in $\mathfrak{X}$ into convergent sequences in $\mathfrak{Y}$." Next,

A.1.2. THEOREM (continuity in terms of open (closed) sets). For sets $\mathfrak{X}$ in $\mathbf{R}^n$ and $\mathfrak{Y}$ in $\mathbf{R}^m$, let f be a point-to-point mapping of $\mathfrak{X}$ into $\mathfrak{Y}$. Then: (a) f is continuous if and only if $f^{-1}(\mathcal{A})$ is open in $\mathfrak{X}$ wherever $\mathcal{A}$ is open in $\mathfrak{Y}$; (b) f is continuous if and only if $f^{-1}(\mathcal{A})$ is closed in $\mathfrak{X}$ whenever $\mathcal{A}$ is closed in $\mathfrak{Y}$.

Hence f is continuous if it "pulls open (closed) sets back to open (closed) sets," *i.e.*, the inverse images of open (closed) sets are open (closed).

The next theorem states that continuous mappings preserve compactness. That is,

A.1.3. THEOREM. Let f be a continuous point-to-point mapping from a set $\mathfrak{X}$ in $\mathbf{R}^n$ into a set $\mathfrak{Y}$ in $\mathbf{R}^m$. If $\mathcal{A}$ is a compact subset of $\mathfrak{X}$, then so is its image $f(\mathcal{A})$.

Two useful concepts which are based upon the notion of a continuous point-to-point function are a "partition of unity subordinate to an open cover" and a "retraction mapping." First, given a set $\mathcal{S}$ in $\mathbf{R}^n$, let $\{\mathcal{E}_i\}$ be an open cover of $\mathcal{S}$. A ***partition of unity subordinate to*** $\{\mathcal{E}_i\}$ is a finite set of point-to-point functions $\{f_1(\mathbf{x}),\ldots,f_k(\mathbf{x})\}$ from $\mathcal{S}$ into the set of nonnegative real numbers $\mathcal{N}$ such that:

(a) $\Sigma_{j=1}^k f_j(\mathbf{x}) \equiv 1$;

(b) for each j there is some $\mathcal{E}_{i_j}$ such that $f_j(\mathbf{x})$ vanished off $\mathcal{E}_{i_j}$.

The following theorem provides us with a constructive method for obtaining the aforementioned partition. Specifically,

A.1.4. THEOREM. Let $\mathcal{S}$ be a compact subset of $\mathbf{R}^n$ and let $\{\mathcal{E}_i\}$ be an open cover of $\mathcal{S}$. Then there exists a partition of order unity subordinate to $\{\mathcal{E}_i\}$.

PROOF. Given that $\mathcal{S}$ is compact, $\{\mathcal{S}_i\}$ has a finite subcover $\mathcal{S}_{i_1},\ldots,\mathcal{S}_{i_k}$. Define a sequence of continuous functions $g_j: \mathcal{S}\rightarrow\mathcal{N}$ such that

$$g_j(\mathbf{x}) = \begin{cases} d(\mathbf{x},\mathbf{y}), & \mathbf{y}\in\mathcal{S}^c_{i_j}, j=1,\ldots,k; \\ 0, & \mathbf{y}\notin\mathcal{S}^c_{i_j}, j=1,\ldots,k. \end{cases}$$

(Note that not all g_j can vanish simultaneously since the $\mathcal{S}_{i_j}$ cover $\mathcal{S}$.) Let $f_j(\mathbf{x}) = g_j(\mathbf{x}) \Big/ \Sigma_{\ell=1}^k g_\ell(\mathbf{x})$. Clearly $\Sigma_{j=1}^k f_j(\mathbf{x}) = 1$. Hence $\{f_1(\mathbf{x}),\ldots,f_k(\mathbf{x})\}$ serves as a partition of unity subordinate to $\{\mathcal{S}_i\}$. Q.E.D.

Next, let $\mathcal{K}$ be a subset of $\mathbf{R}^n$. A continuous point-to-point mapping $g: \mathbf{R}^n\rightarrow\mathcal{K}$ is termed a ***retraction mapping*** on $\mathbf{R}^n$ if $g(\mathbf{x}) = \mathbf{x}$ for all $\mathbf{x}\in\mathcal{K}$. Here $\mathcal{K}$ is called a ***retraction*** of $\mathbf{R}^n$. If $\mathcal{K}$ is contained within an arbitrary subset $\mathcal{S}$ of $\mathbf{R}^n$, then $g: \mathcal{S}\rightarrow\mathcal{K}$ is a retraction of $\mathcal{S}$ onto $\mathcal{K}$ if $g(\mathbf{x}) = \mathbf{x}$ for all $\mathbf{x}\in\mathcal{K}$. Under what conditions will there exist a retraction of $\mathcal{S}$ onto a set $\mathcal{K}$? The answer is provided by

A.1.5. THEOREM [Stoer, Witzgall, 1970]. Let $\mathcal{K}$ be a closed convex subset of an arbitrary set $\mathcal{S}$ in $\mathbf{R}^n$. Then there exists a continuous point-to-point mapping $g: \mathcal{S}\rightarrow\mathcal{K}$ which is a retraction of $\mathcal{S}$ onto $\mathcal{K}$.

Finally, given point-to-point functions $f: \mathcal{X}\rightarrow\mathcal{Y}$ and $g: \mathcal{Y}\rightarrow\mathcal{Z}$, a new point-to-point function $h: \mathcal{X}\rightarrow\mathcal{Z}$ may be defined by setting $h(\mathbf{x}) = g(f(\mathbf{x}))$, $\mathbf{x}\in\mathcal{X}$. The function h is called the ***composition of g with f*** and denoted as $g\circ f$.

A.2. Hemicontinuous Point-to-Set Mappings[1]

Let $2^{\mathcal{Y}}$, called the ***power set of*** $\mathcal{Y}$, represent a collection of nonempty subsets of a set $\mathcal{Y}$ in $\mathbf{R}^m$. Given a set $\mathcal{X}$ in $\mathbf{R}^n$, a ***point-to-set mapping*** (also called a ***transformation*** or ***correspondence*** or ***set-valued function***) $F:\mathcal{X}\to 2^{\mathcal{Y}} \subset \mathcal{Y}$ is a rule which associates with a vector $\mathbf{x}\in\mathcal{X}$ a "nonempty subset" $F(\mathbf{x}) \subset 2^{\mathcal{Y}}$ such that (at least one) $F(\mathbf{x})$ contains more than one element $\mathbf{y}$ or $\mathcal{Y}$. Thus a point-to-set function is a mapping from a set $\mathcal{X}$ into a "set of nonempty subsets" of $\mathcal{Y}$. For $F:\mathcal{X}\to 2^{\mathcal{Y}} \subset \mathcal{Y}$ (hereafter denoted simply as $F:\mathcal{X}\to\mathcal{Y}$) a point-to-set mapping of $\mathcal{X}$ into $\mathcal{Y}$, the ***domain of F*** is the set $\mathcal{D}_F = \{\mathbf{x} \mid \phi \neq F(\mathbf{x}) \subset \mathcal{Y}\}$. Here $F(\mathbf{x})$ is the ***image set*** of $\mathbf{x}\in\mathcal{X}$ while the set $\mathcal{R}_F = \bigcup_{\mathbf{x}\in\mathcal{X}} F(\mathbf{x})$ is the ***range or image of F***. F is termed ***strict*** if all image sets $F(\mathbf{x})$ are nonempty.

Let $F:\mathcal{X}\to\mathcal{Y}$ represent a point-to-set mapping from $\mathcal{X}$ into $\mathcal{Y}$. The subset $\mathcal{G}_F = \{(\mathbf{x},\mathbf{y}) \mid \mathbf{x}\in\mathcal{X}, \mathbf{y}\in F(\mathbf{x}) \subset \mathcal{Y}\}$ of $\mathcal{X}\times\mathcal{Y}$ is called the ***graph of F*** (Figure A.2.1).

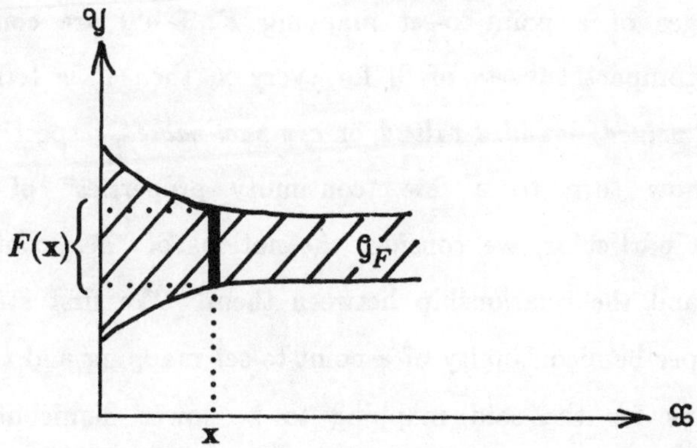

Figure A.2.1

[1] The material presented in this section has been adapted from Berge (1959), Aubin and Frankowska (1990), Nikaido (1972), and Hildenbrand and Kirmen (1976).

For $F: \mathfrak{X} \to \mathcal{Y}$ a point-to-set mapping from the domain of F to its image, the ***inverse mapping*** $F^{-1}: \mathfrak{X} \to \mathcal{Y}$ is a point-to-set mapping from the image of F back to its domain and is defined as $F^{-1}(\mathbf{y}) = \{\mathbf{x} \mid \mathbf{x} \in \mathfrak{X},\ \mathbf{y} \in F(\mathbf{x}),\ (\mathbf{x}, \mathbf{y}) \in \mathcal{G}_F\}$. Thus the domain of F is the image of F^{-1} and the image of F is the domain of F^{-1}. If $\mathcal{B}$ is a subset of $\mathcal{Y}$, there are two ways to define the inverse image of $\mathcal{B}$:

(1) the ***upper (strong) inverse image of $\mathcal{B}$ under F*** (sometimes called the ***core of $\mathcal{B}$ under F***) is the subset $F^{+1}(\mathcal{B}) = \{\mathbf{x} \mid \mathbf{x} \in \mathfrak{X},\ F(\mathbf{x}) \subset \mathcal{B}\}$;

(2) the ***lower (weak) inverse image of $\mathcal{B}$ under F*** is the subset $F^{-1}(\mathcal{B}) = \{\mathbf{x} \mid \mathbf{x} \in \mathfrak{X},\ F(\mathbf{x}) \cap \mathcal{B} \neq \phi\}$.

For $\mathcal{A}$ a subset of $\mathfrak{X}$, the ***restriction of F to $\mathcal{A}$,*** denoted $F \mid \mathcal{A}$, is a point-to-set function defined as

$$F \mid \mathcal{A} = \begin{cases} F(\mathbf{x}), & \mathbf{x} \in \mathcal{A}; \\ \phi, & \mathbf{x} \notin \mathcal{A}. \end{cases}$$

Next, if images of a point-to-set mapping $F: \mathfrak{X} \to \mathcal{Y}$ are convex, closed, bounded, or compact subsets of $\mathcal{Y}$ for every $\mathbf{x}$, then F is termed ***convex-valued***, ***closed-valued***, ***bounded-valued***, or ***compact-valued***, respectively.

We now turn to a few "continuity properties" of point-to-set functions. In particular, we consider the notions of "hemicontinuity" and "closedness" and the relationship between them. We first start with the concept of upper hemicontinuity of a point-to-set mapping and then examine what it means for the said mapping to be lower hemicontinuous. A discussion of closedness then follows.

Upper Hemicontinuity in Terms of Open Sets. The point-to-set function F of $\mathfrak{X}$ into $\mathcal{Y}$ is termed ***upper hemicontinuous (abbreviated uhc) at*** $\mathbf{x}_0 \in \mathfrak{X}$ if for each open set $\mathcal{A}$ in $\mathcal{Y}$ containing the image set $F(\mathbf{x}_0)$ there exists a neighborhood $B(\mathbf{x}_0,\delta)$ of $\mathbf{x}_0$ such that $\mathbf{x} \in B(\mathbf{x}_0,\delta)$ implies $F(\mathbf{x}) \subset \mathcal{A}$ (Figure A.2.2.a).

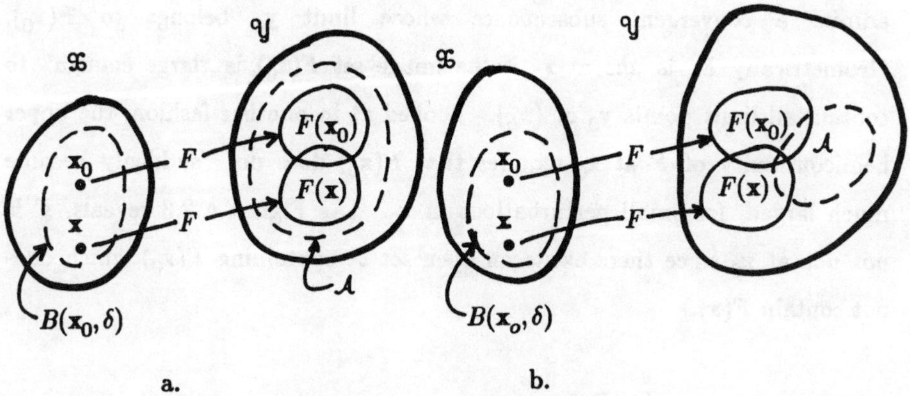

Figure A.2.2

Equivalently, let $\mathfrak{B}$ be an open subset of $\mathfrak{Y}$. Then F is said to be *uhc* at $\mathbf{x}_0$ if $\mathbf{x}_0 \in F^{+1}(\mathfrak{B})$ implies that $B(\mathbf{x}_0,\delta) \subset F^{+1}(\mathfrak{B})$, *i.e.*, a neighborhood of $\mathbf{x}_0$ is in the upper inverse image of $\mathfrak{B}$ wherever $\mathbf{x}_0$ is. In this regard, F is *uhc* if the upper inverse images of open sets are open.

Next, if $F: \mathfrak{X} \rightarrow \mathfrak{Y}$ is compact-valued (the image set $F(\mathbf{x})$ is a compact subset of $\mathfrak{Y}$ for every $\mathbf{x} \in \mathfrak{X}$), then upper hemicontinuity in terms of neighborhoods has an equivalent specification in terms of sequences. That is, a necessary and sufficient condition for F to be *uhc* at a point $\mathbf{x}_0 \in \mathfrak{X}$ is provided by

A.2.1. THEOREM (upper hemicontinuity in terms of convergent sequences). The compact-valued point-to-set function F from $\mathfrak{X}$ into $\mathfrak{Y}$ is *uhc* at $\mathbf{x}_0 \in \mathfrak{X}$ if and only if for every

$\{\mathbf{x}_k\} \rightarrow \mathbf{x}_0$ and every $\{\mathbf{y}_k\}$ with $\mathbf{y}_k \in F(\mathbf{x}_k)$ there exists some $\{\mathbf{y}_{k_j}\} \rightarrow \mathbf{y}_0 \in F(\mathbf{x}_0)$, $j = 1,2,\ldots$.

That is, as $\{\mathbf{x}_k\}$ converges to $\mathbf{x}_0 \in \mathfrak{X}$, the sequence $\{\mathbf{y}_k\}$ with $\mathbf{y}_k \in F(\mathbf{x}_k)$ admits a convergent subsequence whose limit $\mathbf{y}_0$ belongs to $F(\mathbf{x}_0)$. Geometrically F is *uhc* at $\mathbf{x}_0$ if the image set $F(\mathbf{x}_0)$ is "large enough" to contain all limit points $\mathbf{y}_0$ of $\{\mathbf{x}_k\}$. Looked at in another fashion, the upper hemicontinuity of F at $\mathbf{x}_0$ requires that $F(\mathbf{x}_0)$ does not "suddenly become much larger" for small perturbations in $\mathbf{x}_0$. (As Figure A.2.3 reveals, F is not *uhc* at $\mathbf{x}_0$ since there exists an open set $\mathcal{A}$ containing $F(\mathbf{x}_0)$ which does not contain $F(\mathbf{x})$.)

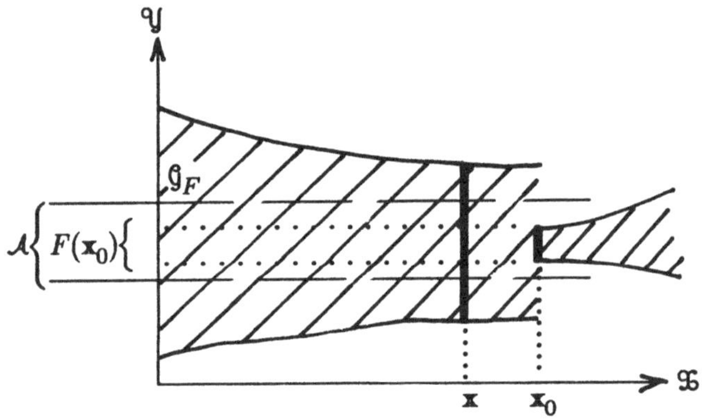

Figure A.2.3

It may be noted further that the set-valued mapping $F: \mathfrak{X} \rightarrow \mathfrak{Y}$ is termed ***uhc*** if it is *uhc* at each point of $\mathfrak{X}$. Additional results pertaining to *uhc* point-to-set functions are:

(a) If the point-to-set function F from $\mathfrak{X}$ into $\mathfrak{Y}$ is *uhc* at $\mathbf{x}_0 \in \mathfrak{X}$, then the ***closure correspondence of F***, denoted $\bar{F}: \mathfrak{X} \rightarrow \mathfrak{Y}$, is also *uhc* at

$\mathbf{x}_0$, where $\overline{F(\mathbf{x}_0)}$ is the closure of $F(\mathbf{x}_0)$. However, the converse assertion does not hold.

(b) If the point-to-set functions $F_i: \mathfrak{X} \to \mathfrak{Y}$, $i=1,\ldots,r$, are all *uhc* at $\mathbf{x}_0 \in \mathfrak{X}$, then their ***union*** $\cup_{i=1}^{r} F_i(\mathbf{x}_0)$ is *uhc* at $\mathbf{x}_0$.

(c) If the point-to-set function $F: \mathfrak{X} \to \mathfrak{Y}$ is compact-valued and *uhc* and $\mathcal{A}$ is a compact subset of $\mathfrak{X}$, then the image set $F(\mathcal{A}) = \cup_{\mathbf{x} \in \mathcal{A}} F(\mathbf{x})$ is compact.

(d) If the point-to-set function $F: \mathfrak{X} \to \mathfrak{Y}$ is compact-valued and *uhc* at $\mathbf{x}_0 \in \mathfrak{X}$, then the ***convex hull correspondence*** $coF: \mathfrak{X} \to \mathfrak{Y}$ is also compact-valued and *uhc*. Here the image set of $\mathbf{x}_0 \in \mathfrak{X}$ is $coF(\mathbf{x}_0)$.

(e) If the point-to-set function $F: \mathfrak{X} \to \mathbf{R}^n$ is *uhc*, then $\mathfrak{D}_F = \{\mathbf{x} \mid \mathbf{x} \in \mathfrak{X}, F(\mathbf{x}) \neq \phi\}$ is a closed subset of $\mathfrak{X}$.

(f) Let the set-valued functions $F, G: \mathfrak{X} \to \mathbf{R}^n$ be *uhc* at $\mathbf{x}$ and closed-valued. Then $\{\mathbf{x} \mid \mathbf{x} \in \mathfrak{X}, F(\mathbf{x}) \cap G(\mathbf{x}) \neq \phi\}$ is a closed subset of $\mathfrak{X}$.

(g) Let F, G be set-valued mappings from $\mathfrak{X}$ in $\mathbf{R}^n$ into $\mathfrak{Y}$ in $\mathbf{R}^m$ and let the ***intersection correspondence*** $H = F \cap G: \mathfrak{X} \to \mathfrak{Y}$ have as its image set $H(\mathbf{x}) = F(\mathbf{x}) \cap G(\mathbf{x}) \neq \phi$ for $\mathbf{x} \in \mathfrak{X}$. If F, G are *uhc* at $\mathbf{x}$ and closed-valued, then H is *uhc* at $\mathbf{x}$. Moreover, if:

(1) F is *uhc* at $\mathbf{x}$ and compact-valued; and

(2) G is closed at $\mathbf{x}$,

then H is *uhc* at $\mathbf{x}$.

(h) Let the set-valued functions F of $\mathfrak{X}$ into $\mathfrak{Y}$ and G of $\mathfrak{Y}$ into $\mathfrak{Z}$ define the ***composition correspondence*** $G \circ F: \mathfrak{X} \to \mathfrak{Z}$ ***with image set*** $(G \circ F)(\mathbf{x}) = G[F(\mathbf{x})] = \cup_{\mathbf{y} \in F(\mathbf{x})} G(\mathbf{y})$. If F, G are *uhc*, then so is $G \circ F$; and if F, G are closed, then $G \circ F$ may or may not be closed.

(i) For the point-to-set functions $F_i: \mathfrak{X} \to \mathfrak{Y}, i=1,\ldots,r$, let us represent the image set of the ***product correspondence*** $F_1 \times \cdots \times F_r$ of $\mathfrak{X}$ into the product space $\mathcal{Y}\times\cdots\times\mathcal{Y}$ as $(F_1\times\cdots\times F_r)(\mathbf{x}) = F_1(\mathbf{x})\times\cdots\times F_r(\mathbf{x})$. If each F_i is *uhc* at $\mathbf{x}$ and compact-valued, then $F_1\times\cdots\times F_r$ is *uhc* at $\mathbf{x}$ and compact-valued. And if each F_i is closed at $\mathbf{x}$, then $F_1\times\cdots\times F_r$ is closed at $\mathbf{x}$.

(j) For the point-to-set functions $F_i: \mathcal{X}\rightarrow R^m, i=1,\ldots,r,$ we may specify the image set of the ***sum correspondence*** $F_1+\cdots+F_r$ of $\mathcal{X}$ into R^m as $(F_1+\cdots+F_r)(\mathbf{x}) = F_1(\mathbf{x})+\cdots+F_r(\mathbf{x})$. If each F_i is *uhc* at $\mathbf{x}$ and compact-valued, then $F_1+\cdots+F_r$ is *uhc* at $\mathbf{x}$ and compact-valued.

We next define

Lower Hemicontinuity in Terms of Open Sets. The point-to-set function F from $\mathcal{X}$ into $\mathcal{Y}$ is termed ***lower hemicontinuous (abbreviated lhc) at*** $\mathbf{x}_0$ if for each open set $\mathcal{A}$ in $\mathcal{Y}$ with $F(\mathbf{x}_0)\cap\mathcal{A}\neq\phi$ there exists a neighborhood $B(\mathbf{x}_0,\delta)$ of $\mathbf{x}_0$ such that $\mathbf{x}\in B(\mathbf{x}_0,\delta)$ implies $F(\mathbf{x}_0)\cap\mathcal{A}\neq\phi$ (Figure A.2.2.b). Equivalently, for $\mathcal{B}$ an open subset of $\mathcal{Y}$, F is said to be *lhc* at $\mathbf{x}_0$ if $\mathbf{x}_0\in F^{-1}(\mathcal{B})$ implies that $B(\mathbf{x}_0,\delta)\subset F^{-1}(\mathcal{B})$, *i.e.*, a neighborhood of $\mathbf{x}_0$ is in the lower inverse image of $\mathcal{B}$ whenever $\mathbf{x}_0$ is. Thus F is *lhc* if the lower inverse images of open sets are open.

For $F:\mathcal{X}\rightarrow\mathcal{Y}$ a set-valued function of $\mathcal{X}$ into $\mathcal{Y}$, lower hemicontinuity in terms of open sets has a parallel representation in terms of sequences. However, unlike the *uhc* case presented earlier, the assumption that F is compact-valued is not needed. To this end we have a necessary and sufficient condition for F to be *lhc* at a point $\mathbf{x}_0\in\mathcal{X}$.

A.2.2. THEOREM (lower hemicontinuity in terms of convergent sequences). The point-to-set function F from $\mathcal{X}$ into $\mathcal{Y}$ is *lhc* at $\mathbf{x}_0\in\mathcal{X}$ if and only if $\{\mathbf{x}_k\}\rightarrow\mathbf{x}_0$, $\{\mathbf{y}_k\}\rightarrow\mathbf{y}_0$, and $\mathbf{y}_k\in F(\mathbf{x}_k)$ for all k imply that $\mathbf{y}_0\in F(\mathbf{x}_0)$.

Here the lower hemicontinuity of F at $\mathbf{x}_0$ requires that $F(\mathbf{x}_0)$ does not "suddenly become much smaller" for small perturbations in $\mathbf{x}_0$. (As exhibited in Figure A.2.4, F is not *lhc* at $\mathbf{x}_0$ since there exists an open set $\mathcal{A}$ having points in common with $F(\mathbf{x}_0)$ but not with $F(\mathbf{x})$. In addition, it is easily seen that F is *uhc* at $\mathbf{x}_0$; F is *lhc* at $\mathbf{x}_0$ in Figure A.2.3).

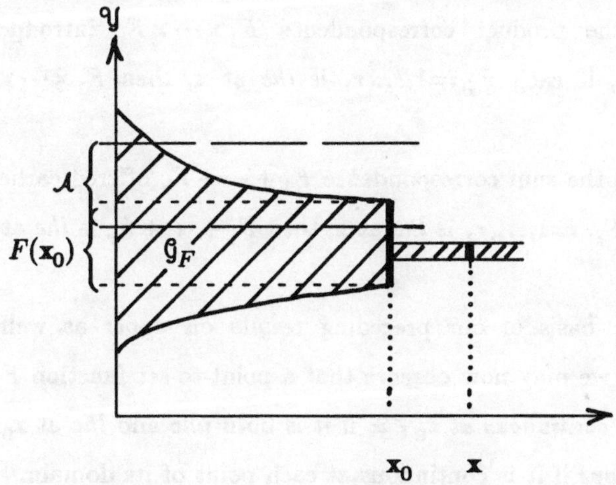

Figure A.2.4

The point-to-set function $F: \mathfrak{X} \rightarrow \mathfrak{Y}$ is said to be ***lhc*** if it is *lhc* at each point of $\mathfrak{X}$. Additional results on point-to-set mappings are:

(a) The point-to-set function F from $\mathfrak{X}$ into $\mathfrak{Y}$ is *lhc* at $\mathbf{x}_0 \in \mathfrak{X}$ if and only if $\bar{F}: \mathfrak{X} \rightarrow \mathfrak{Y}$, the ***closure correspondence of*** F, is *lhc* at $\mathbf{x}_0$. Here $\overline{F(\mathbf{x}_0)}$ depicts the closure of $F(\mathbf{x}_0)$.

(b) If the point-to-set functions $F_i: \mathfrak{X} \rightarrow \mathfrak{Y}$, $i=1,\ldots,r$, are all *lhc* at $\mathbf{x}_0 \in \mathfrak{X}$, then their ***union*** $\cup_{i=1}^{r} F(\mathbf{x}_0)$ is *lhc* at $\mathbf{x}_0$.

(c) If the point-to-set function $F: \mathfrak{X} \rightarrow \mathfrak{Y}$ is *lhc* at $\mathbf{x}_0 \in \mathfrak{X}$, then the ***convex hull correspondence*** $coF: \mathfrak{X} \rightarrow \mathfrak{Y}$ (which renders $coF(\mathbf{x}_0)$ at $\mathbf{x}_0$) is *lhc* at $\mathbf{x}_0$ as well.

(d) If the point-to-set function $F: \mathfrak{X} \rightarrow R^n$ is *lhc*, then $\mathfrak{D}_F = \{\mathbf{x} \mid \mathbf{x} \in \mathfrak{X}, F(\mathbf{x}) \neq \phi\}$ is an open subset of $\mathfrak{X}$.

(e) Given the intersection correspondence $H = F \cap G$ defined in (g) above, if F is *lhc* at $\mathbf{x}$ and G has an open graph, then H is the *lhc* at $\mathbf{x}$.

(f) For the composition correspondence $G \circ F$ defined in (h) above, if F, G are *lhc*, then so is $G \circ F$

(g) For the product correspondence $F_1 \times \cdots \times F_r$ introduced in (i) above, if each $F_i, i=1,\ldots,r$, is *lhc* at $\mathbf{x}$, then $F_1 \times \cdots \times F_r$ is *lhc* at $\mathbf{x}$.

(h) Given the sum correspondence $F_1 + \cdots + F_r$ offered earlier in (j), if each F_i, $i=1,\ldots,r$, is *lhc* at $\mathbf{x}$, then $F_1 + \cdots + F_r$ is *lhc* at $\mathbf{x}$.

On the basis of our preceding results on upper as well as lower hemicontinuity we may now observe that a point-to-set function F of $\mathfrak{X}$ into $\mathfrak{Y}$ is said to be ***continuous at*** $\mathbf{x}_0 \in \mathfrak{X}$ if it is both *uhc* and *lhc* at $\mathbf{x}_0$; and it is termed ***continuous*** if it is continuous at each point of its domain. Moreover, given a set-valued function F from $\mathfrak{X}$ into $\mathfrak{Y}$, if $F(\mathbf{x})$ always consists of a single element $\{\mathbf{y}\} \in \mathfrak{Y}$, then we have, as a special case, a point-to-point function f. In this instance:

(a) The concepts of upper and lower hemicontinuity coincide and are equivalent to the continuity of a point-to-point function f. In fact, either form of hemicontinuity is equivalent to the continuity of f.

(b) The upper inverse image F^{+1} and lower inverse image F^{-1} coincide and simply equal the inverse mapping of f, f^{-1}, when f is a point-to-point function.

A concept closely related to hemicontinuity is that of the "closedness" of a set-valued function F. To explore this notion let $F: \mathfrak{X} \rightarrow \mathfrak{Y}$ be a point-to-set mapping of $\mathfrak{X}$ into $\mathfrak{Y}$. Then F is said to be ***closed at*** $\mathbf{x}_0 \in \mathfrak{X}$

if whenever $\{\mathbf{x}_k\} \rightarrow \mathbf{x}_0$, $\{\mathbf{y}_k\} \rightarrow \mathbf{y}_0$, and $\mathbf{y}_k \in F(\mathbf{x}_k)$ for all k it follows that $\mathbf{y}_0 \in F(\mathbf{x}_0)$. F is termed ***closed*** if it is closed at every $\mathbf{x} \in \mathfrak{D}_F$, *i.e.*, G_F is a closed subset of $\mathfrak{X} \times \mathfrak{Y}$. Clearly a closed point-to-set function is closed-valued (in the sense that $F(\mathbf{x})$ is a closed subset of $\mathfrak{Y}$ for every $\mathbf{x} \in \mathfrak{X}$). It is important to note that, in general, the point-so-set mapping F of $\mathfrak{X}$ into $\mathfrak{Y}$ may be closed but not *uhc* and conversely. However, if F is a closed-valued mapping of $\mathfrak{X}$ into a compact set $\mathfrak{Y}$, then the notions of a closed set-valued function and a *uhc* set-valued function coincide, *i.e.*, by virtue of theorem A.2.1, we can easily see that:

(a) if $F: \mathfrak{X} \rightarrow \mathfrak{Y}$ is closed-valued and *uhc*, then it has a closed graph, *i.e.*, F is closed;

(b) if the range of $F: \mathfrak{X} \rightarrow \mathfrak{Y}$ is compact and F is closed-valued, then F is closed if and only if it is *uhc*.

We note briefly that for $\mathfrak{X}$ in $\mathbf{R}^n$ and $\mathfrak{Y}$ in $\mathbf{R}^m$, the function $F: \mathfrak{X} \rightarrow \mathfrak{Y}$ is termed ***open*** if $\mathcal{G}_F$ is an open set in $\mathfrak{X} \times \mathfrak{Y}$. In this regard, if F is open, then F is *lhc*.

References

Aubin, J.P., H. Frankowska. 1990. *Set-Valued Analysis.* Boston: Birkhaüser.

Begle, E.G. 1950. "A Fixed Point Theorem." *Ann. of Math.*, 51, 544-550.

Berge, C. 1959. "Suruve propriete combinatore des ensembles convexes," *C.R. Academie des Sci, Paris*, 248, 2698-2699.

_______. 1963. *Topological Spaces.* New York: The Macmillan Co.

Border, K. 1985. *Fixed Point Theorems With Applications to Economics and Game Theory.* New York: Cambridge University Press.

Browder, F.E. 1959. "On a Generalization of the Schauder Fixed Point Theorem." *Duke Math. J.*, 26, 291-303.

_______. 1965a. "Another Generalization of the Schauder Fixed Point Theorem." *Duke Math J.*, 32, 399-406.

_______. 1965b. "A Further Generalization of the Schauder Fixed Point Theorem." *Duke Math. J.*, 32, 575-578.

Carathеódory, C. 1907. "Üher den variabilitätsberich der Koeffizienten von Potenzreihen die gegeban Werte Nicht annehmen." *Math. Annalen*, 64, 95-115.

Cronin, J. 1964. *Fixed Points and Topological Degree in Nonlinear Analysis.* Providence, R.I.: Amer. Math. Soc.

Duffin, R.J. 1956. "Infinite Programs." In Kuhn, H.W., A.W. Tucker. 1956.

Eilenberg, S., D. Montgomery. 1946. "Fixed Point Theorems for Multi-Valued Transformations." *Amer. J. Math.*, 68, 214-222.

Fan, K. 1961. "A Generalization of Tychonoff's Fixed Point Theorem." *Math. Annalen*, 142, 305-310.

_______. 1969. "Extensions of Two Fixed Point Theorems of F.E. Browder." *Mathematische Zeitschriff*, 112, 234-240.

Farkas, J. 1902. "Über die Theorie der einfachen Ungleichungen." *Journal für die Reire und Angewandte Math.*, 124, 1-24.

Fenchel, W. 1953. "Convex Cones, Sets and Functions," lecture notes. Princeton: Princeton University.

Forster, W. 1980. "An Application of the Generalized Sperner Lemma to the Computation of Fixed Points in Arbitrary Complexes." In Forster, W. (ed). *Numerical Solution of Highly Nonlinear Problems.* The Netherlands: North-Holland Publishing Co.

Gale, D. 1960. *The Theory of Linear Economic Models.* Now York: McGraw-Hill Book Co.

Gerstenhaber, M. 1951. "Theory of Convex Polyhedral Cones." In Koopmans, T. *Activity Analysis of Production and Allocation.* New York: John Wiley and Sons, Inc.

Goldman, A.J. 1956. "Resolution and Separation Theorems for Polyhedral Convex Sets," In Kuhn, H.W., A.W. Tucker. 1956.

Gordan, P. 1837. "Über die Auflösungen lineara Gleichungen mit reelen coefficienten." *Math. Annalen*, 6, 23-28.

Helly, E. 1921. "Über Mengen Konvexes Körper mit gemeinschafltlichen Punkten." *Jber. Deutsch, Math. Verein*, 32, 175-176.

Hildenbrand, W., A. Kirman. 1976. *Introduction to Equilibrium Analysis.* The Netherlands: North-Holland Publishing Co.

Kakutani, S. 1941. "A Generalization of Brouwer's Fixed Point Theorem." *Duke Math. J.*, 8, 457-459.

Kannai, Y. 1992. "The Cone and Balancedness." In Aumann, R.J., S. Hard (eds). *Handbook of Game Theory With Economic Applications.* The Netherlands: North-Holland Publishing Co.

Klee, V. 1960. "Levay-Schander Theory Without Local Convexity." *Math. Annalen,* 141, 286-296.

Klein, E. 1973. *Mathematical Methods in Theoretical Economics.* New York: Academic Press, Inc.

Knaster, B., K. Kuratowski, S. Mazurkiewicz. 1929. "Ein Beweis des Fixpunktsatzes für n-dimensionale Simplexe." *Fundam. Math.*, 14, 132-137.

Krein, M., D. Milman. 1940. "On Extreme Points of Regularly Convex Sets." *Studia Math.* 9, 133-138.

Kuhn, H.W. 1968. "Simplicial Approximation of Fixed Points." *Proceedings of the National Academy of Sciences*, USA, 61, 1238-1242.

______. 1980. "On the Shapley-Sperner Lemma." In Forster, W. (ed). *Numerical Solution of Highly Nonlinear Problems.* The Netherlands: North-Holland Publishing Co.

Kuhn, H.W., A.W. Tucker. 1956. "Linear Inequalities and Related Systems." *Annals of Mathematics Studies No. 38.* Princeton University Press.

Mangasarian, O. 1969. *Nonlinear Programming.* New York: McGraw-Hill Book Co.

Minkowski, H. 1911. *Geometrie der Zahlen.* Leipzig: Teubner.

Motzkin, J.J. 1936. "Beiträge zur Theorie der Linearen Angleichungen." Inaugural Dissertation, Basel, Jerusalem.

Nikaido, H. 1968. *Convex Structures and Economic Theory.* New York: Academic Press.

Panik. M. 1990. "Farkas' Theorem Under Convex Combination." *Appl. Math. Lett.*, 3, 89-90.

Rockafellar, R.T. 1970. *Convex Analysis.* Princeton: Princeton University Press.

Scarf. H. 1973. *The Computation of Economic Equilibria.* New Haven: Yale University Press.

Schauder, J. 1930. "Der Fixpunktsatz in Funktionalraumen." *Studia Math.*, 2, 171-180.

Shapley, L.S. 1973. *On Balanced Games Without Side Payments.* In Hu, T.C., S.M. Robinson (eds). *Mathematical Programming.* New York: Academic Press.

Shapley, L.S., R. Vohra. 1991. "Onkakutani's Fixed Point Theorem, the K-K-M-S Theorem and the Core of a Balanced Game." *Economic Theory*, 1, 108-116.

Simonnard, M. 1966. *Linear Programming.* New Jersey: Prentice-Hall, Inc.

Slater, M.L. 1951. "A Note on Motzkin's Transposition Theorem." *Econometrica*, 19, 1985-1986.

Sperner, E. 1928. "Neuer Beweis für die Invarianzder Dimensionszahl und des Gebietes." *Abh. Math. Sem. Univ. Hamburg*, 6, 265-272.

______. 1980. "Fifty Years of Further Development of a Combinational Lemma, Parts A, B." In Forster, W. (ed). *Numerical Solution of Highly Nonlinear Problems.* The Netherlands: North-Holland Publishing Co.

Stiemke, E. 1915. "Über positive Lösungen homogener linearer Gleichungen." *Math. Annelen*, 76, 340-342.

Stoer, J., C. Witzgall. 1970. *Convexity and Optimization in Finite Dimensions I.* New York: Springer-Verlag.

Tucker, A.W. 1956. "Dual Systems of Homogeneous Linear Relations." In Kuhn, H.W., A.W. Tucker. 1956.

Tychonoff, A. 1935. "Ein Fixpunktsatz." *Math. Annelen*, 111, 767-776.

von Neumann, J. 1928. "Zur Theorie der Gesellshaftspiele." *Math. Annalen*, 100, 295-320.

Weyl, H. 1950. "The Elementary Theory of Convex Polyhedra." In Kuhn, H.W., A.W. Tucker. *Contributions to the Theory of Games* (Vol. 1). Princeton: Princeton University Press.

Yoseloff, M. 1974. "Topologic Proofs of Some Combinatorial Theorms." *Journal of Combinatorial Theory* (A), 17, 95-111.

Notation Index

$\mathbf{R}$	set of real numbers
$\mathcal{N}$	set of nonnegative real numbers
ϕ	null or empty set
$\mathbf{x} = (x_i)$	the vector whose i^{th} component is x_i
$\mathbf{1}$	the sum vector containing all 1's
$\mathbf{0}$	the null vector/matrix containing all 0's
$\mathbf{I}_n$	identity matrix of order n
$\mathbf{e}_i$	i^{th} unit column vector
$\mathbf{x}'$	transpose of a vector
$\mathbf{x} \cdot \mathbf{y}$	scalar product of vectors $\mathbf{x}$, $\mathbf{y}$
$\Vert \cdot \Vert$	Euclidean norm
$\mathbf{x}/\Vert \mathbf{x} \Vert$	direction of vector $\mathbf{x}$
$\Vert \mathbf{x} - \mathbf{y} \Vert$	distance between points $\mathbf{x}$, $\mathbf{y}$
$\mathbf{A}$	an $(m \times n)$ matrix $\mathbf{A}$
$\rho(\mathbf{A})$	rank of a matrix $\mathbf{A}$
$\vert \mathbf{A} \vert$	determinant of an nth order matrix $\mathbf{A}$
$x \in \mathcal{S}$	x is an element of a set $\mathcal{S}$
$x \notin \mathcal{S}$	x is not an element of a set $\mathcal{S}$
$\mathcal{A} \subseteq \mathcal{B}$	set $\mathcal{A}$ is a subset of set $\mathcal{B}$ (possibly $\mathcal{A} = \mathcal{B}$)
$\mathcal{A} \subset \mathcal{B}$	set $\mathcal{A}$ is a proper subset of set $\mathcal{B}$
$\mathcal{V}_n(\mathcal{F})$	n-dimensional vector space over field $\mathcal{F}$
$\mathcal{V}_n(\mathbf{R}) = \mathbf{R}^n$	n-space or the set of all real n-component vectors
$dim(\mathcal{V})$	dimension of a vector space $\mathcal{V}$
$\mathcal{M}$	subspace or linear manifold
$\ell(\mathcal{M})$	linear hull of subspace $\mathcal{M}$
$\mathcal{M}_1 \oplus \mathcal{M}_2$	direct sum of subspaces $\mathcal{M}_1, \mathcal{M}_2$
$\mathcal{M}^{\perp}$	orthogonal complement or dual space of $\mathcal{M}$

$d(\mathcal{S})$	diameter of a set $\mathcal{S}$
$d(\mathbf{x},\mathcal{S})$	distance between vector $\mathbf{x}$ and set $\mathcal{S}$
$d(\mathcal{S},\mathcal{T})$	distance between sets $\mathcal{S},\mathcal{T}$
$B(\mathbf{x}_0,\delta)$	δ-neighborhood of a point $\mathbf{x}_0 \in \mathbf{R}^n$ (or an open ball or sphere of radius δ about $\mathbf{x}_0$)
$\mathcal{C}$	cone
$\mathcal{C}^+$	polar cone
$\mathcal{C}^*$	dual cone
$(\mathbf{a})$	ray or half-line
$(\mathbf{a})^\perp$	orthogonal cone to $(\mathbf{a})$
$(\mathbf{a})^+$	polar of $(\mathbf{a})$
$(\mathbf{a})^*$	dual of $(\mathbf{a})$
$\mathcal{S}^o$	interior of set $\mathcal{S}$
$\mathcal{S}^c$	complementary set of $\mathcal{S}$
$\bar{\mathcal{S}}$	closure of set $\mathcal{S}$
$\partial(\mathcal{S})$	boundary of set $\mathcal{S}$
$coni(\mathcal{S})$	conical hull of set $\mathcal{S}$
$0^+(\mathcal{S})$	recession cone of set $\mathcal{S}$
$aff(\mathcal{S})$	affine hull of set $\mathcal{S}$
$co(\mathcal{S})$	convex hull of set $\mathcal{S}$
$ri(\mathcal{S})$	relative interior of convex set $\mathcal{S}$
$r\partial(\mathcal{S})$	relative boundary of set $\mathcal{S}$
$\mathcal{S}_1+\mathcal{S}_2$	linear sum of convex sets $\mathcal{S}_1,\mathcal{S}_2$
$cone(\mathcal{S})$	convex cone generated by $\mathcal{S}$
$ray(\mathcal{S})$	union of the origin and the set of recession half-lines
sup	supremum or least upper bound
inf	infimum or greatest lower bound
$\mathcal{A}\times\mathcal{B}$	Cartesian product of sets $\mathcal{A},\mathcal{B}$
$[\mathbf{x}_1,\mathbf{x}_2]$	closed line segment joining points $\mathbf{x}_1,\mathbf{x}_2$

$\{\mathbf{x}_k\}$	sequence of points
$\left\{\mathbf{x}_{k_j}\right\}, j=1,2,$	subsequence of points
$\{\mathbf{x}_k\}_{h\in\mathcal{K}}$	subsequence of points, where $\mathcal{K}$ is a subset of positive integers
$\mathcal{H}$	hyperplane
$[\mathcal{H}^+],[\mathcal{H}^-]$	closed half-planes
$(\mathcal{H}^+),(\mathcal{H}^-)$	open half-planes
σ^k	k-dimensional simplex
$a \equiv b \,(\mathrm{mod}\, m)$	$a-b$ is divisible by m
$f:\mathcal{S}_1 \to \mathcal{S}_2$	point-to-point mapping from $\mathcal{S}_1$ to $\mathcal{S}_2$
$f \circ g$	composition of f with g
$\mathcal{D}_f$	domain of f
$\mathcal{R}_f$	range of f
$\mathcal{G}_f$	graph of f
f^{-1}	single-valued inverse mapping
$2^{\mathcal{Y}}$	power set of $\mathcal{Y}$
$F:\mathcal{S}_1 \to \mathcal{S}_2$	point-to-set mapping from $\mathcal{S}_1$ to $\mathcal{S}_2$
F^{-1}	inverse point-to-set mapping
$F(\mathbf{x})$	image set of $\mathbf{x}$
$F^{+1}(\mathcal{B})$	upper inverse image of $\mathcal{B}$ under F
$F^{-1}(\mathcal{B})$	lower inverse image of $\mathcal{B}$ under F
$G \circ F$	composition correspondence
$F \cap G$	intersection correspondence
$F \times G$	product correspondence
$F + G$	sum correspondence

INDEX

THEORY AND DECISION LIBRARY

SERIES B: MATHEMATICAL AND STATISTICAL METHODS

Editor: H. J. Skala, *University of Paderborn, Germany*

1. D. Rasch and M.L. Tiku (eds.): *Robustness of Statistical Methods and Nonparametric Statistics.* 1984 ISBN 90-277-2076-2
2. J.K. Sengupta: *Stochastic Optimization and Economic Models.* 1986 ISBN 90-277-2301-X
3. J. Aczél: *A Short Course on Functional Equations.* Based upon Recent Applications to the Social Behavioral Sciences. 1987 ISBN Hb 90-277-2376-1; Pb 90-277-2377-X
4. J. Kacprzyk and S.A. Orlovski (eds.): *Optimization Models Using Fuzzy Sets and Possibility Theory.* 1987 ISBN 90-277-2492-X
5. A.K. Gupta (ed.): *Advances in Multivariate Statistical Analysis.* Pillai Memorial Volume. 1987 ISBN 90-277-2531-4
6. R. Kruse and K.D. Meyer: *Statistics with Vague Data.* 1987 ISBN 90-277-2562-4
7. J.K. Sengupta: *Applied Mathematics for Economics.* 1987 ISBN 90-277-2588-8
8. H. Bozdogan and A.K. Gupta (eds.): *Multivariate Statistical Modeling and Data Analysis.* 1987 ISBN 90-277-2592-6
9. B.R. Munier (ed.): *Risk, Decision and Rationality.* 1988 ISBN 90-277-2624-8
10. F. Seo and M. Sakawa: *Multiple Criteria Decision Analysis in Regional Planning.* Concepts, Methods and Applications. 1988 ISBN 90-277-2641-8
11. I. Vajda: *Theory of Statistical Inference and Information.* 1989 ISBN 90-277-2781-3
12. J.K. Sengupta: *Efficiency Analysis by Production Frontiers.* The Non-parametric Approach. 1989 ISBN 0-7923-0028-9
13. A. Chikán (ed.): *Progress in Decision, Utility and Risk Theory.* 1991 ISBN 0-7923-1211-2
14. S.E. Rodabaugh, E.P. Klement and U. Höhle (eds.): *Applications of Category Theory to Fuzzy Subsets.* 1992 ISBN 0-7923-1511-1
15. A. Rapoport: *Decision Theory and Decision Behaviour.* Normative and Descriptive Approaches. 1989 ISBN 0-7923-0297-4
16. A. Chikán (ed.): *Inventory Models.* 1990 ISBN 0-7923-0494-2
17. T. Bromek and E. Pleszczyńska (eds.): *Statistical Inference.* Theory and Practice. 1991 ISBN 0-7923-0718-6

18. J. Kacprzyk and M. Fedrizzi (eds.): *Multiperson Decision Making Models Using Fuzzy Sets and Possibility Theory.* 1990 ISBN 0-7923-0884-0
19. G.L. Gómez M.: *Dynamic Probabilistic Models and Social Structure.* Essays on Socioeconomic Continuity. 1992 ISBN 0-7923-1713-0
20. H. Bandemer and W. Näther: *Fuzzy Data Analysis.* 1992 ISBN 0-7923-1772-6
21. A.G. Sukharev: *Minimax Models in the Theory of Numerical Methods.* 1992 ISBN 0-7923-1821-8
22. J. Geweke (ed.): *Decision Making under Risk and Uncertainty.* New Models and Empirical Findings. 1992 ISBN 0-7923-1904-4
23. T. Kariya: *Quantitative Methods for Portfolio Analysis.* MTV Model Approach. 1993 ISBN 0-7923-2254-1
24. M.J. Panik: *Fundamentals of Convex Analysis.* Duality, Separation, Representation, and Resolution. 1993 ISBN 0-7923-2279-7

KLUWER ACADEMIC PUBLISHERS – DORDRECHT / BOSTON / LONDON

MIX
Papier | Fördert gute Waldnutzung
FSC® C083411

Zeitfracht Medien GmbH
Ferdinand-Jühlke-Straße 7
99095 Erfurt, Deutschland
produktsicherheit@kolibri360.de